Dynamics, Binding and Function of Intrinsically Disordered Proteins

Dynamics, Binding and Function of Intrinsically Disordered Proteins

Editor: Rudyard Ramsey

FOSTER ACADEMICS

www.fosteracademics.com

www.fosteracademics.com

FOSTER ACADEMICS

Cataloging-in-Publication Data

Dynamics, binding and function of intrinsically disordered proteins / edited by Rudyard Ramsey.
p. cm.
Includes bibliographical references and index.
ISBN 978-1-64646-571-2
1. Proteins--Structure. 2. Proteins--Separation. 3. Proteins--Conformation. 4. Proteins--Pathophysiology.
5. Proteins--Metabolism--Disorders. I. Ramsey, Rudyard.
QP551 .I58 2023
572.6--dc23

Foster Academics,
118-35 Queens Blvd., Suite 400,
Forest Hills, NY 11375, USA

ISBN 978-1-64646-571-2 (Hardback)

Contents

Preface

This book was inspired by the evolution of our times; to answer the curiosity of inquisitive minds. Many developments have occurred across the globe in the recent past which has transformed the progress in the field.

Intrinsically disordered proteins (IDPs) are a class of proteins that do not fold into a specific three-dimensional structure, which is a characteristic of proteins. The structural flexibility of IDPs range from fully structured to partially structured, which enables them to complement ordered proteins in order to perform functions that cannot be performed by rigid structures. Several factors such as pH and temperature can affect the shape and level of disorder in proteins. IDPs have been associated with various diseases such as cancer and cardiovascular diseases. The accumulation of an IDP called α-synuclein plays a pathophysiological role in Parkinson's disease (PD). The structural flexibility of the unstructured α-synuclein protein and associated disease mechanisms is affected by several factors such as genetics, oxidative stress, nitrative stress, and mitochondrial impairment. The book aims to shed light on some of the unexplored aspects of intrinsically disordered proteins and the recent researches on them. It will serve as a valuable source of reference for the graduate and postgraduate students.

This book was developed from a mere concept to drafts to chapters and finally compiled together as a complete text to benefit the readers across all nations. To ensure the quality of the content we instilled two significant steps in our procedure. The first was to appoint an editorial team that would verify the data and statistics provided in the book and also select the most appropriate and valuable contributions from the plentiful contributions we received from authors worldwide. The next step was to appoint an expert of the topic as the Editor-in-Chief, who would head the project and finally make the necessary amendments and modifications to make the text reader-friendly. I was then commissioned to examine all the material to present the topics in the most comprehensible and productive format.

I would like to take this opportunity to thank all the contributing authors who were supportive enough to contribute their time and knowledge to this project. I also wish to convey my regards to my family who have been extremely supportive during the entire project.

Editor

1

bHLH–PAS Proteins: Their Structure and Intrinsic Disorder

Marta Kolonko and Beata Greb-Markiewicz *

Department of Biochemistry, Faculty of Chemistry, Wroclaw University of Science and Technology, Wybrzeże Wyspiańskiego 27, 50-370 Wroclaw, Poland
* Correspondence: beata.greb-markiewicz@pwr.edu.pl

Abstract: The basic helix–loop–helix/Per-ARNT-SIM (bHLH–PAS) proteins are a class of transcriptional regulators, commonly occurring in living organisms and highly conserved among vertebrates and invertebrates. These proteins exhibit a relatively well-conserved domain structure: the bHLH domain located at the N-terminus, followed by PAS-A and PAS-B domains. In contrast, their C-terminal fragments present significant variability in their primary structure and are unique for individual proteins. C-termini were shown to be responsible for the specific modulation of protein action. In this review, we present the current state of knowledge, based on NMR and X-ray analysis, concerning the structural properties of bHLH–PAS proteins. It is worth noting that all determined structures comprise only selected domains (bHLH and/or PAS). At the same time, substantial parts of proteins, comprising their long C-termini, have not been structurally characterized to date. Interestingly, these regions appear to be intrinsically disordered (IDRs) and are still a challenge to research. We aim to emphasize the significance of IDRs for the flexibility and function of bHLH–PAS proteins. Finally, we propose modern NMR methods for the structural characterization of the IDRs of bHLH–PAS proteins.

Keywords: bHLH–PAS transcription factor; intrinsically disordered region; IDR; C-terminus

---

1. Introduction to bHLH–PAS Proteins

The basic helix–loop–helix/Per-ARNT-SIM (bHLH–PAS) proteins are a class of transcriptional regulators that commonly occur in living organisms. They play an important role in the regulation of a variety of developmental and physiological events [1]. The maintenance of cellular and systemic oxygen homeostasis is performed by hypoxia-inducible factor 1α (HIF1-α) [2]. In the hypoxia condition, HIF1-α is translocated to the nucleus [3] where it regulates transcription activity related to angiogenesis, cell proliferation/survival, glucose metabolism, and iron metabolism. The incorrect control of the listed processes is fundamental in many diseases, including cancer, strokes, and heart disease [2]. Some bHLH–PAS family members act as receptors for different high and low molecular ligands [1]. The only known small ligand-activated bHLH–PAS protein, aryl hydrocarbon receptor (AHR), is involved in toxin metabolism and binds highly toxic ligands, such as TCDD [4]. The ligated AHR migrates to the nucleus and mediates a wide range of biological responses to poisons. This mediation comprises a wasting syndrome, hepatotoxicity, teratogenesis, and tumor promotion [4]. Overexpression and constitutive activation of the AHR have been observed in various types of tumors [5]. Importantly, the AHR has been described as a critical modulator of host–environment interactions, especially for immune and inflammatory responses [6].

Another interesting example of a bHLH–PAS family member is the single-minded protein (SIM), which plays a significant role during central nerve cord [7] and genital imaginal disc development [8]. As shown, SIM gene mutations contribute to certain dysmorphic features of brain development and also the mental retardation in Down syndrome [9]. Interestingly, SIM overexpression is also associated with breast and prostate cancer [10], which indicates connections between their apparently unrelated signaling pathways.

Members of the bHLH–PAS family were shown to be targets for disease therapy. AHR, highly expressed in multiple organs and tissues, may influence tumorigenesis both by direct effect on the cancer cells and by modulation of the immune system. For this reason, the development of selective AHR modulators active against multiple tumors is a desirable direction of research [11]. Also, targeting of the HIF1-α pathway as a novel cancer therapy is a current project [12]. As AHR was shown to modulate the immune response in the respiratory tract, this protein can be potentially used also as a therapeutic object for the treatment of various inflammatory lung diseases [13,14]. Another member of the family, expressed mainly in the brain, neuronal PAS domain-containing protein 4 (NPAS4) has been proposed as a novel therapeutic target for depression and neurodegenerative diseases [15] and as a component of new stroke therapies [16]. Additionally, NPAS4, whose expression was also detected in the pancreas, was proposed to be a therapeutic target for diabetes [17] and as a treatment during pancreas transplantation [18].

In spite of performing a high diversity of functions, the bHLH–PAS proteins family exhibits a relatively well-conserved domain structure in the N-terminal part of their sequence (Figure 1). The bHLH region contains approximately 60 amino acid (aa) residues and can be divided into two functionally distinctive parts: the basic region responsible for DNA binding (approximately 15 aa), and the neighboring C-terminal HLH region, which takes part in protein dimerization [19]. The PAS domain is located in the central part of the protein and usually comprises about 300 aa residues [1]. It is divided into two structurally conserved regions named PAS-A and PAS-B, which are often connected to a single PAS-associated C-terminal (PAC) motif [20]. The PAS-A and PAS-B regions are separated by a poorly conserved link [1]. The PAS-A region is critical for selecting a dimerization partner and ensuring the specificity of target gene activation [21]. The PAS-B region is usually responsible for sensing diverse exogenous and endogenous signals, and is accompanied by energetic and conformational changes that regulate protein activity [21]. Contrary to conserved domains, the C-termini of bHLH–PAS proteins present significant variability [21] and contain variable transcription activation/repression domains (TAD/RPD) (Figure 1) [22,23]. An example is the mammalian SIM existing in two isoforms: SIM1 and SIM2. Both isoforms present a high amino acid identity in their N-termini (90% identity in the bHLH and PAS regions) and extreme diversity in their C-termini [24]. While SIM1 activates the expression of target genes, SIM2 acts as an inhibitor. Interestingly, the opposite transcriptional effect disappears after the deletion of both SIM1 and SIM2 C-termini, resulting in proteins with a similar activity [25,26]. Moffet and Pelletier [26] demonstrated that a distinct SIM2 C-terminal sequence comprises two repression domains with a high proline/serine and proline/alanine content, respectively. It is a feature of "repressor motifs", which can also be found in a large number of other transcriptional repressors [25,26]. Due to the highly variable amino acid sequence and the lack of predefined domains, C-termini are believed to be responsible for the specific modulation of the functioning of bHLH–PAS proteins and the recognition of partner proteins necessary for their unique action [21].

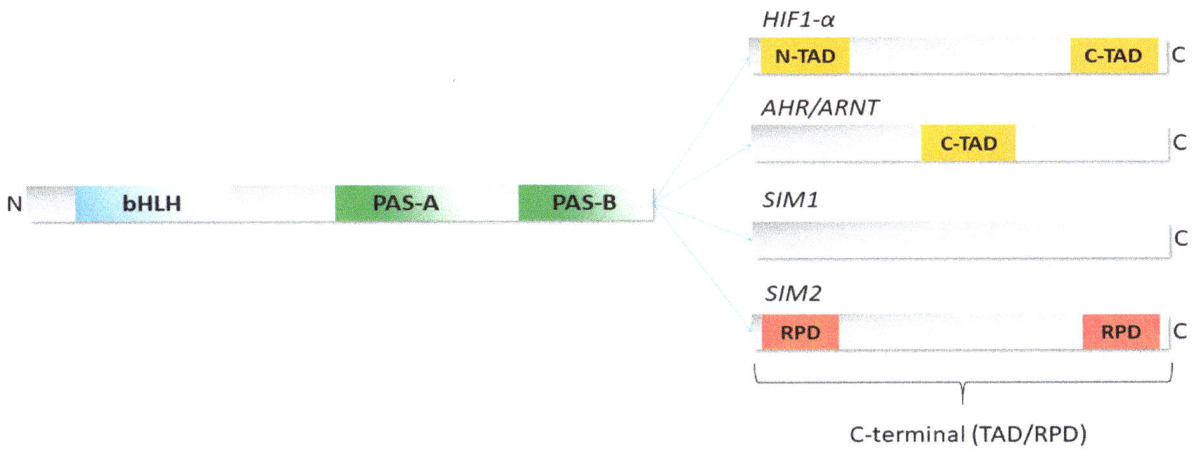

Figure 1. Schematic representation of the bHLH–PAS protein domain structure. The N-terminal part of bHLH–PAS proteins is characterized by the presence of defined domains: bHLH (blue), PAS-A and PAS-B (green). The C-terminal part presents significant diversity and contains variable transactivation/repression domains (TAD/RPD). The C-termini of selected proteins (HIF1-α, AHR/ARNT, SIM1, and SIM2) are presented. Yellow boxes indicate TADs while the red box indicates RPD. Based on [26–29].

Generally, bHLH–PAS proteins can be divided into two classes. While the expression of class I proteins is specifically regulated by diverse physiological states and/or environmental signals [30], class II proteins are expressed continuously and serve as heterodimerization partners for class I members. Only the dimer of the two bHLH–PAS proteins acts as a functional transcription factor complex, regulating the expression of genes under its control [22]. Mammalian bHLH–PAS transcription factors are listed in Table 1.

Table 1. Mammalian class I and class II bHLH–PAS proteins [1,21,30–32].

Class I	Class II	Type of Signal
hypoxia-inducible factors (HIF; HIF1-α, HIF2-α, and HIF3-α)	aryl hydrocarbon receptor nuclear translocator (ARNT), also known as HIF1-β and ARNT2	regulated by hypoxia
aryl hydrocarbon receptor (AHR); aryl hydrocarbon receptor repressor (AHRR)		regulated by xenobiotics
single-minded proteins (SIM1 and SIM2)		developmentally regulated
neuronal PAS domain proteins (NPAS)		developmentally regulated
circadian locomotor output cycles protein kaput (CLOCK)	Circadian rhythm proteins (BMAL1 and BMAL2, also known as ARNTL and ARNTL2)	circadian rhythms

2. bHLH–PAS Protein Conservation between Organisms

bHLH–PAS proteins are highly conserved among different organisms, including vertebrates and invertebrates [33]. Most mammalian representatives possess orthologs in insect species. An example is the *Drosophila melanogaster* TANGO (TGO) protein, which is a homologue of the mammalian class II protein, ARNT [34]. TGO is known as the general dimerization partner for Similar (SIMA), Trachealess (TRH), Single-minded (SIM) protein, Spineless (SS), and Dysfusion (DYS), performing functions equivalent to mammalian ones.

In 2017, the Nobel Prize in Physiology or Medicine was awarded to J. C. Hall, M. Rosbash, and M. W. Young for their discoveries of molecular mechanisms controlling the circadian rhythm in *D. melanogaster*. As shown, the two bHLH–PAS transcription factors CLOCK and CYCLE play a key role as transcriptional activators for *period* (*per*) and *timeless* (*tim*) genes [35,36]. Thanks to the conservation of circadian bHLH–PAS proteins between *D. melanogaster* and mammals [35], the explanation of the fly daily rhythm enabled the understanding of a similar, though much more complicated, process in mammals, controlled by two orthologous to CLOCK/CYCLE heterodimers: CLOCK/BMAL1 and NPAS-2/BMAL1 [22].

In spite of significant similarities, some exceptions between vertebrates and invertebrates can be noticed. The bHLH–PAS transcription factor, Methoprene-tolerant protein (MET), occurs exclusively in insects and to date has no known ortholog in nonarthropod organisms. MET has been recently confirmed as the juvenile hormone (JH) receptor playing a significant role during insect development and maturation [37]. Interestingly, in a few species of insects, like *D. melanogaster* and *Bombyx mori*, there exist the MET paralogs named germ-cell expressed (GCE) and MET2, respectively [38]. MET and GCE participate in modulating JH signaling during *D. melanogaster* development, but their functions are not fully redundant and the proteins exhibit tissue-specific distribution [39]. In turn, the MET2 protein function in *B. mori* is not yet defined [40].

3. Structure of bHLH–PAS Proteins

To date, our knowledge regarding the tertiary structure of bHLH–PAS proteins is limited. All determined structures comprise single isolated domains (PAS-A or PAS-B) or adjacent domains connected with flexible aa chains. C-termini, however, comprising an extensive part of proteins, have not yet been structurally characterized. These regions are not homologous to any described domains and seem to be very disordered. Consequently, it can be seen to be a huge challenge for scientists to determine their structure and combine it with specific protein functions. All bHLH–PAS structures deposited in the Protein Data Bank (PDB) are listed in Table 2 (Nuclear Resonance Magnetism (NMR) structures) and Table 3 (X-ray structures). Most of the listed assemblies correspond to heterodimers.

Table 2. bHLH–PAS protein structures deposited in the PDB obtained with NMR.

Form	Protein	Segment	Organism	PDB ID
monomers	HIF2-α	PAS-B domain	*Homo sapiens*	1P97
	ARNT	PAS-B domain	*Homo sapiens*	1X0O
dimer	HIF-2a:ARNT	PAS-B domains	*Homo sapiens*	2A24

Table 3. bHLH–PAS protein structures deposited in the PDB obtained with X-ray diffraction.

Form	Protein	Segment	Organism	PDB ID
monomers	AHR	PAS-A	*Mus musculus*	4M4X
	ARNT	PAS-B	*Homo sapiens*	2B02
	HIF1-α	PAS-B	*Homo sapiens*	4H6J
dimers	ARNT Homodimer	PAS-B	*Homo sapiens*	4EQ1
	HIF2-α:ARNT	PAS-B	*Homo sapiens*	3F1P
	HIF2-α:ARNT with artificial ligand	PAS-B	*Homo sapiens*	3F1O
	HIF2-α:ARNT	PAS-B	*Homo sapiens*	6D0C
	ARNT/HIF transcription factor/coactivator complex	PAS-B	*Homo sapiens*, *Mus musculus*	4PKY
	ARNT transcription factor/coactivator complex	PAS-B domain	*Homo sapiens*, *Mus musculus*	4LPZ
	HIF2-α:ARNT	bHLH; PAS-A; PAS-B	*Mus musculus*	4ZP4

Table 3. *Cont.*

Form	Protein	Segment	Organism	PDB ID
	HIF2-α:ARNT with HRE DNA	bHLH; PAS-A; PAS-B	*Mus musculus*	4ZPK
	HIF1-α:ARNT with HRE DNA	bHLH; PAS-A; PAS-B	*Mus musculus*	4ZPR
	AHR:ARNT	bHLH; PAS-A	*Homo sapiens*	5NJ8
	AHR:ARNT bound to the dioxin response element (DRE)	bHLH; PAS-A	*Homo sapiens, Mus musculus*	5V0L
	AHRR:ARNT	bHLH; PAS-A; PAS-B	*Homo sapiens, Bos taurus*	5Y7Y
	NPAS1:ARNT	bHLH; PAS-A; PAS-B	*Mus musculus*	5SY5
	NPAS3:ARNT in complex with HRE DNA	bHLH; PAS-A; PAS-B	*Mus musculus*	5SY7
	CLOCK-BMAL1	bHLH	*Homo sapiens*	4H10
	CLOCK:BMAL1	bHLH; PAS-A; PAS-B	*Mus musculus*	4F3L

The first step in determining the structure of bHLH–PAS proteins was the isolation and characterization of PAS-B domains from HIF2-α (Figure 2A) [41] and ARNT (Figure 2B) [42]. Both structures were obtained using the NMR technique and presented a fold characteristic for the PAS domain: a five-stranded antiparallel β-sheet flanked by several α-helices [42]. The next step was the crystallization of the isolated PAS-A domain of AHR (Figure 2C) and the PAS-B domains of ARNT (not shown) and HIF1-α (not shown). Interestingly, the tertiary architecture of all structurally characterized PAS domains is very conserved (Figure 2), despite the fact that their primary sequence is highly divergent (sequence identity lower than 20%) [43].

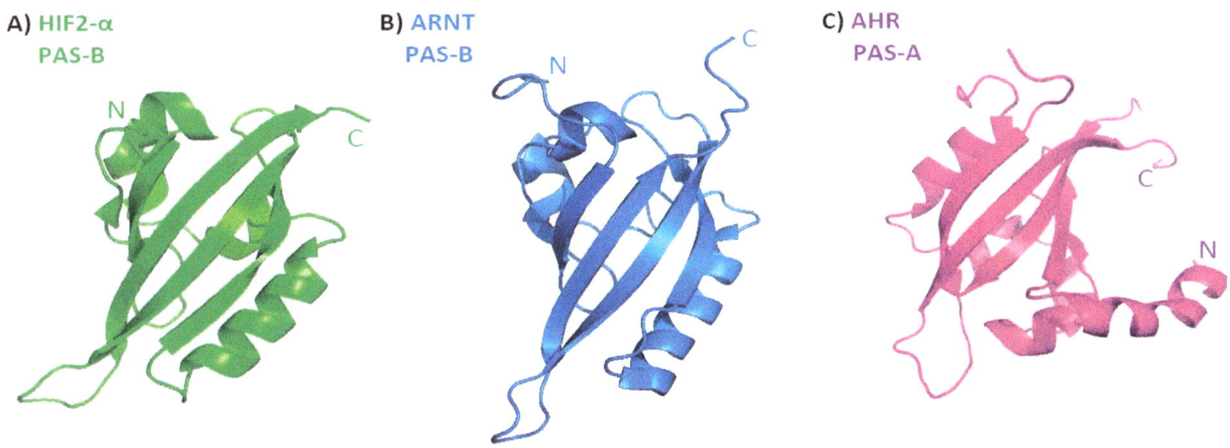

Figure 2. Representation of the PAS fold: a five-stranded antiparallel β-sheet is flanked by several α-helices. (**A**) HIF2-α PAS-B obtained with NMR (PDB 1P97), (**B**) ARNT PAS-B obtained with NMR (PDB 1X0O), (**C**) AHR PAS-A domain obtained with X-ray (PDB 4M4X).

Further experiments led to the cocrystallization of PAS-B domains from the HIF2-α/ARNT heterodimer, which revealed that these two domains form an interaction interface via their β-sheets in an antiparallel form (Figure 3A) [42]. Another measurement covering bHLH domains of BMAL1/CLOCK bound to the DNA defined domain structure and binding properties specifying interactions taking place (Figure 3B) [44]. A typical bHLH domain comprises two long α helices connected by a short loop. The first helix includes the basic domain and interacts with the major groove of the DNA [45]. All presented structures allowed an insight into the organization of bHLH–PAS proteins; however, the structure of the multidomain bHLH–PAS protein was still missing.

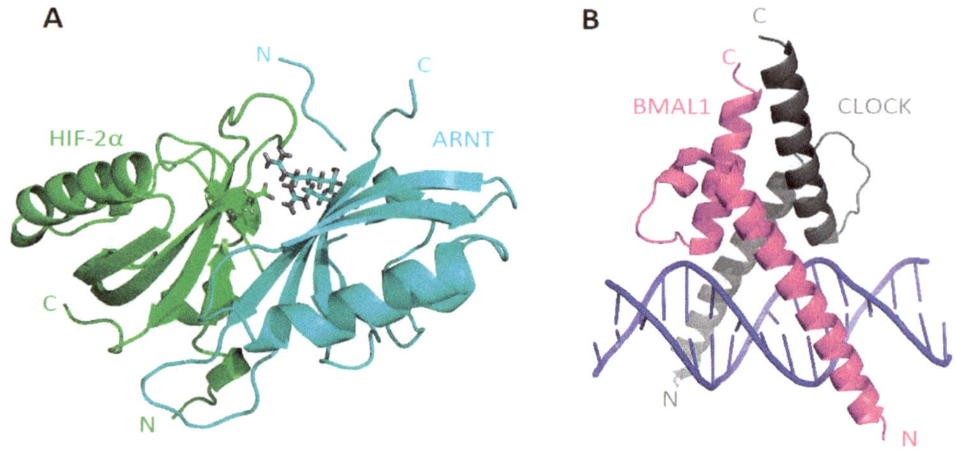

Figure 3. (**A**) HIF2-α PAS-B (green) and ARNT PAS-B (blue) heterodimer (3F1P, [46]). Amino acids creating a salt bridge are marked (HIF2-α E247, ARNT R362, ARNT R379). (**B**) BMAL1 bHLH (magenta) and CLOCK bHLH (grey) domains with E-box DNA (blue) (1H10, [44]).

A turning point was the year 2012, when the first heterodimer comprising the bHLH–PAS-A/PAS-B domains (CLOCK-BMAL1) was crystallized [47] and its structure was resolved (Figure 4A). In 2015, the architecture of two other heterodimers, HIF1-α-ARNT (not shown) and HIF2-α-ARNT (Figure 4B), were obtained [48]. All determined structures present the position of the defined domains in relation to each other in the functional heterodimers. In general, the individual PAS domains are not involved in equal interactions, and the obtained structures are highly asymmetric. Importantly, two groups of heterodimers (based on BMAL-1 or ARNT proteins as a dimerization partner) present separate types of quaternary architecture. All domains in the BMAL-1 group are close spatially to each other (Figure 4A), while ARNT domains do not create intramolecular interactions and can wrap up around a partner protein (Figure 4B) [22,48].

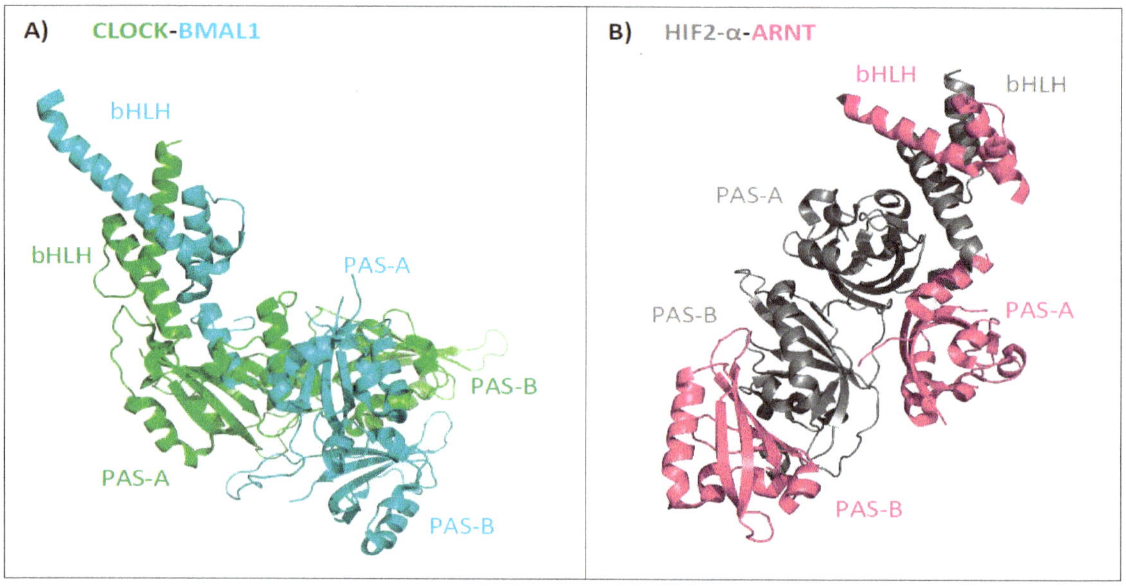

Figure 4. Representatives of the two groups of the bHLH–PAS heterodimers. (**A**) Overall structure of the CLOCK-BMAL1 heterodimer (4f3l, [47]), (**B**) overall structure of the HIF2-α–ARNT heterodimer (4zp4, [48]).

To date, all available structural information concerns mammalian bHLH–PAS proteins. There is almost no information about the structure of proteins derived from other organisms, including invertebrate *D. melanogaster*. It would be interesting to verify evolutionary conservation of the entire bHLH–PAS fold in different organisms. The majority of reported protein structures include defined N-terminal domains, while the structural information about C-terminal regions is still missing and limited to short peptides bound to interacting proteins [22]. An example is a short motif featuring a conserved sequence LIXXL found in *D. melanogaster* MET and GCE, which represents a novel nuclear receptor (NR) box. The Docking models of the MET/GCE NR box associating peptides to the orphan nuclear receptor (FTZ-F1) ligand-binding domain (LBD) revealed their α-helical structure, necessary for hydrophobic interaction [49].

4. Unique Properties of the C-Terminal Domains of bHLH–PAS Proteins as IDRs

While the N-terminal part of bHLH–PAS proteins is responsible for interactions with DNA, ligands/cofactors binding, and heterodimerization, their C-termini are usually responsible for the regulation of the protein and the activity of created complexes [50]. The variability of the amino acid sequence of C-terminal fragments, their transactivation role, and the lack of homology to any described domains prompted us to ask the question about the structural character of these regions and the relationship of their character with the performed function. For a long time, it was believed that spontaneous folding into a well-defined and stable tertiary structure is required for the protein action [51]. However, it is actually known that more than 20–30% of eukaryotic proteins do not have a stable tertiary structure in physiological conditions, but at the same time still perform important biological functions. Such proteins are referred to as intrinsically disordered proteins (IDPs). Simultaneously, over 70% of proteins involved in signal transduction cascades have long intrinsically disordered regions (IDRs). Importantly, the lack of a defined structure is critical for the functionality of IDPs and IDRs [52]. Additionally, the conformational plasticity and elongated shape make them a frequent target of different kinds of post-translational modifications (phosphorylation, acetylation, methylation, and others) that regulate protein activity [53]. IDPs were identified as elements of cellular signaling which control mechanisms and protein interaction networks [54]. IDPs were also shown to take part in disease-related signaling transduction; for example, intrinsically disordered amyloid β-peptides are involved in Alzheimer's disease [55]. Therefore, IDPs can be seen to be targets for drug design strategies.

4.1. *In Silico Analyses of Selected bHLH–PAS Proteins*

To estimate the occurrence of putative IDRs in bHLH–PAS proteins, we performed in silico analyses of the composition, hydropathy, and sequence complexity of amino acid sequences corresponding to selected proteins. We used the previously described human SIM1 and SIM2, as well as their *D. melanogaster* ortholog, SIM (Figure 5A), representing the class I of the family. To obtain a wider spectrum, we studied other human class I members, AHR, HIF1-α, and CLOCK (Figure 5B), which are engaged in different signal transduction pathways. As mentioned previously, class I proteins dimerize with class II proteins to form a functional complex and are crucial for heterodimer specificity. As each bHLH–PAS class II transcription factor is able to interact with different class I members, we found it to be extremely interesting to perform in silico analysis of the structure of class II members. We chose human ARNT, human BMAL1 (Figure 5C), and, additionally, *D. melanogaster* MET (Figure 5C) as a unique protein with an unknown mammalian homolog. MET can be classified as a class II bHLH–PAS family member based on its ability to not only create heterodimers with its paralog GCE, but also homodimers [56].

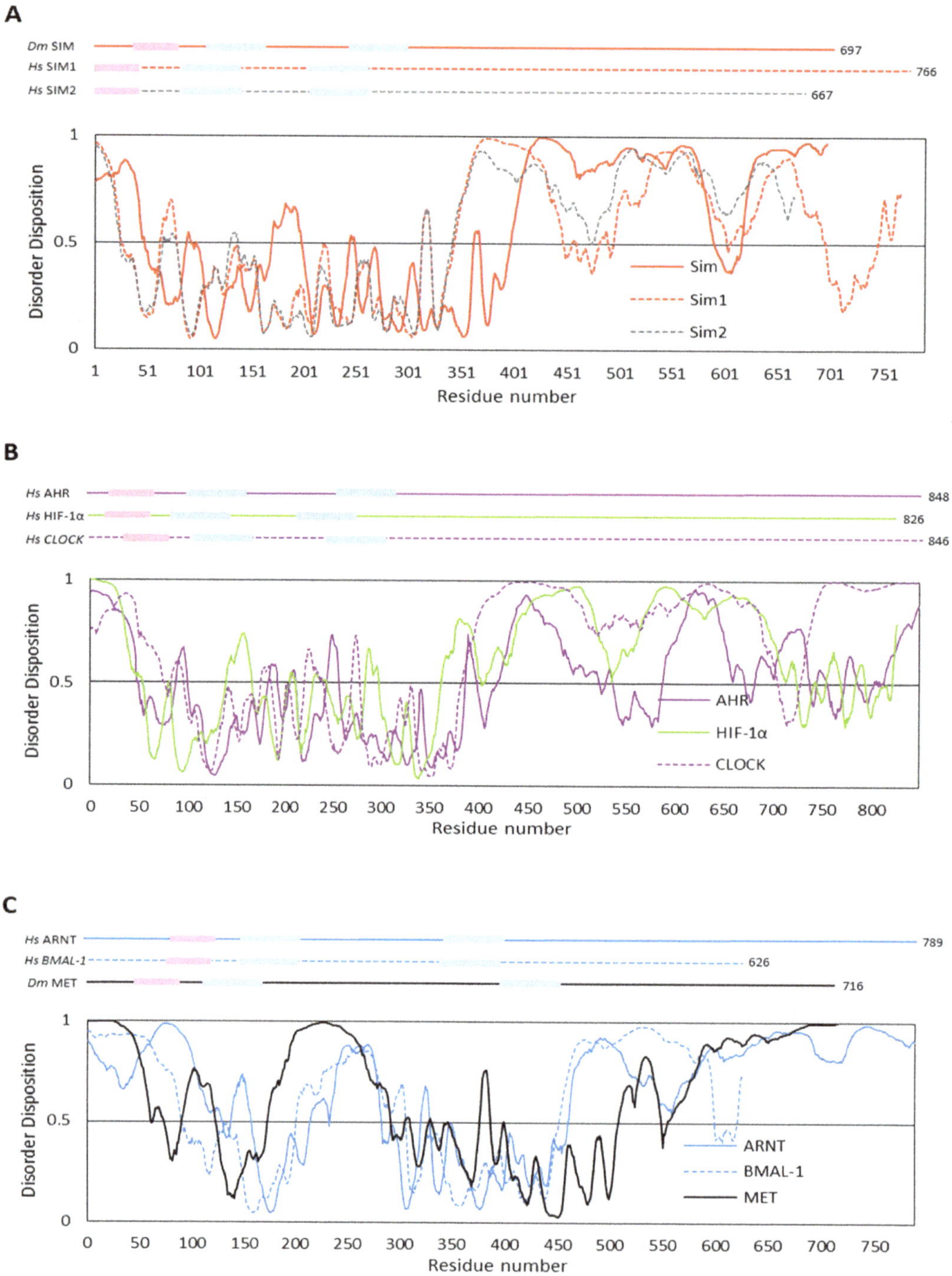

Figure 5. Prediction of intrinsically disordered regions. The top panel presents the domain structure of the analyzed bHLH–PAS proteins. Pink indicates the bHLH domain, whereas blue represents PAS domains. The length of the proteins is marked. The bottom panel presents a prediction of intrinsically disordered regions based on the amino acid sequence of proteins. All calculations were performed using PONDR-VLS2 software [57]. A score over 0.5 indicates disorder. (**A**) The class I proteins: *D. melanogaster* SIM (red line) and its H. sapiens orthologs SIM1 (dashed red line) and SIM2 (dashed grey line). (**B**) The class I proteins: H. sapiens AHR (violet line), HIF1-α (green line), and CLOCK (violet dashed line). (**C**) The class II proteins: H. sapiens ortholog ARNT (blue line), BMAL-1 (blue dashed line), and *D. melanogaster* Met (black line).

We performed in silico analysis using the predictors of intrinsically disordered regions: PONDR-VSL2 [57], PONDR-FIT [58], IUPred [59], and IsUnstruct [60]. Since the results of all the employed predictors were compatible, for the purpose of simplicity, we decided to show only one representative result (PONDR-VSL2) for each protein (Figure 5). All the results (Figure 5) substantiate our hypothesis and indicate the intrinsic character of the long C-termini. It is worth noting that for proteins representing class I (see Figure 5A,B), short ordered fragments in their C-termini are visible. Such fragments are able to act as TAD/RPD or so-called molecular recognition elements (MoREs) [61,62]. The presence of MoREs makes the interactions between partner proteins highly specific and reversible [52]. The presented results revealed some subtle differences in the regions that comprise preserved domains. The structure predicted for class I proteins (Figure 5A,B) is undeniably more ordered, while class II proteins show a marked structure relaxation in their middle part (Figure 5C), which is C-terminally linked to the PAS-A domain responsible for specificity of gene activation by bHLH–PAS proteins [63]. Such a difference explains the ability of class II proteins to serve as an interaction partner for different proteins [22]. The ability of IDRs (and IDPs) to interact with several partners is an undeniable advantage in molecular recognition processes [64]. Importantly, the resulting induced folding may differ depending on the binding partner. For example, a disordered region of p53 protein, a known cell cycle regulator and a tumor suppressor [65], folds into alpha helix or beta strand, depending on the partner protein [66].

4.2. *The Impact of Disordered Regions on Protein Function*

The flexibility and disorder detected in individual C-termini can be related to the ability of individual bHLH–PAS proteins to perform diverse functions. The differences between SIM1 and SIM2 C-termini, regarding their opposite functions (gene activation/repression), have previously been described. C-terminal regions of two other studied proteins, AHR (class I member) and ARNT (class II member), are characterized by the presence of TADs [67], in which functions are mediated by CBP/p300 and RIP140 coactivators. The C-terminal region of ARNT was additionally proposed to be a crucial activator of the estrogen receptor (ER) [68]. Interestingly, the suppression of AHR activity is also connected with the C-terminus and is mediated by the binding of the small peptide inhibitor [69]. Another repressor of the AHR signaling pathway, AHRR, is distinguished from AHR by the presence of three SUMOylation sites in its C-terminus. As shown, SUMOylation is crucial for full suppressive activity of AHRR [70].

Moreover, the C-terminus of another studied protein, HIF1-α, is characterized by the presence of TADs, and it also interacts with the CBP/p300 coactivator. The C-terminus is additionally responsible for protein stability/degradation and contains sequence motifs influencing subcellular localization: nuclear localization signal (NLS) and nuclear export signal (NES) [71,72].

Another remarkable class I bHLH–PAS protein is CLOCK, comprising a domain with histone transacetylase (HAT) activity in the C-terminus. This domain is responsible for histones acetylation, which affects the transcriptional stimulation of clock-controlled genes. Additional acetylation is performed on the R537 residue of the partner protein, BMAL1. R537 residue is located in the C-terminal part of BMAL1 and its modification facilitates the cryptochrome (CRY1)-mediated repression of specific gene transcription [73]. Importantly, CRY1 competes with the CBP/p300 coactivator for BMAL1 TAD binding, and is not able to bind the C-terminus in the paralog protein, BMAL2. Therefore, C-termini distinguish the circadian functions of these two BMAL paralogs [74].

4.3. *Structural Analysis of bHLH–PAS C-Terminal Fragments*

To date, the only structurally characterized C-terminal fragment of the bHLH–PAS protein is the *D. melanogaster* MET C-terminus (MET/C) [75]. It was shown by a series of in vitro analyses that MET/C exhibits a highly disordered character and exists in a solution in extended flexible form with predispositions for conformational changes. It is interesting to note that some short secondary motifs in the structure of MET/C have been predicted. Such short ordered fragments can be important during partner recognition and interactions. It was hypothesized that the intrinsic disorder of the C-terminal fragment was indispensable for the functionality of MET due to it modulating the protein's action in a context-specific way. It enables cross-talk between JH signaling and other signaling pathways during *D. melanogaster* development. Previously, it was shown that Met interacts with FTZ-F1 by its C-terminus [49], thereby modulating stage-specific responses to the hormones during *D. melanogaster* metamorphosis [76].

As all the in vitro analyses results obtained for the MET/C [75] were consistent with the in silico studies presented above (Figure 5C), we hypothesize that the disorder character of the bHLH–PAS proteins subfamily C-terminal fragments can be a more common characteristic and also be very important for their functionality. Previously, the importance of the disordered character of regions flanking the bHLH domain of bHLH transcription factors was shown [77–79].

4.4. Structural Analysis of IDPs

While C-terminal regions of the bHLH–PAS family are considered as IDRs, it can be challenging to detect and characterize them. The reason is that IDPs and IDRs do not adopt a single stable structure and the energetically most favorable conformations can be very distinguished [80,81]. The tiny conformational changes can promote IDPs/IDRs aggregation [82]. Additionally, it was shown that IDPs/IDRs can be highly sensitive to proteolysis [83]. Currently, studies focused on the characterization of IDRs and IDPs are rapidly developing, and techniques enabling the study of proteins in solution are still improving.

There are a number of bioinformatics tools allowing primary recognition of disordered proteins. Since IDPs are characterized by the specific aa composition (a low content of hydrophobic and a high content of charged residues [84]), the Composition Profiler [85] is commonly used to compare aa distribution between the studied protein and IDPs (DisProt3.4 database)/globular proteins (PDB S25 database). Additionally, for IDPs and globular proteins distinguishing, the Uversky diagram plotting mean net charge versus mean hydrophobicity is useful [86]. Disorder predictors (like PONDR-VSL2 [57], PONDR-FIT [58], IUPred [59], and IsUnstruct [60] used in this work) allow determining the probability of IDR occurrence utilizing the neural networks, trained on selected sets of ordered and disordered sequences. Another predictor, DynaMine, provides information about protein backbone flexibility [87,88]. IDPs, once purified, can be identified by various experimental methods. First, the underestimated mobility during SDS-PAGE electrophoresis can indicate the extended and elongated shape of the protein [75]. Hydrodynamic analysis comprising Size Exclusion Chromatography (SEC) [89] and Analytical Ultracentrifugation (AUC) are commonly used to determine hydrodynamic properties, like the Stokes radius (R_S), the sedimentation coefficients (s), and the frictional ratios (f/f_0) [90]. The Circular Dichroism (CD) is useful for secondary structures content calculation [91]. All listed techniques allow obtaining preliminary insight into protein structure properties.

One technique commonly used to study the overall shape and structural transitions of biological macromolecules in solution is small-angle X-ray scattering (SAXS) [92]. However, SAXS only provides limited information about the low-resolution overall shape of the molecule, so it is important to combine it with complementary high-resolution methods like NMR that present the local structure [93]. NMR offers unique opportunities that are based on analyzing the deviations from an idealized random coil devoid of any structural propensity [94]. The random coil exhibits characteristic chemical shifts, which are averages of all the possible conformations that amino acids can adopt in a solution. Therefore, NMR chemical shift deviations from random coil values can be used to evaluate the local transient secondary structure of IDPs [80]. The main problem during spectra assignments of IDPs is spectra overlapping (low chemical shifts dispersion) and a significant proton exchange with bulk water that reduces $^1H^N$ signal intensities, which in turn leads to low signal-to-noise ratios [94]. The exchange with water can be reduced by conducting measurements in low temperature or low pH [95]. Low-resolution spectra require the development of a novel NMR technique. Recently, IDP-dedicated methods such as ^{13}C-direct detected experiments, paramagnetic relaxation enhancements (PREs), or residual dipolar couplings (RDCs) have been described [96].

5. Conclusions

The available structure characterization of bHLH–PAS proteins is limited to the relatively well-conserved domains bHLH, PAS-A, and PAS-B. Importantly, all structures deposited in the Protein Data Bank are obtained for mammalian family members, the majority of them being heterodimers. On the other hand, the important parts of bHLH–PAS factors, which comprise their long C-termini, have not yet been structurally characterized. These fragments perform important functions in the specific modulation of protein action and for the recognition of interacting partners.

Performed in silico analysis revealed that the C-termini of representatives of the class I bHLH–PAS protein family members (SIM, SIM1, SIM2, AHR, HIF1-α, and Clock), and also class II (ARNT, BMAL1, and MET), are predicted as intrinsically disordered regions (IDRs) and are not homologous to any described domains. We discussed the known functions of the presented C-termini proteins according to their disorder character. Moreover, we proposed NMR techniques for intrinsically disordered C-termini characterization [94]. We believe that the structural properties of subsequent IDRs predicted in the sequences of bHLH–PAS transcription factors (mainly C-termini) need to be resolved for a full understanding of the way of bHLH–PAS family transcription factors function.

Acknowledgments: We would like to thank Andrzej Ożyhar (Department of Biochemistry, Faculty of Chemistry, Wroclaw University of Science and Technology) for all valuable comments and suggestions.

Abbreviations

Aa	amino acid
AHR	aryl hydrocarbon receptor
AHRR	aryl hydrocarbon receptor repressor
ARNT	aryl hydrocarbon receptor nuclear translocator
bHLH–PAS	helix–loop–helix/Per-ARNT-SIM
CLOCK	circadian locomotor output cycles protein kaput
CRY1	cryptochrome
DYS	dysfusion protein
ER	estrogen receptor
GCE	germ cell-expressed protein
HAT	histone transacetylase
HIF	hypoxia-inducible factors
IDPs	intrinsically disordered proteins
IDRs	intrinsically disordered regions
JH	juvenile hormone
LBD	ligand-binding domain
MET	methoprene-tolerant protein
MoREs	molecular recognition elements
MET/C	C-terminus of the MET protein
NES	nuclear export signal
NLS	nuclear localization signal
NMR	Nuclear Resonance Magnetism
NPAS	neuronal PAS domain-containing proteins
NR	boxes nuclear receptor boxes
PAC	PAS-associated C-terminal motif
PDB	Protein Data Bank
PREs	paramagnetic relaxation enhancements
RDCs	residual dipolar couplings
SAXS	Small-angle X-ray Scattering
SIM	single-minded proteins

SIMA	similar protein
SS	spineless protein
TAD or TRD	transactivation or repression domain
TGO	TANGO protein
TRH	trachealess protein

References

1. Crews, S.T. Control of cell lineage-specific development and transcription by bHLH-PAS proteins. *Genes Dev.* **1998**, *12*, 607–620. [CrossRef] [PubMed]
2. Lee, J.-W.; Bae, S.-H.; Jeong, J.-W.; Kim, S.-H.; Kim, K.-W. Hypoxia-inducible factor (HIF-1)α: Its protein stability and biological functions. *Exp. Mol. Med.* **2004**, *36*, 1–12. [CrossRef] [PubMed]
3. Ema, M.; Hirota, K.; Mimura, J.; Abe, H.; Yodoi, J.; Sogawa, K.; Poellinger, L.; Fujii-Kuriyama, Y. Molecular mechanisms of transcription activation by HLF and HIF1alpha in response to hypoxia: Their stabilization and redox signal-induced interaction with CBP/p300. *EMBO J.* **1999**, *18*, 1905–1914. [CrossRef] [PubMed]
4. Petrulis, J.R.; Kusnadi, A.; Ramadoss, P.; Hollingshead, B.; Perdew, G.H. The hsp90 Co-chaperone XAP2 Alters Importin β Recognition of the Bipartite Nuclear Localization Signal of the Ah Receptor and Represses Transcriptional Activity. *J. Biol. Chem.* **2003**, *278*, 2677–2685. [CrossRef] [PubMed]
5. Xue, P.; Fu, J.; Zhou, Y. The Aryl Hydrocarbon Receptor and Tumor Immunity. *Front. Immunol.* **2018**, *9*, 286. [CrossRef] [PubMed]
6. Neavin, D.; Liu, D.; Ray, B.; Weinshilboum, R. The Role of the Aryl Hydrocarbon Receptor (AHR) in Immune and Inflammatory Diseases. *Int. J. Mol. Sci.* **2018**, *19*, 3851. [CrossRef]
7. Freer, S.M.; Lau, D.C.; Pearson, J.C.; Talsky, K.B.; Crews, S.T. Molecular and functional analysis of Drosophila single-minded larval central brain expression. *Gene Expr. Patterns* **2011**, *11*, 533–546. [CrossRef] [PubMed]
8. Pielage, J.; Steffes, G.; Lau, D.C.; Parente, B.A.; Crews, S.T.; Strauss, R.; Klämbt, C. Novel behavioral and developmental defects associated with Drosophila single-minded. *Dev. Biol.* **2002**, *249*, 283–299. [CrossRef]
9. Chen, H.; Chrast, R.; Rossier, C.; Gos, A.; Antonarakis, S.E.; Kudoh, J.; Yamaki, A.; Shindoh, N.; Maeda, H.; Minoshima, S.; et al. Single–minded and Down syndrome? *Nat. Genet.* **1995**, *10*, 9–10. [CrossRef]
10. Bersten, D.C.; Sullivan, A.E.; Peet, D.J.; Whitelaw, M.L. bHLH–PAS proteins in cancer. *Nat. Rev. Cancer* **2013**, *13*, 827–841. [CrossRef]
11. Safe, S.; Lee, S.O.; Jin, U.H. Role of the aryl hydrocarbon receptor in carcinogenesis and potential as a drug target. *Toxicol. Sci.* **2013**, 1–16. [CrossRef] [PubMed]
12. Masoud, G.N.; Li, W. HIF-1α pathway: Role, regulation and intervention for cancer therapy. *Acta Pharm. Sin. B* **2015**, *5*, 378–389. [CrossRef] [PubMed]
13. Beamer, C.A.; Shepherd, D.M. Role of the aryl hydrocarbon receptor (AhR) in lung inflammation. *Semin. Immunopathol.* **2013**, *35*, 693–704. [CrossRef] [PubMed]
14. Puccetti, M.; Paolicelli, G.; Oikonomou, V.; De Luca, A.; Renga, G.; Borghi, M.; Pariano, M.; Stincardini, C.; Scaringi, L.; Giovagnoli, S.; et al. Towards Targeting the Aryl Hydrocarbon Receptor in Cystic Fibrosis. *Mediat. Inflamm.* **2018**, *2018*, 1601486. [CrossRef] [PubMed]
15. Zhang, Z.; Fei, P.; Mu, J.; Li, W.; Song, J. Hippocampal expression of aryl hydrocarbon receptor nuclear translocator 2 and neuronal PAS domain protein 4 in a rat model of depression. *Neurol. Sci.* **2014**, *35*, 277–282. [CrossRef] [PubMed]
16. Choy, F.C.; Klarić, T.S.; Koblar, S.A.; Lewis, M.D. The Role of the Neuroprotective Factor Npas4 in Cerebral Ischemia. *Int. J. Mol. Sci.* **2015**, *16*, 29011–29028. [CrossRef] [PubMed]
17. Sabatini, P.V.; Lynn, F.C. All-encomPASsing regulation of β-cells: PAS domain proteins in β-cell dysfunction and diabetes. *Trends Endocrinol. Metab.* **2015**, *26*, 49–57. [CrossRef]
18. Speckmann, T.; Sabatini, P.V.; Nian, C.; Smith, R.G.; Lynn, F.C. Npas4 Transcription Factor Expression Is Regulated by Calcium Signaling Pathways and Prevents Tacrolimus-induced Cytotoxicity in Pancreatic Beta Cells. *J. Biol. Chem.* **2016**, *291*, 2682–2695. [CrossRef]
19. Li, X.; Duan, X.; Jiang, H.; Sun, Y.; Tang, Y.; Yuan, Z.; Guo, J.; Liang, W.; Chen, L.; Yin, J.; et al. Genome-Wide Analysis of Basic/Helix-Loop-Helix Transcription Factor Family in Rice and Arabidopsis. *Plant Physiol.* **2006**, *141*, 1167–1184. [CrossRef]

20. Ponting, C.P.; Aravind, L. PAS: A multifunctional domain family comes to light. *Curr. Biol.* **1997**, *7*, R674–R677. [CrossRef]
21. Kewley, R.J.; Whitelaw, M.L.; Chapman-Smith, A. The mammalian basic helix–loop–helix/PAS family of transcriptional regulators. *Int. J. Biochem. Cell Biol.* **2004**, *36*, 189–204. [CrossRef]
22. Wu, D.; Rastinejad, F. Structural characterization of mammalian bHLH-PAS transcription factors. *Curr. Opin. Struct. Biol.* **2017**, *43*, 1–9. [CrossRef] [PubMed]
23. Partch, C.L.; Gardner, K.H. Coactivator recruitment, a new role for PAS domains in transcriptional regulation by the bHLH-PAS family. *J. Cell. Physiol.* **2010**, *223*, 553–557. [CrossRef] [PubMed]
24. Woods, S.L.; Whitelaw, M.L. Differential Activities of Murine Single Minded 1 (SIM1) and SIM2 on a Hypoxic Response Element. *J. Biol. Chem.* **2002**, *277*, 10236–10243. [CrossRef] [PubMed]
25. Moffett, P.; Reece, M.; Pelletier, J. The murine Sim-2 gene product inhibits transcription by active repression and functional interference. *Mol. Cell. Biol.* **1997**, *17*, 4933–4947. [CrossRef] [PubMed]
26. Moffett, P.; Pelletier, J. Different transcriptional properties of mSim-1 and mSim-2. *FEBS Lett.* **2000**, *466*, 80–86. [CrossRef]
27. Vorrink, S.U.; Domann, F.E. Regulatory crosstalk and interference between the xenobiotic and hypoxia sensing pathways at the AhR-ARNT-HIF1α signaling node. *Chem. Biol. Interact.* **2014**, *218*, 82–88. [CrossRef] [PubMed]
28. Tal, R.; Shaish, A.; Bangio, L.; Peled, M.; Breitbart, E.; Harats, D. Activation of C-transactivation domain is essential for optimal HIF-1α-mediated transcriptional and angiogenic effects. *Microvasc. Res.* **2008**, *76*, 1–6. [CrossRef] [PubMed]
29. Fukunaga, B.N.; Probst, M.R.; Reisz-Porszasz, S.; Hankinson, O. Identification of functional domains of the aryl hydrocarbon receptor. *J. Biol Chem.* **1995**, *270*, 29270–29278. [CrossRef]
30. Fribourgh, J.L.; Partch, C.L. Assembly and function of bHLH-PAS complexes. *Proc. Natl. Acad. Sci. USA* **2017**, *114*, 5330–5332. [CrossRef]
31. Hoffman, E.C.; Reyes, H.; Chu, F.F.; Sander, F.; Conley, L.H.; Brooks, B.A.; Hankinson, O. Cloning of a factor required for activity of the Ah (dioxin) receptor. *Science* **1991**, *252*, 954–958. [CrossRef] [PubMed]
32. Andreasen, E.A.; Spitsbergen, J.M.; Tanguay, R.L.; Stegeman, J.J.; Heideman, W.; Peterson, R.E. Tissue-Specific Expression of AHR2, ARNT2, and CYP1A in Zebrafish Embryos and Larvae: Effects of Developmental Stage and 2,3,7,8-Tetrachlorodibenzo-p-dioxin Exposure. *Toxicol. Sci.* **2002**, *68*, 403–419. [CrossRef] [PubMed]
33. Crews, S.T.; Fan, C.-M. Remembrance of things PAS: Regulation of development by bHLH–PAS proteins. *Curr. Opin. Genet. Dev.* **1999**, *9*, 580–587. [CrossRef]
34. Sonnenfeld, M.; Ward, M.; Nystrom, G.; Mosher, J.; Stahl, S.; Crews, S. The Drosophila tango gene encodes a bHLH-PAS protein that is orthologous to mammalian Arnt and controls CNS midline and tracheal development. *Development* **1997**, *124*, 4571–4582. [PubMed]
35. Panda, S.; Hogenesch, J.B.; Kay, S.A. Circadian rhythms from flies to human. *Nature* **2002**, *417*, 329–335. [CrossRef] [PubMed]
36. Paranjpe, D.A.; Sharma, V.K. Evolution of temporal order in living organisms. *J. Circadian Rhythm.* **2005**, *3*, 7. [CrossRef]
37. Li, K.-L.; Lu, T.-M.; Yu, J.-K. Genome-wide survey and expression analysis of the bHLH-PAS genes in the amphioxus Branchiostoma floridae reveal both conserved and diverged expression patterns between cephalochordates and vertebrates. *Evodevo* **2014**, *5*, 20. [CrossRef] [PubMed]
38. Cheng, D.; Meng, M.; Peng, J.; Qian, W.; Kang, L.; Xia, Q. Genome-wide comparison of genes involved in the biosynthesis, metabolism, and signaling of juvenile hormone between silkworm and other insects. *Genet. Mol. Biol.* **2014**, *37*, 444–459. [CrossRef] [PubMed]
39. Abdou, M.; Peng, C.; Huang, J.; Zyaan, O.; Wang, S.; Li, S.; Wang, J. Wnt Signaling Cross-Talks with JH Signaling by Suppressing Met and gce Expression. *PLoS ONE* **2011**, *6*, e26772. [CrossRef]
40. Jindra, M.; Bellés, X.; Shinoda, T. Molecular basis of juvenile hormone signaling. *Curr. Opin. Insect. Sci.* **2015**, *11*, 39–46. [CrossRef]
41. Erbel, P.J.A.; Card, P.B.; Karakuzu, O.; Bruick, R.K.; Gardner, K.H. Structural basis for PAS domain heterodimerization in the basic helix-loop-helix-PAS transcription factor hypoxia-inducible factor. *Proc. Natl. Acad. Sci. USA* **2003**, *100*, 15504–15509. [CrossRef] [PubMed]
42. Card, P.B.; Erbel, P.J.A.; Gardner, K.H. Structural Basis of ARNT PAS-B Dimerization: Use of a Common

Beta-sheet Interface for Hetero- and Homodimerization. *J. Mol. Biol.* **2005**, *353*, 664–677. [CrossRef] [PubMed]
43. Henry, J.T.; Crosson, S. Ligand-binding PAS domains in a genomic, cellular, and structural context. *Annu. Rev. Microbiol.* **2011**, *65*, 261–286. [CrossRef] [PubMed]
44. Wang, Z.; Wu, Y.; Li, L.; Su, X.-D. Intermolecular recognition revealed by the complex structure of human CLOCK-BMAL1 basic helix-loop-helix domains with E-box DNA. *Cell Res.* **2013**, *23*, 213. [CrossRef] [PubMed]
45. Jones, S. An overview of the basic helix-loop-helix proteins. *Genome Biol.* **2004**, *5*, 226. [CrossRef] [PubMed]
46. Scheuermann, T.H.; Tomchick, D.R.; Machius, M.; Guo, Y.; Bruick, R.K.; Gardner, K.H. Artificial ligand binding within the HIF2 PAS-B domain of the HIF2 transcription factor. *Proc. Natl. Acad. Sci. USA* **2009**, *106*, 450–455. [CrossRef] [PubMed]
47. Huang, N.; Chelliah, Y.; Shan, Y.; Taylor, C.A.; Yoo, S.-H.; Partch, C.; Green, C.B.; Zhang, H.; Takahashi, J.S. Crystal structure of the heterodimeric CLOCK:BMAL1 transcriptional activator complex. *Science* **2012**, *337*, 189–194. [CrossRef] [PubMed]
48. Wu, D.; Potluri, N.; Lu, J.; Kim, Y.; Rastinejad, F. Structural integration in hypoxia-inducible factors. *Nature* **2015**, *524*, 303–308. [CrossRef]
49. Bernardo, T.J.; Dubrovsky, E.B. The Drosophila juvenile hormone receptor candidates methoprene-tolerant (MET) and germ cell-expressed (GCE) utilize a conserved LIXXL motif to bind the FTZ-F1 nuclear receptor. *J. Biol. Chem.* **2012**, *287*, 7821–7833. [CrossRef]
50. Furness, S.G.B.; Lees, M.J.; Whitelaw, M.L. The dioxin (aryl hydrocarbon) receptor as a model for adaptive responses of bHLH/PAS transcription factors. *FEBS Lett.* **2007**, *581*, 3616–3625. [CrossRef]
51. Mirsky, A.E.; Pauling, L. On the Structure of Native, Denatured, and Coagulated Proteins. *Proc. Natl. Acad. Sci. USA* **1936**, *22*, 439–447. [CrossRef] [PubMed]
52. Tompa, P. Intrinsically unstructured proteins. *Trends Biochem. Sci.* **2002**, *27*, 527–533. [CrossRef]
53. Wright, P.E.; Dyson, H.J. Intrinsically disordered proteins in cellular signalling and regulation. *Nat. Rev. Mol. Cell Biol.* **2014**, *16*, 18–29. [CrossRef] [PubMed]
54. Uversky, V.N. Flexible Nets of Malleable Guardians, Intrinsically Disordered Chaperones in Neurodegenerative Diseases. *Chem. Rev.* **2011**, *111*, 1134–1166. [CrossRef] [PubMed]
55. Ball, K.A.; Wemmer, D.E.; Head-Gordon, T. Comparison of Structure Determination Methods for Intrinsically Disordered Amyloid-β Peptides. *J. Phys. Chem. B* **2014**, *118*, 6405–6416. [CrossRef] [PubMed]
56. Godlewski, J.; Wang, S.; Wilson, T.G. Interaction of bHLH-PAS proteins involved in juvenile hormone reception in Drosophila. *Biochem. Biophys. Res. Commun.* **2006**, *342*, 1305–1311. [CrossRef] [PubMed]
57. Li, X.; Romero, P.; Rani, M.; Dunker, A.K.; Obradovic, Z. Predicting Protein Disorder for N-, C-, and Internal Regions. *Genome Inform.* **1999**, *10*, 30–40.
58. Xue, B.; Dunbrack, R.L.; Williams, R.W.; Dunker, A.K.; Uversky, V.N. PONDR-FIT: A meta-predictor of intrinsically disordered amino acids. *Biochim. Biophys. Acta* **2010**, *1804*, 996–1010. [CrossRef]
59. Dosztanyi, Z.; Csizmok, V.; Tompa, P.; Simon, I. IUPred: Web server for the prediction of intrinsically unstructured regions of proteins based on estimated energy content. *Bioinformatics* **2005**, *21*, 3433–3434. [CrossRef]
60. Lobanov, M.Y.; Sokolovskiy, I.V.; Galzitskaya, O.V. IsUnstruct: Prediction of the residue status to be ordered or disordered in the protein chain by a method based on the Ising model. *J. Biomol. Struct. Dyn.* **2013**, *31*, 1034–1043. [CrossRef]
61. Dziedzic-Letka, A.; Ozyhar, A. Intrinsically disordered proteins. *Postepy Biochem.* **2012**, *58*, 100–109. [PubMed]
62. Vacic, V.; Oldfield, C.J.; Mohan, A.; Radivojac, P.; Cortese, M.S.; Uversky, V.N.; Dunker, A.K. Characterization of Molecular Recognition Features, MoRFs, and Their Binding Partners. *J. Proteome Res.* **2007**, *6*, 2351–2366. [CrossRef] [PubMed]
63. Zelzer, E.; Wappner, P.; Shilo, B.-Z. The PAS domain confers target gene specificity of Drosophila bHLH/PAS proteins. *Genes Dev.* **1997**, *11*, 2079–2089. [CrossRef] [PubMed]
64. Uversky, V.N. Natively unfolded proteins: A point where biology waits for physics. *Protein Sci.* **2002**, *11*, 739–756. [CrossRef] [PubMed]
65. Bell, S.; Klein, C.; Müller, L.; Hansen, S.; Buchner, J. p53 contains large unstructured regions in its native state. *J. Mol. Biol.* **2002**, *322*, 917–927. [CrossRef]

66. Riley, T.; Sontag, E.; Chen, P.; Levine, A. Transcriptional control of human p53-regulated genes. *Nat. Rev. Mol. Cell Biol.* **2008**, *9*, 402–412. [CrossRef] [PubMed]
67. Jain, S.; Dolwick, K.M.; Schmidt, J.V.; Bradfield, C.A. Potent transactivation domains of the Ah receptor and the Ah receptor nuclear translocator map to their carboxyl termini. *J. Biol. Chem.* **1994**, *269*, 31518–31524. [PubMed]
68. Brunnberg, S.; Pettersson, K.; Rydin, E.; Matthews, J.; Hanberg, A.; Pongratz, I. The basic helix-loop-helix-PAS protein ARNT functions as a potent coactivator of estrogen receptor-dependent transcription. *Proc. Natl. Acad. Sci. USA* **2003**, *100*, 6517–6522. [CrossRef]
69. Ren, L.; Thompson, J.D.; Cheung, M.; Ngo, K.; Sung, S.; Leong, S.; Chan, W.K. Selective suppression of the human aryl hydrocarbon receptor function can be mediated through binding interference at the C-terminal half of the receptor. *Biochem. Pharmacol.* **2016**, *107*, 91–100. [CrossRef]
70. Kawajiri, K.; Fujii-Kuriyama, Y. The aryl hydrocarbon receptor: A multifunctional chemical sensor for host defense and homeostatic maintenance. *Exp. Anim.* **2017**, *66*, 75–89. [CrossRef]
71. Kallio, P.J.; Okamoto, K.; Brien, S.O.; Carrero, P.; Makino, Y.; Tanaka, H.; Poellinger, L. Signal transduction in hypoxic cells: Inducible nuclear translocation and recruitment of the CBP/p300 coactivator by the hypoxia-inducible factor-1 α. *EMBO J.* **1998**, *17*, 6573–6586. [CrossRef] [PubMed]
72. Mylonis, I.; Chachami, G.; Paraskeva, E.; Simos, G. Atypical CRM1-dependent nuclear export signal mediates regulation of hypoxia-inducible factor-1alpha by MAPK. *J. Biol. Chem.* **2008**, *283*, 27620–27627. [CrossRef] [PubMed]
73. Doi, M.; Hirayama, J.; Sassone-Corsi, P. Circadian regulator CLOCK is a histone acetyltransferase. *Cell* **2006**, *125*, 497–508. [CrossRef] [PubMed]
74. Xu, H.; Gustafson, C.L.; Sammons, P.J.; Khan, S.K.; Parsley, N.C.; Ramanathan, C.; Lee, H.-W.; Liu, A.C.; Partch, C.L. Cryptochrome 1 regulates the circadian clock through dynamic interactions with the BMAL1 C terminus. *Nat. Struct. Mol. Biol.* **2015**, *22*, 476–484. [CrossRef] [PubMed]
75. Kolonko, M.; Ożga, K.; Hołubowicz, R.; Taube, M.; Kozak, M.; Ożyhar, A.; Greb-Markiewicz, B. Intrinsic Disorder of the C-Terminal Domain of Drosophila Methoprene-Tolerant Protein. *PLoS ONE* **2016**, *11*, e0162950. [CrossRef] [PubMed]
76. Broadus, J.; McCabe, J.R.; Endrizzi, B.; Thummel, C.S.; Woodard, C.T. The Drosophila beta FTZ-F1 orphan nuclear receptor provides competence for stage-specific responses to the steroid hormone ecdysone. *Mol. Cell* **1999**, *3*, 143–149. [CrossRef]
77. Fuxreiter, M.; Simon, I.; Bondos, S. Dynamic protein–DNA recognition: Beyond what can be seen. *Trends Biochem. Sci.* **2011**, *36*, 415–423. [CrossRef]
78. Guo, X.; Bulyk, M.L.; Hartemink, A.J. Intrinsic disorder within and flanking the DNA-binding domains of human transcription factors. *Biocomputing* **2012**, 104–115. [CrossRef]
79. Roschger, C.; Schubert, M.; Regl, C.; Andosch, A.; Marquez, A.; Berger, T.; Huber, C.G.; Lütz-Meindl, U.; Cabrele, C. The Recombinant Inhibitor of DNA Binding Id2 Forms Multimeric Structures via the Helix-Loop-Helix Domain and the Nuclear Export Signal. *Int. J. Mol. Sci.* **2018**, *19*, 1105. [CrossRef]
80. Eliezer, D. Biophysical characterization of intrinsically disordered proteins. *Curr. Opin. Struct. Biol.* **2009**, *19*, 23–30. [CrossRef]
81. Jensen, M.R.; Zweckstetter, M.; Huang, J.; Blackledge, M. Exploring Free-Energy Landscapes of Intrinsically Disordered Proteins at Atomic Resolution Using NMR Spectroscopy. *Chem. Rev.* **2014**, *114*, 6632–6660. [CrossRef] [PubMed]
82. Levine, Z.A.; Larini, L.; LaPointe, N.E.; Feinstein, S.C.; Shea, J.-E. Regulation and aggregation of intrinsically disordered peptides. *Proc. Natl. Acad. Sci. USA* **2015**, *112*, 2758–2763. [CrossRef] [PubMed]
83. Tolkatchev, D.; Plamondon, J.; Gingras, R.; Su, Z.; Ni, F. Recombinant Production of Intrinsically Disordered Proteins for Biophysical and Structural Characterization. In *Instrumental Analysis of Intrinsically Disordered Proteins*; John Wiley & Sons, Inc.: Hoboken, NJ, USA, 2010; pp. 653–670.
84. Dunker, A.K.; Lawson, J.D.; Brown, C.J.; Williams, R.M.; Romero, P.; Oh, J.S.; Oldfield, C.J.; Campen, A.M.; Ratliff, C.M.; Hipps, K.W.; et al. Intrinsically disordered protein. *J. Mol. Graph. Model.* **2001**, *19*, 26–59. [CrossRef]
85. Vacic, V.; Uversky, V.N.; Dunker, A.K.; Lonardi, S. Composition Profiler: A tool for discovery and visualization of amino acid composition differences. *BMC Bioinform.* **2007**, *8*, 211. [CrossRef] [PubMed]

86. Uversky, V.N.; Gillespie, J.R.; Fink, A.L. Why are "natively unfolded" proteins unstructured under physiologic conditions? *Proteins* **2000**, *41*, 415–427.
87. Cilia, E.; Pancsa, R.; Tompa, P.; Lenaerts, T.; Vranken, W.F. From protein sequence to dynamics and disorder with DynaMine. *Nat. Commun.* **2013**, *4*. [CrossRef] [PubMed]
88. Cilia, E.; Pancsa, R.; Tompa, P.; Lenaerts, T.; Vranken, W.F. The DynaMine webserver: Predicting protein dynamics from sequence. *Nucleic Acids Res.* **2014**, *42*, W264–W270. [CrossRef]
89. Uversky, V.N. Size-exclusion chromatography in structural analysis of intrinsically disordered proteins. *Methods Mol. Biol.* **2012**, *896*, 179–194. [CrossRef]
90. Schuck, P. Size-distribution analysis of macromolecules by sedimentation velocity ultracentrifugation and lamm equation modeling. *Biophys. J.* **2000**, *78*, 1606–1619. [CrossRef]
91. Greenfield, N.; Fasman, G.D. Computed circular dichroism spectra for the evaluation of protein conformation. *Biochemistry* **1969**, *8*, 4108–4116. [CrossRef]
92. Kikhney, A.G.; Svergun, D.I. A practical guide to small angle X-ray scattering (SAXS) of flexible and intrinsically disordered proteins. *FEBS Lett.* **2015**, *589*, 2570–2577. [CrossRef] [PubMed]
93. Kachala, M.; Valentini, E.; Svergun, D.I. Application of SAXS for the Structural Characterization of IDPs. In *Intrinsically Disordered Proteins Studied by NMR Spectroscopy*; Springer: Cham, Switzerland, 2015; pp. 261–289.
94. Konrat, R. NMR contributions to structural dynamics studies of intrinsically disordered proteins. *J. Magn. Reson.* **2014**, *241*, 74–85. [CrossRef] [PubMed]
95. Solyom, Z.; Schwarten, M.; Geist, L.; Konrat, R.; Willbold, D.; Brutscher, B. BEST-TROSY experiments for time-efficient sequential resonance assignment of large disordered proteins. *J. Biomol. NMR* **2013**, *55*, 311–321. [CrossRef] [PubMed]
96. Kosol, S.; Contreras-Martos, S.; Cedeño, C.; Tompa, P. Structural Characterization of Intrinsically Disordered Proteins by NMR Spectroscopy. *Molecules* **2013**, *18*, 10802–10828. [CrossRef] [PubMed]

2

Modeling of Disordered Protein Structures using Monte Carlo Simulations and Knowledge-Based Statistical Force Fields

Maciej Pawel Ciemny [1,2], Aleksandra Elzbieta Badaczewska-Dawid [1], Monika Pikuzinska [1], Andrzej Kolinski [1] and Sebastian Kmiecik [1,*]

1 Faculty of Chemistry, Biological and Chemical Research Center, University of Warsaw, Pasteura 1, 02-093 Warsaw, Poland; maciej.ciemny@fuw.edu.pl (M.P.C.); adawid@chem.uw.edu.pl (A.E.B.-D.); m.pikuzinska@student.uw.edu.pl (M.P.); kolinski@chem.uw.edu.pl (A.K.)

2 Faculty of Physics, University of Warsaw, Pasteura 5, 02-093 Warsaw, Poland

* Correspondence: sekmi@chem.uw.edu.pl

Abstract: The description of protein disordered states is important for understanding protein folding mechanisms and their functions. In this short review, we briefly describe a simulation approach to modeling protein interactions, which involve disordered peptide partners or intrinsically disordered protein regions, and unfolded states of globular proteins. It is based on the CABS coarse-grained protein model that uses a Monte Carlo (MC) sampling scheme and a knowledge-based statistical force field. We review several case studies showing that description of protein disordered states resulting from CABS simulations is consistent with experimental data. The case studies comprise investigations of protein–peptide binding and protein folding processes. The CABS model has been recently made available as the simulation engine of multiscale modeling tools enabling studies of protein–peptide docking and protein flexibility. Those tools offer customization of the modeling process, driving the conformational search using distance restraints, reconstruction of selected models to all-atom resolution, and simulation of large protein systems in a reasonable computational time. Therefore, CABS can be combined in integrative modeling pipelines incorporating experimental data and other modeling tools of various resolution.

Keywords: coarse-grained; CABS model; MC simulations; statistical force fields; disordered protein; protein structure

1. Introduction

There is a growing body of evidence that some proteins act in multiple structural states [1]. It has been demonstrated that the ability of these proteins to switch between distinct structural states may be crucial for their function and regulation [1]. Additionally, a number of key biological functions have been proven to be performed by disordered or partially unstructured proteins [2]. Some proteins fold and obtain their structure only upon binding to their partners, while others form so called "fuzzy complexes" in which both proteins retain a certain degree of disorder [3]. These discoveries modified the core biochemistry principle of "structure determines function". As for now, a consensus has been reached that protein function may be a result of an interplay between protein structure and its dynamics [4,5].

Internal protein motions may be studied both experimentally and with computational methods [6,7]. For example, nuclear magnetic resonance (NMR) spectroscopy is one of the richest sources of information on protein structure and dynamics, especially when accompanied with assisting methods that enhance resolution or provide an additional insight into the dynamics of structures [8]. This approach, however, results in an averaged image of the structural ensemble.

A variety of computational techniques have been developed to assist these challenging experimental studies [7,9]. In the last decades, molecular modeling was dominated by structure-based models or Go-like models (approaches that are biased toward known folded conformations [10,11]). These indeed lead to significant speedup of simulations but may result for example in an unrealistic picture of protein folding, which in reality may also depend on non-native interactions [12–14].

Recent works show that methods combining experimental data and computational approaches may produce the most promising pictures of protein equilibrium dynamics [15,16]. However, the development of these methods poses a number of challenges—both in terms of the validity of the approach and its computationally efficient implementation [17].

Molecular dynamics (MD) has been so far the most widespread computational method for the investigation of protein motions [18]. However, standard all-atom MD implementations are limited to sub-microsecond timescales and may suffer from limited sampling despite recent significant advances in code optimization and hardware [19]. To overcome this problem various MD extensions have been proposed. These extensions include for example replica-exchange MD, meta-dynamics, Markov state models and simulated annealing algorithms [6,20–23].

A number of non-MD sampling methods have also been developed to provide a comprehensive image of protein dynamics using limited computational resources. Of these, Monte Carlo (MC) is perhaps the most commonly used and generally applicable sampling method [11]. Monte Carlo randomly generates conformations and uses an energy-based acceptance criterion that promotes pseudo-trajectory convergence to an energetic minimum. On the expense of losing a direct image of the timescales or kinetics of the ensemble, MC manages to overcome some of the major limitations of MD [24].

Aside from the sampling method, a further extension of effective timescales is possible by using a simplified representation of protein structures to reduce the number of a system's degrees of freedom. The accuracy of the available coarse-grained (CG) models may vary from detailed, almost atomistic representations (Primo [25], Rosetta [26]), medium resolution models (in which a single amino acid is represented by three to five beads: UNRES [27], CABS [28], AWSEM [29], MARTINI [30], PaLaCe [31]), and Scorpion [32]) to significantly simplified models like SURPASS [33,34]. Applications and implementations of these and other CG models are described in detail in a recent review [11].

In addition to the representation and sampling method, the choice of the force field to perform the simulation determines the success of modeling. Traditionally, force fields are divided into two main groups: physics-based, which involve (usually pairwise) interaction terms [35], and those employing a statistical approach; however, most of the successful approaches are usually a mixture of the two. A statistical force field is constructed using the probability of a chosen observable (or a set of observables) in a given ensemble of structures [36]. Early attempts focused on straightforward pairwise contacts [37]; however, with further development, more complex observables were analyzed. This resulted in a generation of knowledge-based force fields, or scores, for various representations, coarse-grained and all-atom: CABS [28], Rosetta [38], DOPE [39], GOAP [40], QUARK [41], Bcl::Score [42] or BACH [36]. Newly developed approaches go a step further and improve the results by combining these methods with experimental data [43,44]. An example of such approach is RosettaEPR [45], which includes distance data from site-directed spin labeling electron paramagnetic resonance experiments. It is generally agreed that statistical force fields frequently allow more accurate scoring than physics-based potentials [11]. The combination of knowledge-based force fields or scores with effective sampling schemes seems to be a promising approach to a number of problems [11], such as protein structure prediction [43,44,46,47], investigation of protein interactions [48] or studies of protein dynamics [17,49–51].

This review briefly describes one of these approaches: an MC-based and knowledge-based interaction scheme for modeling protein–peptide interactions and unfolded states of globular proteins using the CABS coarse-grained protein model. Firstly, the main features of the CABS method will be described, with a focus on their applicability for modeling disordered or unfolded proteins or their fragments. Subsequently, representative case studies will be discussed to provide detailed insights into the modeling results obtained for systems characterized by a varying level of disorder.

2. CABS Dynamics and Interaction Model

Since its development, the CABS model (C-alpha, C-beta and Side chain model) has been applied to a variety of modeling problems, such as protein folding mechanisms [49,50,52–57], protein structure prediction [58–61], protein–peptide docking including large-scale conformational flexibility [62–68] and simulations of near-native fluctuations of globular proteins [69–73]. When combined with careful bioinformatics selection of the generated models, CABS proved to be one of the two most accurate structure prediction tools evaluated in the CASP (Critical Assessment of protein Structure Prediction) experiment [60]. The CABS model uses up to four atoms or pseudo-atoms per residue (see the description below), but outputs protein systems in C-alpha representation only. Therefore, for practical applications, the obtained models need to be reconstructed to all-atom representation. In various multiscale modeling tools discussed below, CABS has been integrated with the MODELLER-based reconstruction procedure [74]. Other reconstruction scenarios are also possible to ensure the best possible quality of local protein structure. This can be realized by combination of different tools for protein backbone reconstruction from the C-alpha trace and side chain reconstruction, like BBQ [75] or SCWRL [76] for example, and optionally further refinement [77].

In this review, we discuss the applicability of the CABS CG model and its knowledge-based statistical force field [28] to the modeling of disordered or unfolded protein states. In the CABS model the polypeptide chain representation is reduced to up to four unified atoms per residue (see Figure 1). These interaction centers represent lattice-confined C-alpha atoms, C-beta atoms, the united side chain pseudo-atom, and additionally, pseudo-atoms representing geometrical centers of peptide bonds needed to define the hydrogen pseudo-bond. An example of a polypeptide chain in CABS representation is presented in Figure 1b. Even though the restriction of the C-alpha trace to the underlying low spacing (0.61 Å [28]) cubic lattice may appear to be a drastic simplification, it is not. Allowing small fluctuations of the C-alpha, C-alpha distance enables hundreds of possible orientations of this pseudo bond, and thereby the resulting model chains do not show any noticeable directional biases. Furthermore, the averaged resolution of the C-alpha traces is acceptable and below 0.5 Å [28]. Additionally, the lattice representation enables pre-calculation of local moves and corresponding changes of interactions, leading to a few times faster simulations in comparison with otherwise equivalent continuous space CG models [11].

The CABS model uses a knowledge-based statistical force field that consists of generic, sequence-independent interaction terms that favor protein-like conformations, and sequence-dependent interaction terms that determine some structural details [11,28,78]. The generic force field terms are derived from general features of polypeptide chains that result in protein-like behavior of the model chains. They account for properties of protein chains such as local stiffness, their biases toward secondary structures and packing compactness. The residue–residue interaction terms are derived from contact geometry statistics derived from folded globular proteins (illustrated in Figure 2a). Nevertheless, the local packing regularities in unfolded states appear to be very similar to that observed in native structures [11,28,33]. Thereby, CABS simulations provided correct pictures of protein folding [49,52–56,60] and flexibility of globular proteins [70,71].

The resulting force field takes a form of a precomputed matrix of contact pseudo-energies, presented schematically in Figure 2b. Additionally, to allow successful modeling of membrane proteins the CABS force field can be extended by introducing effective dielectric constant terms [79].

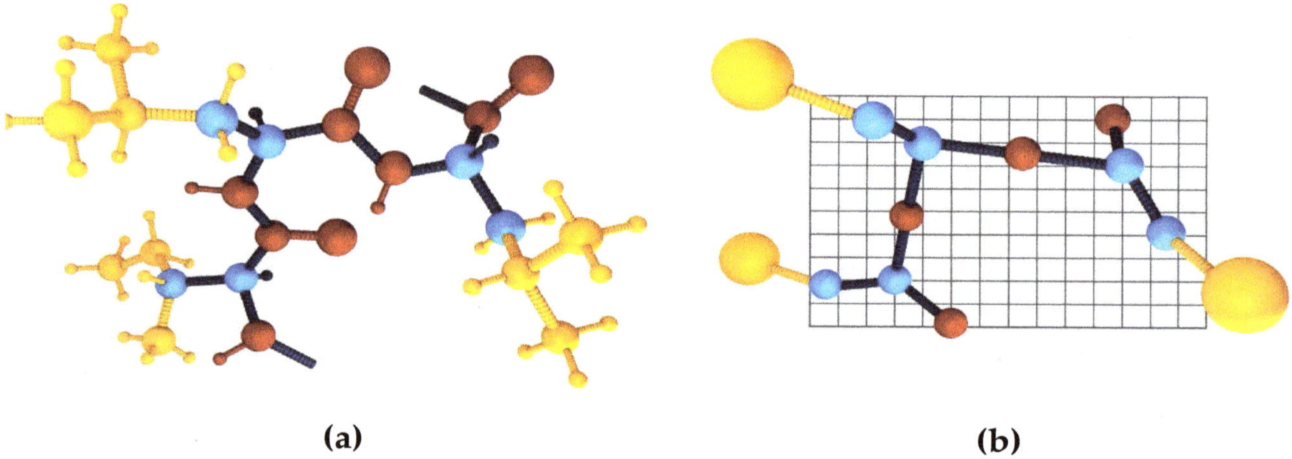

Figure 1. A three-residue protein fragment in: all-atom (**a**) and CABS model (**b**) representation. The spheres represent atoms: blue, C-alpha and C-beta atoms (the same in both representations); yellow, side chain atoms (one pseudo-atom in CABS); red, atoms involved in the peptide bond (one pseudo-atom in CABS placed in the geometric center of the peptide bond. A single slice (layer) of the lattice that confines the C-alpha trace in the CABS model is also presented.

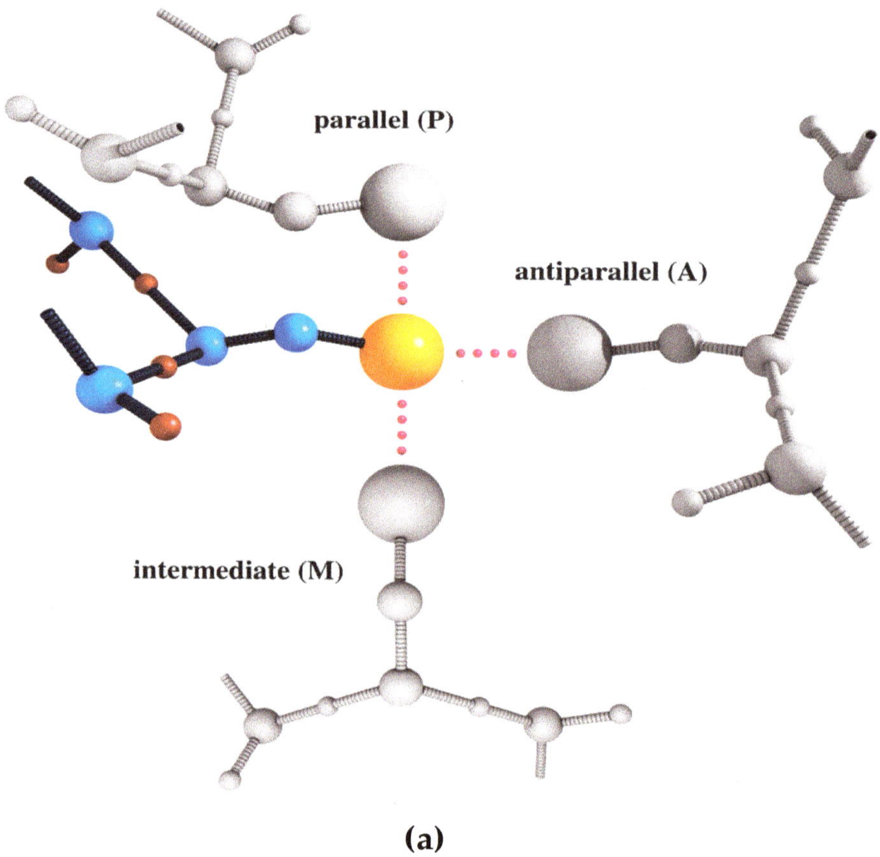

Figure 2. *Cont.*

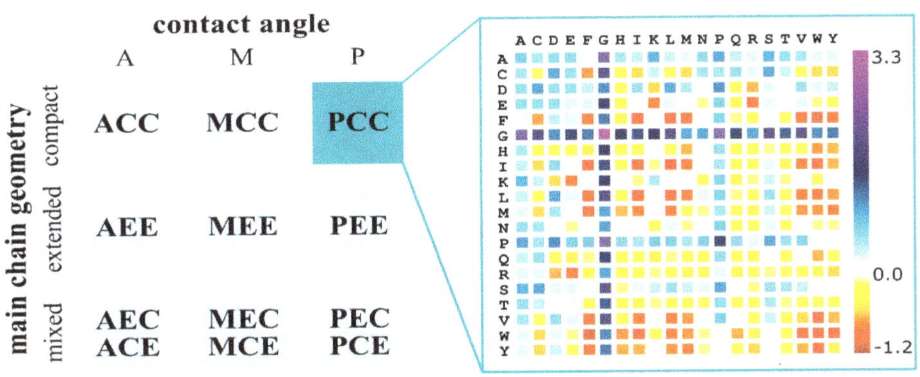

(b)

Figure 2. Key elements of a residue–residue interaction term in the CABS model force field. Panel (**a**) shows three examples of contact geometries in CABS representation: parallel (P), antiparallel (A), and intermediate (M), used to derive contact statistics from experimentally-derived structures of folded globular proteins. Panel (**b**) shows an example matrix of contact energies which depend on the geometry of the contacting pair, main chain geometry (compact (C) or extended (E)) for both amino acids (left part of the panel), and also on the amino acid identities (right part of the panel, the amino acids are represented using the one-letter code). The PCC matrix is presented which shows interaction energies between residues being in parallel orientation (P), where one residue belongs to a compact type of structure (C) and the second one as well (C).

The main difference between CABS and other statistical force fields used in CG models of similar resolution [11] is the context and orientation dependence of side chain interaction pseudo-energy that encodes characteristic patterns observed in globular proteins. For instance, the oppositely charged side chains in single globules mostly contact in an almost parallel fashion (usually on the surface of a globule), while the antiparallel contacts (usually in the buried regions of the protein globule) are very rare. Therefore, in the context dependent force field these antiparallel contacts of oppositely charged residues are treated as repulsive. This way, the CABS force field implicitly incorporates information on the complicated interaction patterns with the solvent (via contact statistics) and its entropic contribution to system thermodynamics [11,28].

Using the mean-force force field derived from folded proteins to simulations of less-structured systems raises justified questions about the validity of this approach in studies of the disordered protein regions. The folding events observed in simulations performed using the CABS force field are consistent with both the experimental data and all-atom MD simulations [49,52,80,81]. Thus, it is hypothesized that unstructured (unfolded, partially unfolded or intrinsically disordered) proteins to a significant extent share similar stabilizing interaction patterns with the patterns observed for their well-structured counterparts [82,83].

The CABS method uses the MC asymmetric Metropolis sampling scheme that governs a set of local motions as well as multi-residue, small distance moves of the C-alpha atoms (see Figure 3). The method uses a replica exchange algorithm with simulated annealing to enhance the sampling of conformational states. The simulation is organized as a set of nested loops, in which the s number of MC steps are organized into the y number of MC cycles, and these in the a number of annealing cycles. Each of the MC steps consists of a per-set number of attempts to perform each of the five standard precomputed moves. The available motions and the details of implementation of the sampling scheme are presented in Figure 3.

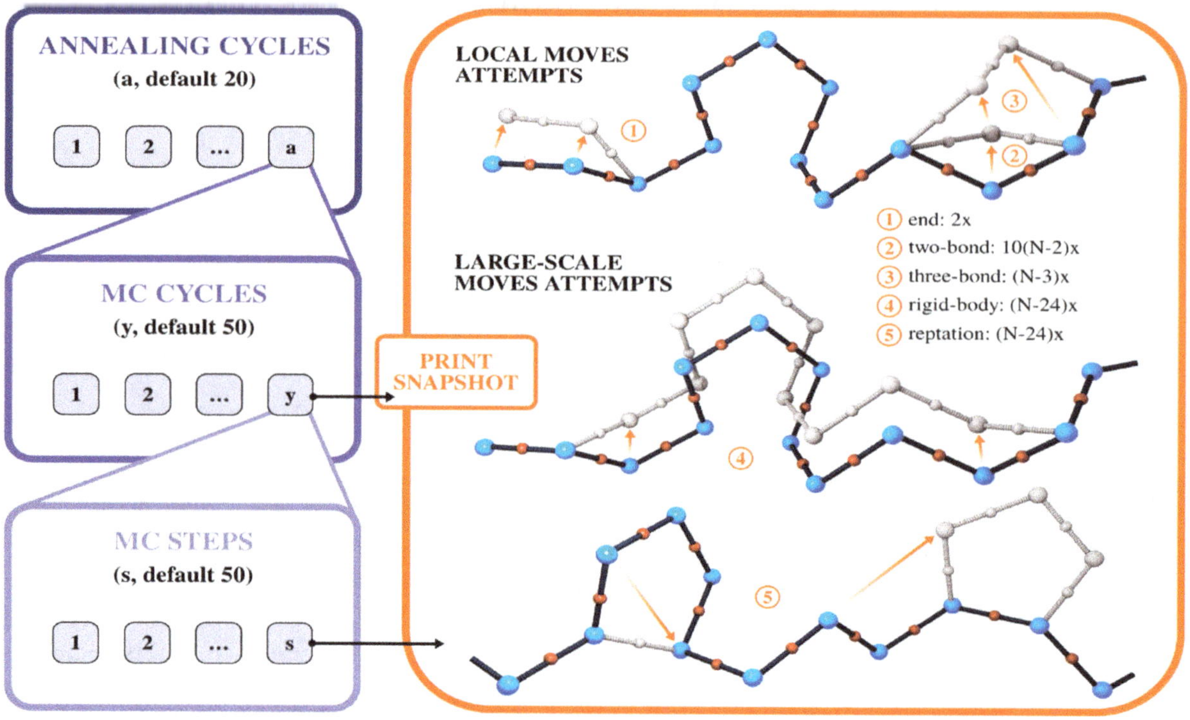

Figure 3. Sampling scheme of the CABS model. Blue panels show implementation details of Monte Carlo (MC) iterations (loops). The orange panel shows all motions that may be performed in a single MC step. The simulation is organized as a set of nested loops, in which the s number of MC steps is organized into the y number of cycles, and these in a annealing cycles (number of a, y or s cycles can be controlled by the user in CABS-flex and CABS-dock standalone packages [72]). In the orange panel, numbers 1 to 5 denote the available moves, presented together with the number of attempts to perform a move in each of the MC steps. The resulting trajectory is comprised of simulation snapshots saved at the end of each MC cycle.

The combination of the key features of CABS—its representation, force field and the scale of the movements used in the MC scheme—makes it suitable for the investigation of protein pseudo-dynamics. As mentioned above, the fine-grained lattice improves sampling efficiency, achieving effective timescales of milliseconds. As compared with MD, this is a considerably broader time range (in the study of flexibility of folded proteins [71] the CABS dynamics was estimated to be around 6×10^3 cheaper in terms of computational cost than the classical MD). The chosen micro-motions allow (via accumulation over simulation steps) cooperative, large-scale motions. The ensemble of structures produced by the CABS method resembles a dynamic ensemble averaged over the effective timescale. Due to the nature of the method, the picture of local dynamics is distorted (on the level of local moves); however, it may be argued (based on the works mentioned above that compared our simulations with experimental data) that the long-time pseudo-dynamics recovers the realistic picture of protein motions averaged over time.

The timescale of the CABS simulations is not a priori defined and depends on the CABS simulation temperature, due to hidden entropic contributions in the force field, accounting for implicit solvent effects and multi-body interactions encoded in the statistical force field. Nevertheless, the effective timescale of MC dynamics can be approximately identified by comparison with MD trajectories from sufficiently long simulations. This comparison was thoroughly discussed previously, and the results were compared to MD results [69] and NMR ensembles [71].

The CABS model is presently used as a simulation engine of a few multiscale modeling tools that merge CABS with models reconstruction to all-atom resolution. Those include the CABS-dock method for flexible protein-peptide docking (available as a web server [62] at http://biocomp.chem.uw.edu.pl/CABSdock and a standalone application [84] at https://bitbucket.org/lcbio/cabsdock/)

(accessed on 30 January 2019). In comparison to other protein–peptide docking tools, reviewed recently [85], CABS-dock offers a unique opportunity for modeling large-scale rearrangements of protein receptor structure during on-the-fly docking of fully flexible peptides. Another CABS-based tool, CABS-flex, enables fast simulations of protein flexibility (available as a web server [73] at http://biocomp.chem.uw.edu.pl/CABSflex and a standalone application [72] at https://bitbucket.org/lcbio/cabsflex/, accessed on 30 January 2019). This approach has been also incorporated as the module in the Aggrescan3D method for prediction of protein aggregation properties (available as a web server [86] at http://biocomp.chem.uw.edu.pl/A3D and a standalone application at https://bitbucket.org/lcbio/aggrescan3D, accessed on 30 January 2019). By using CABS-flex predictions, Aggrescan3D enables predicting the impact of protein conformational fluctuations on aggregation properties. Finally, the CABS model is used in the CABS-fold method for protein structure prediction: in the de novo fashion (from an amino acid sequence only), guided by user-provided templates or user-provided distance restraints (available as a web server [58] at http://biocomp.chem.uw.edu.pl/CABSfold/, accessed on 30 January 2019). The access to CABS-based tools, together with the tools description, is also available from websites of the laboratories: http://biocomp.chem.uw.edu.pl/ and http://lcbio.pl/ (accessed on 30 January 2019).

3. CABS Applications to Simulation of Disordered or Unfolded Proteins

In this section, we review CABS applications to simulations of protein–peptide binding (Section 3.1) and folding of globular proteins (Section 3.2). We briefly discuss modeling results for the binding of three protein–peptide systems and protein folding of one protein system. Figure 4 shows native conformations of these systems determined by X-ray crystallography or NMR. In the figure, they are arranged according to the size of a fully flexible fragment of the modeled system, effective timescales required for a meaningful simulation of their motions, and thus the modeling difficulty: (1) modeling of FxxLF motif peptide docking to an androgen receptor (AR), (2) investigation of binding and folding of an unstructured pKID protein to KIX protein, (3) modeling of p53-derived peptide docking to the MDM2 protein receptor with partially unstructured regions, and (4) simulation of the de novo folding of barnase. The simulations were performed using the CABS-dock method for protein–peptide docking [62] and CABS-flex methodology [72,73] that enable running de novo folding simulations.

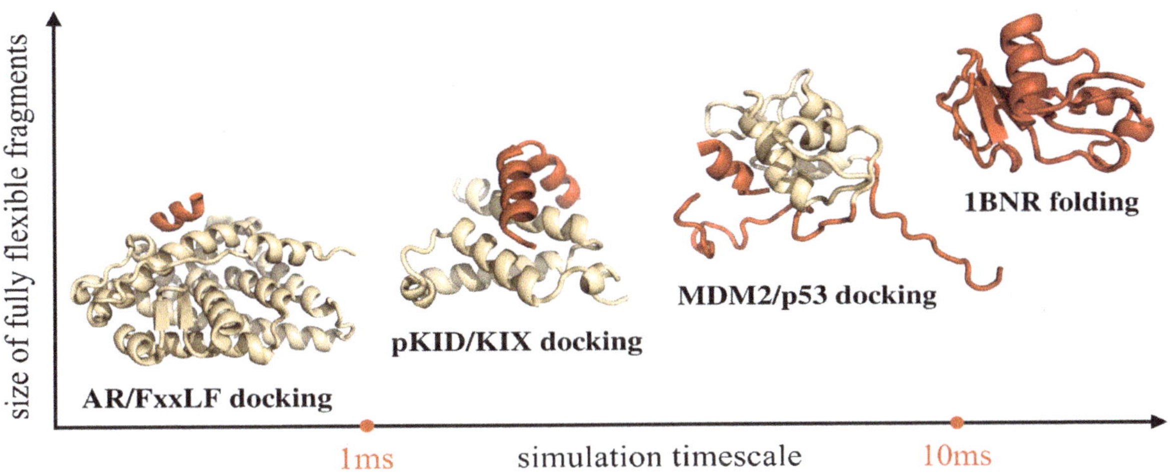

Figure 4. Presentation of the modeling cases discussed in this work. The modeled systems are arranged according to the size of the fully flexible fragment of the modeled system and the effective timescales required to observe their motions. The regions of the systems that were modeled as fully flexible are marked with red, while the regions in which backbone fluctuations were limited to 1 Å RMSD with beige. The presented millisecond values are approximated up to the order of magnitude.

3.1. Protein–Peptide Binding

The CABS-dock method has been extensively tested using the PeptiDB benchmark set of protein–peptide complexes [62,65,87]. One of the benchmark cases is the androgen receptor ligand binding domain (AR) in complex with a peptide with the FxxLF motif [88] (PDB code: 1T7R). To further analyze the interaction details of this complex, we performed blind global docking (using no knowledge about the binding site and peptide conformation) using CABS-dock [62]. As the input we used information on peptide sequence (incorporating the FxxLF motif: SSRFESLFAGEKESR), peptide secondary structure information assigned by the DSSP method [89] and the structure of the AR protein receptor. In this docking study, the peptide structure was simulated as fully flexible, while fluctuations of the protein receptor were limited to small backbone movements around the input structure (around 1 Å). The docking simulation started from random peptide conformations placed in random positions around the receptor structure. During simulation, the peptide remained unstructured until it was bound to the receptor binding site (Figure 5a). The docking simulations provided a set of high-quality models—the best model was characterized by a peptide-RMSD (root-mean-square deviation) value of 1.97 Å—and contact maps in strong agreement with the experimental data. As expected from the experimentally obtained structures and sequence analysis [88] the FxxLF interaction motif residues were most frequently involved in stabilizing hydrophobic interactions with the receptor. These high-frequency contacts are clearly visible in Figure 5a.

The study of the pKID/KIX system [63] involved performing a folding simulation of an intrinsically disordered protein (pKID) and its binding to a well-structured KIX receptor (Figure 5b). According to the experimental studies, the pKID structure is disordered in its unbound form with a slight propensity toward a helix (for detailed description on how one-dimensional secondary structure information is used in the CABS model see [78]). In the complex with the KIX protein, pKID adopts a characteristic conformation of two perpendicular helices that wrap around the receptor. However, most simulation results for the coupled folding and binding of this system published prior to the CABS-based study used models which biased pKID toward its native conformation (see the discussion in [63]). Using our method for studying this system enabled fully flexible treatment of the pKID protein. The obtained results [63] suggested the binding mechanism that involve two encounter complexes and were in well agreement with the available NMR experimental data. The predicted models presented high fractions of native contacts and allowed identification of residues essential for the binding and stabilization of the complex.

In the simulation of MDM2/p53 binding [64], the most challenging task was to adequately model the flexibility of the relatively long, unstructured regions of the protein receptor in addition to the fully flexible peptide [64,90] (Figure 5c). To provide a detailed insight into MDM2/p53 binding, we performed CABS-dock simulations and captured system behavior in agreement with the experimental data [64]. During the simulation, the flexible N- and C- terminal MDM2 fragments remained significantly disordered. The best resulting model was characterized by a peptide-RMSD value of 2.76 Å and 54% of the native contacts while the top ranked model by 3.74 Å and 60%, respectively. During simulations, we observed ensembles of models in which the peptide adopted different conformations loosely bound to the binding site and models in which the N-terminal highly flexible MDM2 fragment was interacting with the binding site. These findings are in agreement with the experimental data suggesting that p53-MDM2 binding is affected by significant rearrangements of the N-terminal MDM2 fragment (see discussion in [64]).

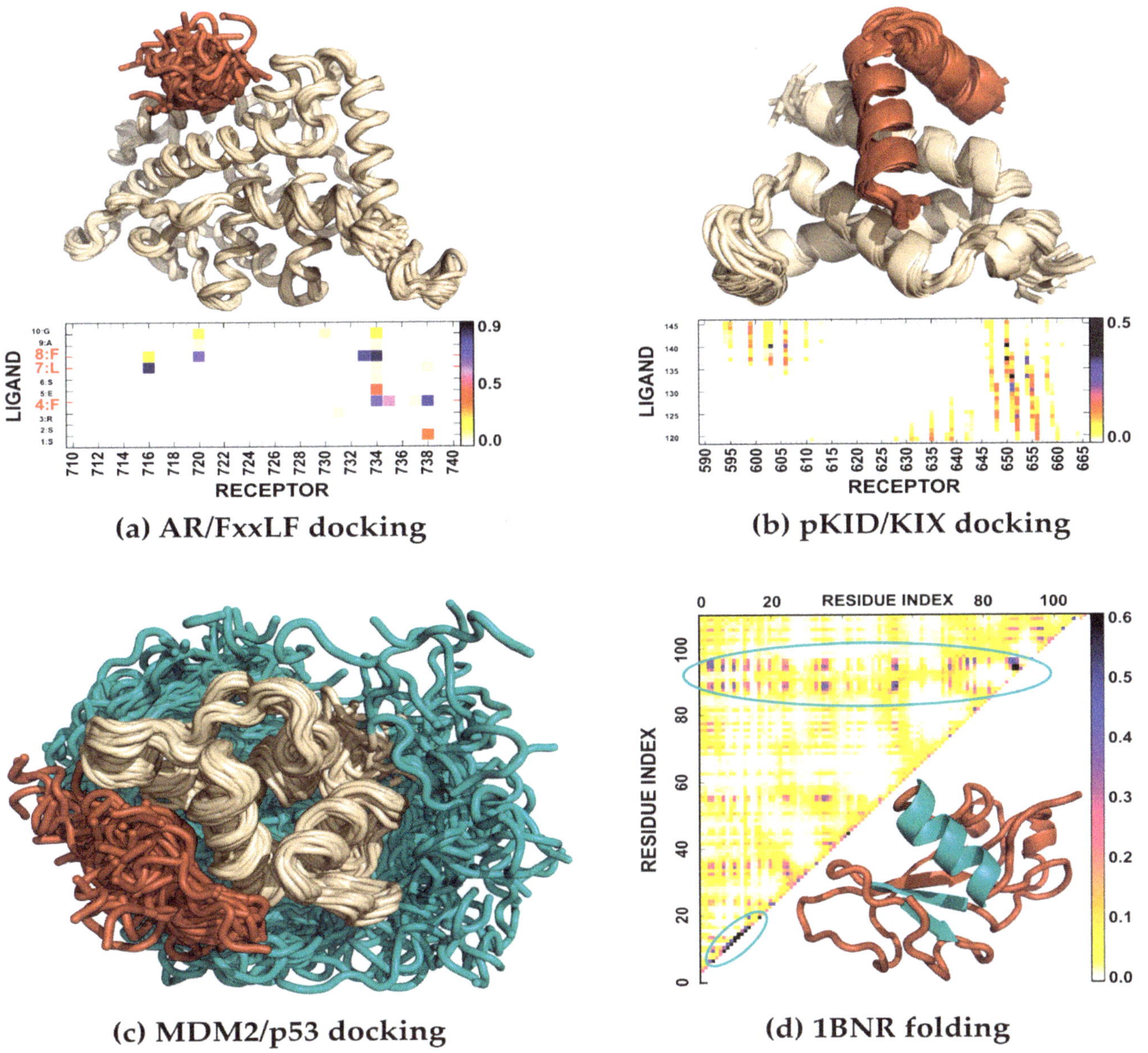

Figure 5. Case studies of modeling disordered or unfolded structures of proteins with CABS-based tools. In the figures, red or cyan marks structure fragments simulated as fully flexible (cyan was used to mark regions of interest discussed in the text), while beige marks regions whose motions were confined to small backbone movements (around 1 Å from the input structure). (**a**) Modeling of the dynamics of a flexible peptide representing the FxxLF motif in the proximity of the binding site of AR protein together with an averaged contact map showing frequency of residue–residue contacts during the docking simulation. (**b**) Modeling of coupled folding and binding of the disordered pKID to the KIX domain [63]; the map presents the frequency of contacts of near-native conformations obtained in the simulation. (**c**) Modeling of p53 peptide binding to the MDM2 receptor [64], which includes fully-flexible regions of the protein receptor (shown in cyan) interacting with a fully-flexible peptide (shown in red). (**d**) Modeling of barnase folding [52] in the de novo fashion (using no knowledge about the structure); the map is a residue–residue contact map showing relative contact frequencies in denaturing conditions; the protein fragments that form the folding nucleation site are colored in cyan in the presented folded structure of barnase.

3.2. Folding and Flexibility of Globular Proteins

The CABS model has been applied to de novo simulations of protein folding (using no knowledge about the protein structure) for several model systems that have been extensively studied by experiment and simulation tools. Those studies include barnase [50,52], chymotrypsin inhibitor [50,52], B1 domain of protein G [49,50], B domain of protein A [53], and others [50,54]. The CABS modeling protocol was also extended to enable studies of the chaperonin effect on the folding mechanism [55]. In these works, various parameters have been studied, including residue–residue contact frequency, radius of gyration, residual secondary structure and others. The obtained pictures, which covered protein dynamics from highly denatured states to ensembles close to the folded states, agreed well with available experimental data.

For example, simulation of barnase folding resulted in the adequate reproduction of the folding pathway in strong agreement with NMR data for denatured states and phi-value analysis [52]. The performed simulations show that barnase folding starts with developing a folding nucleation site that consists of protein fragments corresponding to two strands of a beta sheet and one of the helices in the folded structure (presented in Figure 5d). In addition, the characteristic patterns of hydrophobic interactions that are crucial for the initiation and sustenance of folding are in agreement with the experimental data (see discussion in Reference [52], the contact map resulting from these simulations is presented in Figure 5d).

4. Conclusions

The presented case studies review the applications of the CABS model in simulations of disordered or unfolded protein states. As discussed, the method succeeded in capturing the experimentally determined features of the investigated systems, such as binding site localization, key contacts, peptide hot-spot areas, distinctive conformational states of the system, transient encounter complexes and intermediate states in protein folding [49,52,63,64]. Additionally, CABS enables an investigation of fluctuations of globular proteins around the native (input) structure [69–73].

There is a number of tools commonly used for sampling of disordered protein states, which predictions agree with the experimental studies [91–95]. The CABS method is complementary to these and provides a unique approach allowing for effective modeling both ordered and disordered elements of the system. As observed in many previous studies, these features of CABS method allow for providing accurate pictures of folding pathways [49,52–56,60] and near-native dynamics [70,71]. Obviously, due to its coarse-graining, the geometric details are missed, and their reconstructions is approximate [11,28]. The main distinctive feature of CABS method as compared to the available tools is that the ensemble generation is (pseudo-)energy driven and thus may provide some information on the dynamics on the system. This is not the case in the above-mentioned examples of methods based on random-walk [91,92,95].

On the other hand, CABS force field side-chain interactions escape a clear interpretation, which may be a disadvantage compared to physics-based approaches that allow for straightforward and detailed description of each of the terms [93,94].

It is, however, noteworthy that statistical force fields suffer from inherent limitations, depending on the chosen method of derivation. The most commonly discussed challenges include the transferability, solvent interactions and integration of experimental data. Here, we briefly summarize these topics, a detailed discussion of the limitations of this approach, and possible workarounds may be found in review works [11,17]. The transferability of statistical force fields may be limited as they are applicable always to a certain subset of proteins. Therefore, the performance of knowledge-based approaches may be poor for rare or atypical structures, for which appropriate statistics of contact patterns could not be collected. It should also be noted that interactions with solvent are averaged and treated implicitly, which may lead to significant discrepancies if the method is applied to non-standard solvent conditions (such as extreme pH values). The CABS force field is derived assuming averaged effect solvent conditions for folded globular proteins. Therefore, a subtle effect of small molecules, such as pH, cannot be simulated in a strict fashion, although averaged effects (see modeling the chaperonin effect [55]) can be approximately taken into considerations.

One of the most challenging tasks in modeling protein systems is the effective incorporation of sparse experimental data to drive the modeling procedure. In the CABS model, the experimental data may be readily introduced into the simulation as geometry distance restraints and weighted according to their certainty. A thorough discussion of this possibility is presented in the documentation of CABS-based tools for the fast modeling of protein flexibility and protein–peptide docking [66,72,73]. On a similar basis, CABS simulations can be guided by computational predictions from other sources or integrated with other modeling tools of various resolution. Therefore, the CABS model can be incorporated into integrative modeling pipelines that would benefit from its effective sampling scheme. The recently published standalone application and web server tools are available for integration with external pipelines (access links are presented in the last paragraph of Section 2).

Author Contributions: S.K. and A.K. conceptualized this review. M.P. performed the simulations and analyzed the results for the AR/FxxLF system. The review was written by M.P.C., A.E.B-D., A.K. and S.K.

Abbreviations

CABS	Cα, Cβ, Side chain model
MC	Monte Carlo
NMR	nuclear magnetic resonance
MD	molecular dynamics
CG	coarse-grained
AR	androgen receptor
DSSP	dictionary of protein secondary structure
RMSD	root-mean-square deviation of atomic positions
PDB	Protein Data Bank
CASP	Critical Assessment of protein Structure Prediction

References

1. Dishman, A.F.; Volkman, B.F. Unfolding the Mysteries of Protein Metamorphosis. *ACS Chem. Biol.* **2018**, *13*, 1438–1446. [CrossRef] [PubMed]
2. Uversky, V.N. Dancing protein clouds: The strange biology and chaotic physics of intrinsically disordered proteins. *J. Biol. Chem.* **2016**, *291*, 6681–6688. [CrossRef] [PubMed]
3. Wright, P.E.; Dyson, H.J. Intrinsically disordered proteins in cellular signalling and regulation. *Nat. Rev. Mol. Cell Biol.* **2015**, *16*, 18–29. [CrossRef] [PubMed]
4. Henzler-Wildman, K.; Kern, D. Dynamic personalities of proteins. *Nature* **2007**, *450*, 964–972. [CrossRef] [PubMed]
5. Vendruscolo, M.; Dobson, C.M. Dynamic visions of enzymatic reactions. *Science* **2006**, *313*, 1586–1587. [CrossRef] [PubMed]
6. Wei, G.; Xi, W.; Nussinov, R.; Ma, B. Protein Ensembles: How Does Nature Harness Thermodynamic Fluctuations for Life? the Diverse Functional Roles of Conformational Ensembles in the Cell. *Chem. Rev.* **2016**, *116*, 6516–6551. [CrossRef]
7. Best, R.B. Computational and theoretical advances in studies of intrinsically disordered proteins. *Curr. Opin. Struct. Biol.* **2017**, *42*, 147–154. [CrossRef]
8. Kay, L.E. NMR studies of protein structure and dynamics. *J. Magn. Reson.* **2011**, *213*, 477–491. [CrossRef]
9. Robustelli, P.; Piana, S.; Shaw, D.E. Developing a molecular dynamics force field for both folded and disordered protein states. *Proc. Natl. Acad. Sci. USA* **2018**, *115*, E4758–E4766. [CrossRef]
10. Bowman, G.R.; Voelz, V.A.; Pande, V.S. Taming the complexity of protein folding. *Curr. Opin. Struct. Biol.* **2011**, *21*, 4–11. [CrossRef]

11. Kmiecik, S.; Gront, D.; Kolinski, M.; Wieteska, L.; Dawid, A.E.; Kolinski, A. Coarse-Grained Protein Models and Their Applications. *Chem. Rev.* **2016**, *116*, 7898–7936. [CrossRef] [PubMed]
12. Zhang, Z.; Chan, H.S. Competition between native topology and nonnative interactions in simple and complex folding kinetics of natural and designed proteins. *Proc. Natl. Acad. Sci. USA* **2010**, *107*, 2920–2925. [CrossRef] [PubMed]
13. Shan, B.; Eliezer, D.; Raleigh, D. The unfolded state of the C-terminal domain of the ribosomal protein L9 contains both native and non-native structure. *Biochemistry* **2009**, *48*, 4707–4719. [CrossRef] [PubMed]
14. Rothwarf, D.M.; Scheraga, H.A. Role of non-native aromatic and hydrophobic interactions in the folding of hen egg white lysozyme. *Biochemistry* **1996**, *35*, 13797–13807. [CrossRef] [PubMed]
15. Cavalli, A.; Montalvao, R.W.; Vendruscolo, M. Using chemical shifts to determine structural changes in proteins upon complex formation. *J. Phys. Chem. B* **2011**, *115*, 9491–9494. [CrossRef] [PubMed]
16. Fu, B.; Kukic, P.; Camilloni, C.; Vendruscolo, M. MD Simulations of Intrinsically Disordered Proteins with Replica-Averaged Chemical Shift Restraints. *Biophys. J.* **2014**, *106*, 481a. [CrossRef]
17. Kar, P.; Feig, M. Recent advances in transferable coarse-grained modeling of proteins. *Adv. Protein Chem. Struct. Biol.* **2014**, *96*, 143–180. [CrossRef]
18. Greener, J.G.; Filippis, I.; Sternberg, M.J.E. Predicting Protein Dynamics and Allostery Using Multi-Protein Atomic Distance Constraints. *Structure* **2017**, *25*, 546–558. [CrossRef]
19. Klepeis, J.L.; Lindorff-Larsen, K.; Dror, R.O.; Shaw, D.E. Long-timescale molecular dynamics simulations of protein structure and function. *Curr. Opin. Struct. Biol.* **2009**, *19*, 120–127. [CrossRef]
20. Bernardi, R.C.; Melo, M.C.R.; Schulten, K. Enhanced sampling techniques in molecular dynamics simulations of biological systems. *Biochim. Biophys. Acta - Gen. Subj.* **2015**, *1850*, 872–877. [CrossRef]
21. Shukla, D.; Hernández, C.X.; Weber, J.K.; Pande, V.S. Markov state models provide insights into dynamic modulation of protein function. *Acc. Chem. Res.* **2015**, *48*, 414–422. [CrossRef] [PubMed]
22. Kolinski, A. Toward more efficient simulations of slow processes in large biomolecular systems: Comment on "Ligand diffusion in proteins via enhanced sampling in molecular dynamics" by Jakub Rydzewski and Wieslaw Nowak. *Phys. Life Rev.* **2017**, *22–23*, 75–76. [CrossRef] [PubMed]
23. Rydzewski, J.; Nowak, W. Ligand diffusion in proteins via enhanced sampling in molecular dynamics. *Phys. Life Rev.* **2017**, *22–23*, 82–84. [CrossRef] [PubMed]
24. Maximova, T.; Moffatt, R.; Ma, B.; Nussinov, R.; Shehu, A. Principles and Overview of Sampling Methods for Modeling Macromolecular Structure and Dynamics. *PLoS Comput. Biol.* **2016**, *12*, e1004619. [CrossRef] [PubMed]
25. Hatherley, R.; Brown, D.K.; Glenister, M.; Bishop, Ö.T. PRIMO: An interactive homology modeling pipeline. *PLoS ONE* **2016**, *11*, e0166698. [CrossRef] [PubMed]
26. Das, R.; Baker, D. Macromolecular Modeling with Rosetta. *Annu. Rev. Biochem.* **2008**, *77*, 363–382. [CrossRef] [PubMed]
27. Czaplewski, C.; Karczyńska, A.; Sieradzan, A.K.; Liwo, A. UNRES server for physics-based coarse-grained simulations and prediction of protein structure, dynamics and thermodynamics. *Nucleic Acids Res.* **2018**, *46*, W304–W309. [CrossRef] [PubMed]
28. Kolinski, A. Protein modeling and structure prediction with a reduced representation. *Acta Biochim. Pol.* **2004**, *51*, 349–371.
29. Davtyan, A.; Schafer, N.P.; Zheng, W.; Clementi, C.; Wolynes, P.G.; Papoian, G.A. AWSEM-MD: Protein structure prediction using coarse-grained physical potentials and bioinformatically based local structure biasing. *J. Phys. Chem. B* **2012**, *116*, 8494–8503. [CrossRef]
30. Marrink, S.J.; Tieleman, D.P. Perspective on the Martini model. *Chem. Soc. Rev.* **2013**, *42*, 6801. [CrossRef]
31. Pasi, M.; Lavery, R.; Ceres, N. PaLaCe: A coarse-grain protein model for studying mechanical properties. *J. Chem. Theory Comput.* **2013**, *9*, 785–793. [CrossRef] [PubMed]
32. Basdevant, N.; Borgis, D.; Ha-Duong, T. Modeling protein-protein recognition in solution using the coarse-grained force field SCORPION. *J. Chem. Theory Comput.* **2013**, *9*, 803–813. [CrossRef] [PubMed]
33. Dawid, A.E.; Gront, D.; Kolinski, A. SURPASS Low-Resolution Coarse-Grained Protein Modeling. *J. Chem. Theory Comput.* **2017**, *13*, 5766–5779. [CrossRef] [PubMed]
34. Dawid, A.E.; Gront, D.; Kolinski, A. Coarse-Grained Modeling of the Interplay between Secondary Structure Propensities and Protein Fold Assembly. *J. Chem. Theory Comput.* **2018**, *14*, 2277–2287. [CrossRef] [PubMed]

35. Lopes, P.E.M.; Guvench, O.; MacKerell, A.D. Current Status of Protein Force Fields for Molecular Dynamics Simulations. In *Molecular Modeling of Proteins*; Humana Press: New York, NY, USA, 2015; pp. 47–71.
36. Cossio, P.; Granata, D.; Laio, A.; Seno, F.; Trovato, A. A simple and efficient statistical potential for scoring ensembles of protein structures. *Sci. Rep.* **2012**, *2*, 351. [CrossRef]
37. Tanaka, S.; Scheraga, H.A. Medium- and Long-Range Interaction Parameters between Amino Acids for Predicting Three-Dimensional Structures of Proteins. *Macromolecules* **1976**, *9*, 945–950. [CrossRef] [PubMed]
38. Tsai, J.; Bonneau, R.; Morozov, A.V.; Kuhlman, B.; Rohl, C.A.; Baker, D. An improved protein decoy set for testing energy functions for protein structure prediction. *Proteins Struct. Funct. Genet.* **2003**, *53*, 76–87. [CrossRef] [PubMed]
39. Shen, M.; Sali, A. Statistical potential for assessment and prediction of protein structures. *Protein Sci.* **2006**, *15*, 2507–2524. [CrossRef]
40. Zhou, H.; Skolnick, J. GOAP: A Generalized Orientation-Dependent, All-Atom Statistical Potential for Protein Structure Prediction. *Biophys. J.* **2011**, *101*, 2043–2052. [CrossRef]
41. Xu, D.; Zhang, Y. Ab initio protein structure assembly using continuous structure fragments and optimized knowledge-based force field. *Proteins Struct. Funct. Bioinforma.* **2012**, *80*, 1715–1735. [CrossRef]
42. Woetzel, N.; Karakaş, M.; Staritzbichler, R.; Müller, R.; Weiner, B.E.; Meiler, J. BCL::Score—Knowledge Based Energy Potentials for Ranking Protein Models Represented by Idealized Secondary Structure Elements. *PLoS ONE* **2012**, *7*, e49242. [CrossRef] [PubMed]
43. Ovchinnikov, S.; Park, H.; Kim, D.E.; Liu, Y.; Wang, R.Y.-R.; Baker, D. Structure prediction using sparse simulated NOE restraints with Rosetta in CASP11. *Proteins Struct. Funct. Bioinforma.* **2016**, *84*, 181–188. [CrossRef] [PubMed]
44. Ovchinnikov, S.; Kim, D.E.; Wang, R.Y.-R.; Liu, Y.; DiMaio, F.; Baker, D. Improved de novo structure prediction in CASP11 by incorporating coevolution information into Rosetta. *Proteins Struct. Funct. Bioinforma.* **2016**, *84*, 67–75. [CrossRef] [PubMed]
45. Hirst, S.J.; Alexander, N.; Mchaourab, H.S.; Meiler, J. RosettaEPR: An integrated tool for protein structure determination from sparse EPR data. *J. Struct. Biol.* **2011**, *173*, 506–514. [CrossRef]
46. Yang, J.; Zhang, W.; He, B.; Walker, S.E.; Zhang, H.; Govindarajoo, B.; Virtanen, J.; Xue, Z.; Shen, H.B.; Zhang, Y. Template-based protein structure prediction in CASP11 and retrospect of I-TASSER in the last decade. *Proteins* **2016**, *84*, 233–246. [CrossRef]
47. Russel, D.; Lasker, K.; Webb, B.; Velázquez-Muriel, J.; Tjioe, E.; Schneidman-Duhovny, D.; Peterson, B.; Sali, A. Putting the Pieces Together: Integrative Modeling Platform Software for Structure Determination of Macromolecular Assemblies. *PLoS Biol.* **2012**, *10*, e1001244. [CrossRef]
48. Rodrigues, J.P.G.L.M.; Bonvin, A.M.J.J. Integrative computational modeling of protein interactions. *FEBS J.* **2014**, *281*, 1988–2003. [CrossRef]
49. Kmiecik, S.; Kolinski, A. Folding pathway of the B1 domain of protein G explored by multiscale modeling. *Biophys. J.* **2008**, *94*, 726–736. [CrossRef]
50. Kolinski, A. Multiscale approaches to protein modeling: Structure prediction, dynamics, thermodynamics and macromolecular assemblies. In *Multiscale Approaches to Protein Modeling: Structure Prediction, Dynamics, Thermodynamics and Macromolecular Assemblies*; Kolinski, A., Ed.; Springer: New York, NY, USA, 2011; pp. 1–355. ISBN 9781441968890.
51. Kmiecik, S.; Kouza, M.; Badaczewska-Dawid, A.E.; Kloczkowski, A.; Kolinski, A. Modeling of Protein Structural Flexibility and Large-Scale Dynamics: Coarse-Grained Simulations and Elastic Network Models. *Int. J. Mol. Sci.* **2018**, *19*, 3496. [CrossRef]
52. Kmiecik, S.; Kolinski, A. Characterization of protein-folding pathways by reduced-space modeling. *Proc. Natl. Acad. Sci. USA* **2007**, *104*, 12330–12335. [CrossRef]
53. Kmiecik, S.; Gront, D.; Kouza, M.; Kolinski, A. From coarse-grained to atomic-level characterization of protein dynamics: Transition state for the folding of B domain of protein A. *J. Phys. Chem. B* **2012**, *116*, 7026–7032. [CrossRef] [PubMed]
54. Kmiecik, S.; Kurcinski, M.; Rutkowska, A.; Gront, D.; Kolinski, A. Denatured proteins and early folding intermediates simulated in a reduced conformational space. *Acta Biochim. Pol.* **2006**, *53*, 131–143. [CrossRef] [PubMed]
55. Kmiecik, S.; Kolinski, A. Simulation of chaperonin effect on protein folding: A shift from nucleation - Condensation to framework mechanism. *J. Am. Chem. Soc.* **2011**, *133*, 10283–10289. [CrossRef] [PubMed]

56. Jamroz, M.; Kolinski, A.; Kmiecik, S. Protocols for efficient simulations of long-time protein dynamics using coarse-grained CABS model. *Methods Mol. Biol.* **2014**, *1137*, 235–250. [CrossRef] [PubMed]
57. Wabik, J.; Kmiecik, S.; Gront, D.; Kouza, M.; Koliński, A. Combining coarse-grained protein models with replica-exchange all-atom molecular dynamics. *Int. J. Mol. Sci.* **2013**, *14*, 9893–9905. [CrossRef] [PubMed]
58. Blaszczyk, M.; Jamroz, M.; Kmiecik, S.; Kolinski, A. CABS-fold: Server for the de novo and consensus-based prediction of protein structure. *Nucleic Acids Res.* **2013**, *41*, W406–W411. [CrossRef] [PubMed]
59. Kmiecik, S.; Jamroz, M.; Kolinski, M. Structure prediction of the second extracellular loop in G-protein-coupled receptors. *Biophys. J.* **2014**, *106*, 2408–2416. [CrossRef]
60. Koliński, A.; Bujnicki, J.M. Generalized protein structure prediction based on combination of fold-recognition with de novo folding and evaluation of models. *Proteins Struct. Funct. Genet.* **2005**, *61*, 84–90. [CrossRef]
61. Jamroz, M.; Kolinski, A. Modeling of loops in proteins: A multi-method approach. *BMC Struct. Biol.* **2010**, *10*. [CrossRef]
62. Kurcinski, M.; Jamroz, M.; Blaszczyk, M.; Kolinski, A.; Kmiecik, S. CABS-dock web server for the flexible docking of peptides to proteins without prior knowledge of the binding site. *Nucleic Acids Res.* **2015**, *43*, W419–W424. [CrossRef]
63. Kurcinski, M.; Kolinski, A.; Kmiecik, S. Mechanism of folding and binding of an intrinsically disordered protein as revealed by ab initio simulations. *J. Chem. Theory Comput.* **2014**, *10*, 2224–2231. [CrossRef] [PubMed]
64. Ciemny, M.P.; Debinski, A.; Paczkowska, M.; Kolinski, A.; Kurcinski, M.; Kmiecik, S. Protein-peptide molecular docking with large-scale conformational changes: The p53-MDM2 interaction. *Sci. Rep.* **2016**, *6*. [CrossRef] [PubMed]
65. Blaszczyk, M.; Kurcinski, M.; Kouza, M.; Wieteska, L.; Debinski, A.; Kolinski, A.; Kmiecik, S. Modeling of protein-peptide interactions using the CABS-dock web server for binding site search and flexible docking. *Methods* **2016**, *93*, 72–83. [CrossRef] [PubMed]
66. Ciemny, M.; Kurcinski, M.; Kozak, K.; Kolinski, A.; Kmiecik, S. Highly flexible protein-peptide docking using cabs-dock. *Methods Mol. Biol.* **2017**, *1561*, 69–94. [CrossRef]
67. Blaszczyk, M.; Ciemny, M.P.; Kolinski, A.; Kurcinski, M.; Kmiecik, S. Protein–peptide docking using CABS-dock and contact information. *Brief. Bioinform.* **2018**, *bby080*. [CrossRef]
68. Ciemny, M.P.; Kurcinski, M.; Blaszczyk, M.; Kolinski, A.; Kmiecik, S. Modeling EphB4-EphrinB2 protein-protein interaction using flexible docking of a short linear motif. *Biomed. Eng. Online* **2017**, *16*, 71. [CrossRef]
69. Jamroz, M.; Orozco, M.; Kolinski, A.; Kmiecik, S. Consistent view of protein fluctuations from all-atom molecular dynamics and coarse-grained dynamics with knowledge-based force-field. *J. Chem. Theory Comput.* **2013**, *9*, 119–125. [CrossRef]
70. Jamroz, M.; Kolinski, A.; Kmiecik, S. CABS-flex: Server for fast simulation of protein structure fluctuations. *Nucleic Acids Res.* **2013**, *41*, W427–W431. [CrossRef] [PubMed]
71. Jamroz, M.; Kolinski, A.; Kmiecik, S. CABS-flex predictions of protein flexibility compared with NMR ensembles. *Bioinformatics* **2014**, *30*, 2150–2154. [CrossRef] [PubMed]
72. Kurcinski, M.; Oleniecki, T.; Ciemny, P.M.; Kuriata, A.; Kolinski, A.; Kmiecik, S. CABS-flex standalone: A simulation environment for fast modeling of protein flexibility. *Bioinformatics* **2018**, *bty685*. [CrossRef] [PubMed]
73. Kuriata, A.; Gierut, A.M.; Oleniecki, T.; Ciemny, M.P.; Kolinski, A.; Kurcinski, M.; Kmiecik, S. CABS-flex 2.0: A web server for fast simulations of flexibility of protein structures. *Nucleic Acids Res.* **2018**, *46*, W338–W343. [CrossRef] [PubMed]
74. Eswar, N.; John, B.; Mirkovic, N.; Fiser, A.; Ilyin, V.A.; Pieper, U.; Stuart, A.C.; Marti-Renom, M.A.; Madhusudhan, M.S.; Yerkovich, B.; Sali, A. Tools for comparative protein structure modeling and analysis. *Nucleic Acids Res.* **2003**, *31*, 3375–3380. [CrossRef] [PubMed]
75. Gront, D.; Kmiecik, S.; Kolinski, A. Backbone building from quadrilaterals: A fast and accurate algorithm for protein backbone reconstruction from alpha carbon coordinates. *J. Comput. Chem.* **2007**, *28*, 1593–1597. [CrossRef] [PubMed]
76. Canutescu, A.A.; Shelenkov, A.A.; Dunbrack, R.L. A graph-theory algorithm for rapid protein side-chain prediction. *Protein Sci.* **2003**, *12*, 2001–2014. [CrossRef] [PubMed]

77. Gront, D.; Kmiecik, S.; Blaszczyk, M.; Ekonomiuk, D.; Koliński, A. Optimization of protein models. *Wiley Interdiscip. Rev. Comput. Mol. Sci.* **2012**, *2*, 479–493. [CrossRef]
78. Kmiecik, S.; Kolinski, A. One-dimensional structural properties of proteins in the coarse-grained cabs model. *Methods Mol. Biol.* **2017**, *1484*, 83–113. [CrossRef] [PubMed]
79. Pulawski, W.; Jamroz, M.; Kolinski, M.; Kolinski, A.; Kmiecik, S. Coarse-grained simulations of membrane insertion and folding of small helical proteins using the CABS model. *J. Chem. Inf. Model.* **2016**, *56*, 2207–2215. [CrossRef]
80. Adhikari, A.N.; Freed, K.F.; Sosnick, T.R. De novo prediction of protein folding pathways and structure using the principle of sequential stabilization. *Proc. Natl. Acad. Sci. USA* **2012**, *109*, 17442–17447. [CrossRef]
81. Adhikari, A.N.; Freed, K.F.; Sosnick, T.R. Simplified protein models: Predicting folding pathways and structure using amino acid sequences. *Phys. Rev. Lett.* **2013**, *111*, 028103. [CrossRef]
82. Konrat, R. NMR contributions to structural dynamics studies of intrinsically disordered proteins. *J. Magn. Reson.* **2014**, *241*, 74–85. [CrossRef]
83. Kmiecik, S.; Wabik, J.; Kolinski, M.; Kouza, M.; Kolinski, A. Coarse-Grained Modeling of Protein Dynamics. In *Computational Methods to Study the Structure and Dynamics of Biomolecules*; Springer: Berlin/Heidelberg, Germany, 2014; Volume 1, pp. 55–79. ISBN 978-3-642-28553-0.
84. Kurcinski, M.; Ciemny, M.P.; Oleniecki, T.; Kuriata, A.; Badaczewska-Dawid, A.E.; Kolinski, A.; Kmiecik, S. CABS-dock standalone: A toolbox for flexible protein-peptide docking. *Bioinformatics* **2019**. submitted.
85. Ciemny, M.; Kurcinski, M.; Kamel, K.; Kolinski, A.; Alam, N.; Schueler-Furman, O.; Kmiecik, S. Protein–peptide docking: Opportunities and challenges. *Drug Discov. Today* **2018**, *23*, 1530–1537. [CrossRef] [PubMed]
86. Zambrano, R.; Jamroz, M.; Szczasiuk, A.; Pujols, J.; Kmiecik, S.; Ventura, S. AGGRESCAN3D (A3D): Server for prediction of aggregation properties of protein structures. *Nucleic Acids Res.* **2015**, *43*, W306–W313. [CrossRef]
87. London, N.; Movshovitz-Attias, D.; Schueler-Furman, O. The Structural Basis of Peptide-Protein Binding Strategies. *Structure* **2010**, *18*, 188–199. [CrossRef] [PubMed]
88. Hur, E.; Pfaff, S.J.; Sturgis Payne, E.; Grøn, H.; Buehrer, B.M.; Fletterick, R.J. Recognition and accommodation at the androgen receptor coactivator binding interface. *PLoS Biol.* **2004**, *2*, E274. [CrossRef] [PubMed]
89. Kabsch, W.; Sander, C. Dictionary of protein secondary structure: Pattern recognition of hydrogen-bonded and geometrical features. *Biopolymers* **1983**, *22*, 2577–2637. [CrossRef] [PubMed]
90. Kussie, P.H.; Gorina, S.; Marechal, V.; Elenbaas, B.; Moreau, J.; Levine, A.J.; Pavletich, N.P. Structure of the MDM2 oncoprotein bound to the p53 tumor suppressor transactivation domain. *Science* **1996**, *274*, 948–953. [CrossRef] [PubMed]
91. Ozenne, V.; Bauer, F.; Salmon, L.; Huang, J.R.; Jensen, M.R.; Segard, S.; Bernadó, P.; Charavay, C.; Blackledge, M. Flexible-meccano: A tool for the generation of explicit ensemble descriptions of intrinsically disordered proteins and their associated experimental observables. *Bioinformatics* **2012**, *28*, 1463–1470. [CrossRef] [PubMed]
92. Feldman, H.J.; Hogue, C.W.V. Probabilistic sampling of protein conformations: New hope for brute force? *Proteins Struct. Funct. Genet.* **2002**, *46*, 8–23. [CrossRef] [PubMed]
93. Vitalis, A.; Pappu, R.V. ABSINTH: A new continuum solvation model for simulations of polypeptides in aqueous solutions. *J. Comput. Chem.* **2009**, *30*, 673–699. [CrossRef] [PubMed]
94. Baul, U.; Chakraborty, D.; Mugnai, M.L.; Straub, J.E.; Thirumalai, D. Sequence effects on size, shape, and structural heterogeneity in Intrinsically Disordered Proteins. *bioRxiv* **2018**, 427476. [CrossRef]
95. Estaña, A.; Sibille, N.; Delaforge, E.; Vaisset, M.; Cortés, J.; Bernadó, P. Realistic Ensemble Models of Intrinsically Disordered Proteins Using a Structure-Encoding Coil Database. *Structure* **2018**. [CrossRef] [PubMed]

3

Co-Evolution of Intrinsically Disordered Proteins with Folded Partners Witnessed by Evolutionary Couplings

Rita Pancsa [1,*,†], Fruzsina Zsolyomi [2,†] and Peter Tompa [1,3,*]

1 Research Centre for Natural Sciences of the Hungarian Academy of Sciences, Institute of Enzymology, 1117 Budapest, Hungary
2 Department of Biochemistry and Molecular Biology, Faculty of Medicine, University of Debrecen, 4032 Debrecen, Hungary; zsolyomi.fruzsina@med.unideb.hu
3 Center for Structural Biology, Flanders Institute for Biotechnology (VIB), Vrije Universiteit Brussel, 1050 Brussels, Belgium
* Correspondence: pancsa.rita@ttk.mta.hu (R.P.); peter.tompa@vub.be (P.T.)

† These authors contributed equally to this work.

Abstract: Although improved strategies for the detection and analysis of evolutionary couplings (ECs) between protein residues already enable the prediction of protein structures and interactions, they are mostly restricted to conserved and well-folded proteins. Whereas intrinsically disordered proteins (IDPs) are central to cellular interaction networks, due to the lack of strict structural constraints, they undergo faster evolutionary changes than folded domains. This makes the reliable identification and alignment of IDP homologs difficult, which led to IDPs being omitted in most large-scale residue co-variation analyses. By preforming a dedicated analysis of phylogenetically widespread bacterial IDP–partner interactions, here we demonstrate that partner binding imposes constraints on IDP sequences that manifest in detectable interprotein ECs. These ECs were not detected for interactions mediated by short motifs, rather for those with larger IDP–partner interfaces. Most identified coupled residue pairs reside close (<10 Å) to each other on the interface, with a third of them forming multiple direct atomic contacts. EC-carrying interfaces of IDPs are enriched in negatively charged residues, and the EC residues of both IDPs and partners preferentially reside in helices. Our analysis brings hope that IDP–partner interactions difficult to study could soon be successfully dissected through residue co-variation analysis.

Keywords: intrinsically disordered; disordered protein; structural disorder; correlated mutations; co-evolution; evolutionary couplings; residue co-variation; interaction surface; residue contact network

1. Introduction

Protein sequences provide rich information on structural and functional constraints in the form of residue co-variation in evolution. With the rapid expansion of available sequence data and computational power, and improvements in global statistical approaches [1], the problem of transitive residue correlations (false positive correlations observed for residues that do not actually contact each other in space) could be largely overcome and the analysis of evolutionary couplings (ECs) between protein residues has achieved important breakthroughs [2,3]. Several groups have demonstrated that the analysis of sequence co-variation can be efficiently used for predicting protein structures [2–10], including transmembrane proteins [11,12], defining evolutionary units within proteins [13], and identifying contacting residues of interaction partners [14,15], and interacting subunits of larger complexes [14]. The reason why co-variation analysis does not (yet) provide the ultimate solution to the sequence-based prediction of protein structures is that it requires large, good-quality alignments of sufficiently diverse sequences [16], restricting its applicability to phylogenetically widespread and reasonably conserved proteins.

Intrinsically disordered proteins/regions (IDPs/IDRs) lack well-defined 3D structures, rather, they exist and function as ensembles of rapidly interconverting conformers [17–20]. The conformational variability and adaptability, extended interaction surface, various embedded interaction motifs [21,22], and post-translational modification sites [23] of IDPs make structural disorder indispensable in regulatory [24], complex-assembly [25], and scaffolding [26,27] functions. IDPs are central to cellular interaction networks [28,29] and are frequently associated with human diseases [30].

IDPs mostly interact with their partners through eukaryotic/short linear motifs (ELMs/SLiMs) comprised of a few specificity-determining residues embedded into a disordered sequence environment that ensures the right positioning of their mostly hydrophobic, crucial interaction residues [21,22,31–33]. SLiM-mediated interactions are frequently switched on and off by alternative splicing [31,34–36], and are frequently rewired in evolution [31]. As IDPs lack a well-defined tertiary structure and their functional modules are restricted to a few critical residues, most of their sequences are under limited structural and/or functional constraints. As compared to folded domains, this inherent freedom leads to increased rates [37] and altered types [38] of residue changes in evolution, hampering both the identification and correct alignment of homologous IDPs. Despite their indisputable importance, true IDPs (with the exception of ribosomal proteins [15]) and their interactions were not subject to high-scale residue co-variation analyses [6,7,9,11,12,14,15] until very recently. Although disordered regions have less co-varying residues than folded domains [39], Toth-Petroczy and colleagues have recently refined their method, EVfold, to predict potential structured states of IDPs through detecting ECs within their chains [40]. The prevalent phenomenon of induced folding or disorder-to-order transition [41] in IDP-partner recognition, however, suggests that the (partly) structured states of IDPs are encoded not only in their own sequence but also in that of the partner. It follows then that co-evolutionary signals in the interface of the IDP and its partner may actually be just as pronounced as those within a protein fold or between two folded interaction partners.

Therefore, to see if functional constraints originating from partner binding resulted in detectable interprotein ECs of (at least for certain phylogenetically wide-spread) IDPs and to characterize the respective protein pairs, interfaces, and their co-evolved residue pairs, we performed a targeted screen to identify such cases.

2. Results

Even though bacterial proteomes are relatively poor in IDPs/IDRs [42–44], we restricted our search to bacterial IDP–partner interactions to ensure a sufficient number and diversity of orthologous sequences for residue co-variation analysis. Although there are already a few hundred eukaryotic genomes available, they are not evenly distributed among phylogenetic groups (i.e., a large fraction of them are from mammals) and thus do not show enough sequence diversity on the level of proteins. This is well supported by the fact that recent large-scale residue co-variation analyses have all focused on bacterial protein complexes [9,14,15]. Therefore, we analyzed the 42 bacterial IDPs bound to their folded partners available in the Database of disordered binding sites (DIBS) [45] because there the structural states of the constituent protein chains are backed by experimental evidence. After getting rid of 4 redundant structures and one where the IDP chain was only four residues long, the phylogenetically fairly wide-spread 19 complexes (those with >130 sequences for the IDP in Pfam database 31 [46] full alignments) were selected for co-variation analysis (Supplementary Tables S1 and S2). This way, species–specific complexes, such as virulence factors, certain toxin–antitoxin, effector–chaperone and effector–immunity protein pairs could be counter-selected instead of needlessly occupying co-evolution analysis servers. The sequences were trimmed for the interacting regions/domains or extended to reach the minimum length of 30 residues required for Gremlin analysis as indicated in Table 1, and the Gremlin [15] and EVcomplex [14] servers were used to identify interprotein ECs (See Methods for further details). These methods perform co-variation analyses along similar lines. They first prepare a so-called paired alignment for the protein pair provided, which means that they detect the closest homolog of each of the two provided proteins in each analyzed proteome, and build an alignment wherein the interacting sequences are linked together and filtered for similarity. If the resulting paired alignment has enough sequences, they proceed with detecting co-varying residue pairs (evolutionary couplings, ECs) therein. They use somewhat different approaches to score the pairs of residues based on evolutionary co-variation. From the outputs we exclusively take interprotein ECs, where the coupled residues come from the two different proteins because those provide clues on the interaction. Gremlin could be successfully run on 13 complexes (Table 1), while it stopped in 6 cases due to the alignments being insufficient for analysis. For 7 of the 13 successfully analyzed complexes, Gremlin detected coupled interprotein residue pairs with a scaled score $\geq$1.3 and probability in the top 12% ($p > 0.88$), hereafter referred to as ECs. For EVcomplex, interprotein ECs with an EVcomplex score >0.9 have been accepted as ECs. EVcomplex also identified ECs for 7/19 complexes (Supplementary Table S1). Both the identified complexes and the detected ECs showed a good overlap between the two methods. Since Gremlin ran the sequence search on a more up-to-date sequence database, it obtained better sequence coverage values and consequently identified more ECs than EVcomplex for almost all the analyzed protein pairs. Based on their residue–residue distances, almost all Gremlin EC pairs fell spatially close, implying that they could be correctly identified co-varying pairs. Thus, we decided to continue the analysis and show the results for Gremlin ECs. The ECs were then checked by PDBe PISA [47] in order to see if they are at the interface (IF) and if they engage in physical interaction (hydrogen bonds or salt bridges; see Table 1).

2.1. ECs Were Detected Outside the Sequence Ranges with PDB Coordinates, but Visible ECs Reside on the Interfaces

Gremlin identified 31 ECs in 7 complexes (Supplementary Tables S1 and S3). Intriguingly, for 9 of these co-varying residue pairs, at least one of the constituent residues (mostly the IDP residue) fell outside the segments with PDB coordinates (invisible ECs; marked with N/A in Supplementary Table S3). These invisible ECs make up a remarkably high percentage of all identified ECs. Based on their Gremlin scores and probabilities, they do not seem to be mistakenly predicted pairs as they often have higher scores and probabilities than interface residue pairs with several atomic contacts. Of the 10 invisible EC residues, 5 were present in the respective PDB sequences but had no atomic coordinates, while the other 5 were not included in the constructs used for structure determination. High occurrence of these invisible EC residues suggests that the IDP segments used for structure determination and interaction analyses are often too short, lacking residues that could still form important contacts with the partner. There are actually several instances where the contribution to binding of segments that are disordered in the partner-bound state has been proven, such as binding of CREB to CBP [48] and binding of SF1 to U2Af65 [49], which forms the basis of the concept of fuzzy interactions [50].

All except one of the 22 visible IDP EC residues (Table 1, Figure 1) are interface residues according to PDBe PISA, and 12 of them are assigned as forming a direct H-bond or salt bridge (or both) with the partner (Table 1), although not necessarily with their corresponding EC pair (Supplementary Table S3). These values already support the validity of the predicted ECs, nevertheless, we decided to investigate them in more detail.

Table 1. Residue co-variation analysis of IDP–partner interactions with broad phylogenetic spread.

Complex		Folded Partner		IDP			Gremlin Analysis Results		
DIBS ID-PDB ID (Chains Partner(s) | IDP)	IF Area (Å^2)	Gene Name	Uniprot AC_region	Gene Name	Uniprot AC_Region	Length	# of Seq. in PFAM 31 Full	Coverage (seq/res)	ECs by Gremlin/by EVComplex/ Gremlin IDP ECs on IF/in Bonds [a]
DI4200001-3B1K (BG | D)	1016	*gap2*	Q9R6W2_78-215	*cp12*	Q6BBK3_46-75	30	507	0.76	1/0/1/1
DI2200001-3HPW (AB | C)	1483	*ccdB*	P62554_1-101	*ccdA*	P62552_37-72	36	395	4.06	7/0/6(1 inv [b])/2
DI2200002-5CQX (AB | C)	1128	*mazF*	P0AE70_1-111	*mazE*	P0AE72_53-82	30	4349	3.00	9/1/5(4 inv)/5
DI2200004-3M91 (AC | B)	1065	*mpa*	P9WQN5_46-96	*pup*	P9WHN5_21-64	44	511	2.19	5/0/2(2 inv)/1
DI2200006-3M4W (AC | E)	2321	*rseB*	P0AFX9_220-318	*rseA*	P0AFX7_125-195	71	253	1.67	2/0/2/1
DI1200004-1SC5 (A | B)	1641	*fliA*	O67268_1-236	*flgM*	O66683_1-88	88	1450	3.43	3/10/3/1
DI1210003-2A7U (B | A)	611	*atpH*	P0ABA5_1-134	*atpA*	P0ABB3_1-30	30	14,854	8.34	4/3/2(2 inv)/1
DI2200005-1SUY (AB | C)	1588	*kaiA*	Q79V62_177-283	*kaiC*	Q79V60_485-518	34	2739	0.85	0/0/0/0
DI1200003-1QFN (A | B)	732	*grxA*	P68688_1-85	*nrdA*	P00452_732-761	30	7563	1.59	0/0/0/0
DI1200001-1R1R (A | D)	851	*nrdA*	P00452_335-729	*nrdB*	P69924_347-376	30	4425	0.78	0/1/0/0
DI1200006-5D0O (D | C)	2072	*bamD*	P0AC02_1-245	*bamC*	P0A903_30-85	56	726	0.38	0/0/0/0
DI1210006-4Z0U (B | E)	358	*rnhA*	A7ZHV1_1-155	*ssb*	P0AGE0_149-178	30	8735	0.31	0/0/0/0
DI1210004-3C94 (A | B)	320	*sbcB*	P04995_13-355	*ssb*	A0A0H3GL04_145-174	30	8735	0.11	0/0/0/0

[a] In this column, bonds were assigned to the IDP EC residues based on PDBe PISA H-bond and salt bridge annotations. [b] The locations, bonds, and distances of invisible EC pairs could not be analyzed as they reside outside the sequence ranges with PDB coordinates. The numbers of such invisible EC pairs (inv) are indicated in brackets in the last column along with interface (IF) pairs, since they provide the explanation for the difference between the number of identified ECs and interface ECs.

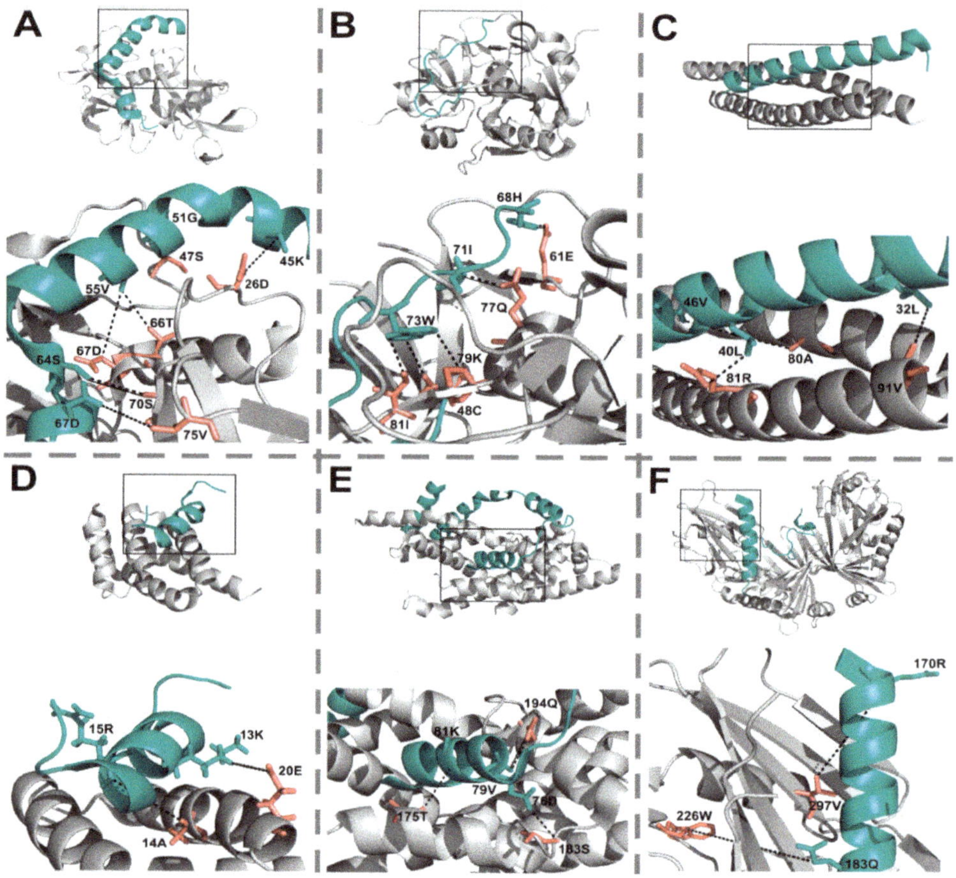

Figure 1. High-scoring interprotein ECs indicating IDP-partner co-evolution. The six protein complexes with multiple high-scoring interprotein ECs predicted by Gremlin are introduced: (**A**) PDB complex 3HPW; (**B**) 5CQX; (**C**) 3M91; (**D**) 2A7U; (**E**) 1SC5; (**F**) 3M4W. IDPs are depicted in teal cartoon style, while the partners are in light grey. Rectangles indicate the part of the complexes which are depicted as detailed interaction maps below. In the interaction maps, the EC side chains are depicted as sticks colored teal in the IDP and salmon in the partner. The interprotein ECs have their respective residue side chains labelled and are connected by dashed lines. For the few pairs forming direct H-bonds or salt bridges according to PISA, the connecting lines are doubled. The 183Q-226W IDP-partner EC pair on the F panel had a distance 17.5Å, thus it was handled as an outlier. None of the depicted complexes represent SLiM-mediated interactions, as we could not identify any ECs for those (see Section 2.5 for further details).

2.2. EC Pairs Have More Atomic Contacts between Them than Other Pairs

Interchain residue–residue atomic contacts have been obtained from the Protein Contacts Atlas (PCA) [51] portal for the 34 X-Ray structures. Among EC-containing complexes, 2A7U is an NMR structure so it was not included in the following residue-contact analysis. From the contact lists between the IDP chain and the relevant partner chains, the numbers of atomic contacts between each contacting residue pair have been obtained, the residue pairs have been grouped as ECs, EC residue/no-EC residue pairs and no-EC residue/no-EC residue pairs, and the numbers of atomic contacts for these three groups have been compared. Although only 7 of the 20 visible ECs of X-ray structures had atomic contacts in PCA, with a median of 5, ECs had significantly more contacts between them than contacting residue pairs in the other two groups (Figure 2A). A likely reason behind EC/no-EC pairs having significantly less contacts than no-EC/no-EC pairs is that in the EC/no-EC dataset the contacting residues of EC residues were skimmed. Their respective EC pairs were not taken into account in this dataset; still, they had less space to make contacts with other, no-EC residues. We also wanted to see how many contacts EC residues have, regardless of their partner residue. To this end, we added up the total number of atomic contacts (with all contacting partner residues) for all residues in the contact lists for the IDPs and partners separately. These values were then compared between EC and non-EC residues. We did not find significant differences neither for IDPs nor for their partners (Supplementary Figure S1). Thus, we can claim that although the individual residues of IDP-partner ECs do not have more atomic contacts than other residues, ECs taken as pairs do have more atomic contacts with each other than other residue pairs.

2.3. ECs Are Significantly Closer to Each Other than Randomly Paired IDP–Partner Interface Residues

We were also interested in the distance distribution of ECs. For the pairs with PCA atomic contacts, we took the shortest atomic contact as residue–residue distance, while for the rest of the visible ECs (including the ones in the NMR structure 2A7U) we calculated the minimal residue–residue distances using PyMOL (Supplementary Table S3). We compared the distance distribution so obtained to an equivalent reference distance distribution of randomly selected and subsequently paired IDP and partner residues picked from the same interfaces and found that ECs are highly significantly closer in space than the randomly paired interchain interface residue pairs (Figure 2B). The sampling of randomly paired interface residues was carried out 100 times and their distances were consistently significantly larger than those of ECs ($p < 0.01$). The descriptive distribution features (minimum, 1st quantile, mean, median, 3rd quantile, maximum) of the 100 samples have shown an average standard deviation of 2.31 which we interpret as low variation among the random samples. 18/22 visible ECs were closer than 8 Å, two were between 8 and 10 Å far, and only two had >10 Å distances (Figure 2C). The latter two were assigned as outliers by R, they probably represent mistakenly identified pairs (or have larger distances due to other reasons [52]) and thus have been handled as outliers and excluded from distance comparisons (Figure 2B). One of the outliers is the 183Q-226W IDP-partner EC pair in the RseA/RseB (PDB: 3M4W) complex as indicated in Figure 1. The other outlier is the sole EC pair (74D-133Y) identified in the CP12/Glyceraldehyde-3-phosphate dehydrogenase (PDB: 3B1K) complex. Therefore, the latter complex is left without any reliable ECs, and thus it is not shown in Figure 1.

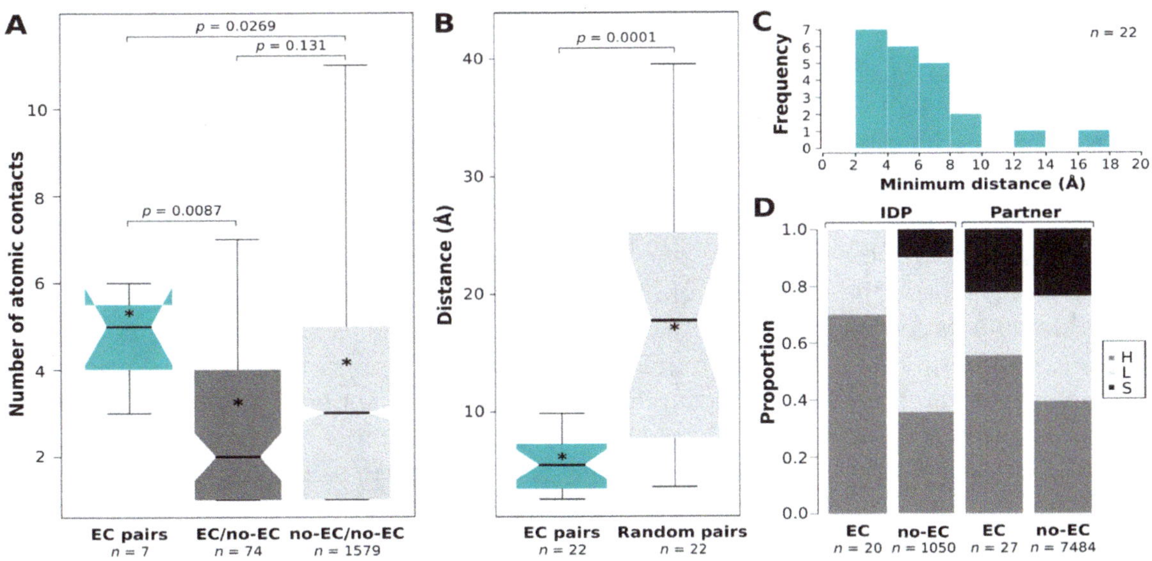

Figure 2. Atomic contacts and distances of ECs. (**A**) The numbers of atomic contacts connecting EC pairs according to PCA compared to those of other contacting residue pairs. Pairwise comparisons were done with Mann–Whitney U test with the corresponding p-values indicated; (**B**) the minimum distances between visible EC pairs are compared to those of randomly selected and paired IDP–partner interface residues; (**C**) histogram showing the distribution of minimum distances between visible ECs; (**D**) the proportions of main secondary structure element types (H—helix, S—strand, L—loop) are compared between EC and no-EC residues for IDPs and partners separately. In the boxplots stars (*) indicate the average values of the datasets.

2.4. EC Residues Preferentially Occur in Helices in Both IDPs and Partners

To assess the structural preferences of EC residues, we have computed DSSP secondary structure assignments for the complexes. The EC residues preferentially resided in helices for both IDPs ($p = 0.003$) and partners ($p = 0.039$) compared to other residues (Figure 2D). Also, the detected interprotein ECs typically cluster in one alpha helix of the IDPs, although in most cases the studied chains did not have more: in anti-sigma-28 factor FlgM they cluster in helix4 (Figure 1E), while in RseA in the longer helix of the two (Figure 1F). Furthermore, ECs have also been detected in the toxin-antitoxin MazF-MazE pair, where the IDP MazE does not fold into a regular secondary structure on the surface of the MazF dimer (Figure 1B).

2.5. General Trends Observed for the Protein Complexes with High-Scoring Interprotein ECs

The seven protein pairs yielding high-scoring ECs mostly had >2 Gremlin sequence/residue alignment coverage values, while the ones that stopped running or did not provide high-scoring ECs tended to have values <1 (Table 1). The coverage values did not correlate with the number of IDP homologs in PFAM. Longer-interacting IDP regions with wider phylogenetic spread (for the corresponding phylogenetic groups see Supplementary Table S1) had a higher chance for sufficient coverage values, while short IDP chains that required significant extensions for reaching the minimum length of 30 residues did not have a good chance for >1 coverage values even if mediating a phylogenetically widely conserved interaction. Therefore, the complexes where the IDPs interact with their partners in the typical IDP manner—through ELMs/SLiMs or molecular recognition features (MoRFs)—namely the enolase- and PNPase-interacting motifs of RNase E [53], the SspB-interacting region of the N-terminal part of RseA, and the C-terminal interaction motif of single-stranded DNA-binding protein (SSB) recruiting diverse partner proteins [54,55], did not have enough PFAM sequences for analysis or, despite a vast amount of sequences in PFAM, did not show any high-scoring ECs. The lack of a sufficient number or diversity of detectable homologs for these important interaction motifs/regions could be due to their short length, the relatively small fraction of the actual specificity-determining residues and the fast evolutionary turnover of the surrounding other residues [31].

In the case of the interactions mediated by the SSB C-terminal motif, that are among the phylogenetically most widespread ones, the lack of detected ECs could be attributed to different reasons. The C-terminal motif is 9 residues in length, so it had to be extended by 21 residues from the poorly conserved SSB linker region to reach a total length of 30 residues. The lack of conservation in the linker segment dilutes the information in the motif, whereas the multitude of interaction partners simultaneously restricting the evolution of the motif [55,56] leads to a complete lack of sequence variation in most of its residues, which could both contribute to the lack of detected ECs.

While EC residues showed a strong preference for interface helices, only 58.8% of ELM instances form secondary structure elements, with only 16.2% of them being mostly helical and 7.6% being partially helical according to a large-scale analysis of the eukaryotic linear motif (ELM) database [31]. Although our dataset only contains a few bacterial short linear motifs, which are not part of the ELM database, they show a similar distribution among secondary structure types as proposed for their eukaryotic counterparts. The SSB C-terminal motif does not form a secondary structure with any of its 5 partners, while the SspB binding motif of RseA forms two very short helices. The PNPase- and enolase-binding motifs of RNase E were not subjected to EC analysis because they did not have the sufficient amount of PFAM sequences, but the former binds through beta sheet augmentation, while the latter forms a short helix. Thus, the identified SLiMs do not show a preference for helical conformation, and by mostly spanning only 4–9 residues, even the helix-forming ones are not long enough to form extended helical structures on the surface of binding partners, which could also contribute to the lack of detected ECs. In all, the complete lack of ECs for SLiM-mediated interactions regardless of their phylogenetic spread definitely represents a major limitation of applying residue co-variation-based approaches for the analysis of IDPs.

Although multisubunit enzymes often contain predicted disordered chains that occupy a completely extended conformation in the complex, and such cases would be perfect candidates for IDP-partner co-evolution analysis, the structural features of these subunits are rarely analyzed on their own. Due to this reason, our DIBS-derived dataset only has 4/37 permanent protein complexes, while most of them depict transient interactions (Supplementary Table S1). The seven complexes with identified ECs contained both permanent (1) and transient (6) complexes in similar fractions as seen for the overall dataset, so permanence of the interactions does not seem to largely affect their co-evolution patterns at first glance. However, it is interesting to note that all the complexes with identified ECs had IDP–partner interface areas >1000 Å^2, except for the sole complex with the permanent interaction of two subunits of the ATP synthase that only had 610.5 Å^2. This might imply that, in permanent protein complexes, which are phylogenetically widely distributed, co-evolution of interacting subunits is so prominent that it can be detected even if the corresponding interaction surfaces are relatively small.

2.6. ECs Could Not Be Identified for Very Small Interfaces with no H-Bonds or Salt Bridges

By comparing the interfaces of the complexes with and without ECs (the latter group comprised all the 30 non-redundant complexes that did not provide high-scoring ECs by Gremlin) for their interface areas, H-bond and salt bridge densities, we did not find any significant difference (Figure 3). However, it became evident that ECs could not be detected for interfaces <600 Å^2 (for transient complexes this threshold rather seems as 1000 Å^2) in size, not even for exceptionally widely conserved interactions. ECs could also not be identified for interfaces without any assigned H-bonds or salt bridges (Figure 3C).

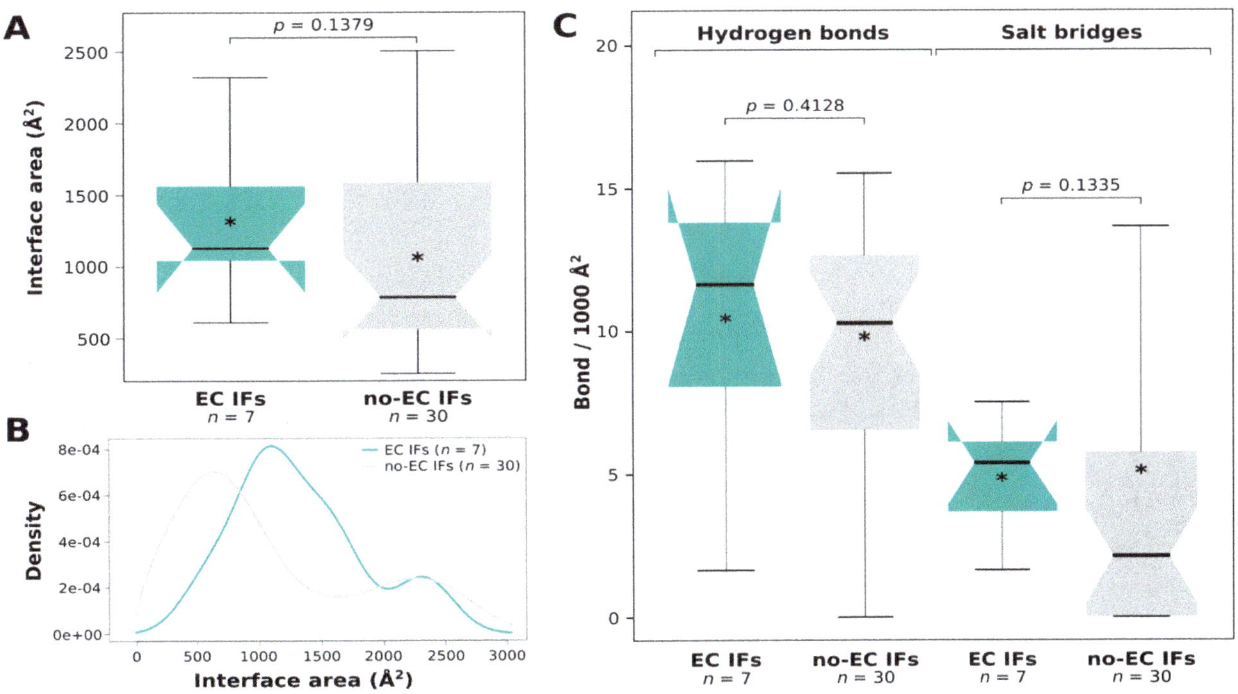

Figure 3. Comparison of interface areas and bonds for EC and no-EC interfaces. (**A**) Comparison of EC and non-EC interface areas with Mann–Whitney U test. (**B**) Density plot of the interfaces in the two groups. (**C**) Comparison of the normalized numbers of H-bonds and salt bridges between EC and non-EC interfaces with Mann–Whitney U test, with the corresponding *p*-values indicated. In the boxplots, stars (*) indicate the average values of the datasets.

2.7. EC Interfaces Are Enriched in Negatively Charged Residues

We have also calculated the amino acid and amino acid group compositions of the trimmed protein segments (analyzed sequence ranges), all interfaces (all IFs), interfaces without any detectable EC pairs

(No-EC IFs), EC-carrying interfaces (EC IFs) and EC residues for the IDPs and partners separately. Comparing EC IFs and ECs to all IFs, we found no significant differences either for IDPs (Figure 4, Supplementary Figure S2) or their partners (Supplementary Figure S3), although the enrichment of the EC-carrying interfaces of IDPs in negative residues was nearly significant ($p = 0.0519$). We have to note though that we had only 28 IDP- and 30 partner EC residues, which is suboptimal for rigorous statistical analysis. Due to the small sample size, even seemingly very large differences were not statistically significant, for instance, IDP EC residues were not found to be enriched in positive amino acids compared to all IDP interface residues, despite having almost two times as high fraction of positives (28.6% vs. 14.5%; Figure 4). When comparing the compositions of EC interfaces to those of no-EC interfaces, we found that the EC-carrying interfaces of IDPs harbor significantly more negatively charged residues ($p = 0.0101$), in particular glutamates ($p = 0.0240$), than the IDP interfaces with no ECs (Figure 4, Supplementary Figure S2). For the partner interfaces no significant differences were found (Supplementary Figure S3).

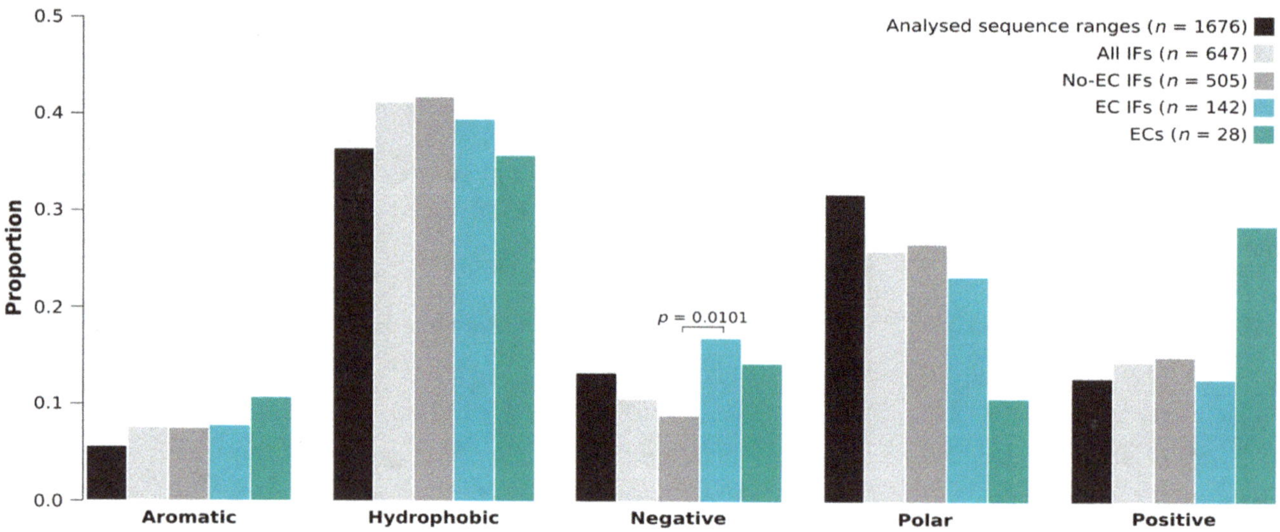

Figure 4. The amino acid group compositions of IDP EC-carrying interfaces and ECs. The amino acid group compositions of trimmed protein segments (analyzed sequence ranges), all interfaces (all IFs), interfaces with no ECs (no-EC IFs), EC-carrying interfaces (EC IFs) and EC residues of IDPs. Both EC-carrying interfaces and EC residues have been compared to all IFs. Also, EC interfaces were compared to no-EC interfaces. P-values are only indicated for amino acid group proportion differences that were found significant by the built-in test of equal proportions of R.

3. Discussion

Due to increased substitution rates [37], IDP homologs are difficult to identify and correctly align. Also, their extended, floppy conformational states manifest in less contacting residue pairs per unit length compared to folded proteins/domains, also leaving less room for residue co-evolution [39]. Furthermore, IDPs excel in moonlighting [57], meaning that they can interact with multiple partners (often through the very same sequence region) and thus their evolution might well be simultaneously constrained by many different partners. However, these multiple different interactions are only poorly represented in structural databases, and residue co-variation analysis can also only be performed for one pair of sequences at a time. Another important factor is that the ensembles and interactions of IDRs are frequently modulated by post-translational modifications (although these are largely limited to eukaryotic proteins), which currently cannot be taken into account by the available residue co-variation analysis methods. Therefore, we need to acknowledge that, at the moment, it is not feasible to comprehensively cover the full range of structural and interaction plasticity of disordered proteins by such analyses. Due to these reasons, IDPs and their interactions have hardly been investigated for the occurrence of residue co-variation [6,7,9,11,12,14,15]. Although a recent analysis suggested structured states of IDPs based on observable intrachain ECs [40], the co-evolution of IDPs with folded partners remained to be elucidated. To possibly demonstrate and understand IDP–partner co-evolution, we performed a comprehensive analysis of IDP–partner interprotein evolutionary couplings on the available bacterial complexes of the DIBS database using dedicated methods [14,15]. To ensure sufficient diversity and depth of sequence coverage by homologs, we restricted our analysis to bacterial IDPs and their cognate partners. Furthermore, to improve internal consistency of the predictions, we applied two different algorithms, Gremlin and EVcomplex. We could identify high-scoring interprotein ECs for seven interacting protein pairs, typically the ones with the best alignment coverage. Alignment coverage is, however, not strictly correlated with phylogenetic spread; it also depends on the length of interacting protein regions. For complexes where the IDP contacts the partner over a relatively long sequence bit, like, for instance, for the FlgM-FliA complex (Figure 1E), we had a good chance to identify some ECs. However, although SLiM-mediated interactions are highly preferred by IDPs and some of them are exceptionally widely conserved, they did not show high-scoring interprotein ECs, probably due to their short length, wherein the evolutionary information on a relatively small fraction of specificity-determining residues is diluted by many more surrounding other residues undergoing fast evolutionary turnover [31]. The absence of detected ECs for SLiM-mediated interactions could be a major limitation of performing residue co-variation analyses on IDPs, as SLiMs are crucial and widely applied interaction units of IDPs that mediate contacts with many of their partners.

Supporting the validity of the predicted ECs, i.e., that they represent true physical contacts between the IDP and its partner, or at least could be spatially restrained by each other, the majority of the visible ECs fall within 8 Å. Residue contact networks from PCA also showed that several EC pairs had multiple atomic contacts; moreover, ECs had more atomic contacts than other contacting IDP–partner residue pairs. Regarding their structural preferences, EC residues tend to be located in helices in both IDPs and folded partners. Although EC-carrying interfaces did not significantly differ in size or in the surface-normalized number of H-bonds and salt bridges, it is important to note that no ECs could be identified in interfaces that are small (<600 Å^2) or have no hydrogen bonds or salt bridges, no matter how widely conserved the interaction is. Since for the partners we did not find enrichment in any residue type, we do not claim that the enrichment of EC-carrying IDP interfaces in negative residues would imply the employment of increased electrostatic attraction compared to other IDP–partner interfaces. Therefore, we cannot ascertain the biological relevance of the noted preference for negatively charged residues.

Importantly, ECs could be identified for both permanent and transient IDP–partner interactions. This shows that not only obligatory, permanent associations, such as the subunits of the ribosome [15] or heteromeric enzymes, but also well-established signaling relationships, such as regulatory antisigma factors, have a trace of detectable co-evolution between the IDP and its interaction partner. Regarding the evolutionary history and genomic proximity of the genes encoding the protein pairs with high-scoring ECs, the CcdA-CcdB (PDB: 3HPW) and MazE-MazF (5CQX) toxin–antitoxin pairs, the alpha and delta subunits of the ATP synthase (2A7U) and the Sigma-E/Anti-Sigma-E regulatory factors RseB and RseA (3M4W), are each encoded on common operons. This suggests that these protein pairs are co-expressed and their co-occurrence is strongly preferred in evolution. To the contrary, the other two pairs, FliA with its anti-sigma factor FlgM (1SC5) and proteasome-associated ATPase (product of *mpa* gene) with prokaryotic ubiquitin-like protein Pup (3M91) are located in different operons and are thus substantially less strictly associated both in gene-regulatory and evolutionary terms. Therefore, based on our findings, a phylogenetically widely preserved protein–protein interaction that buries an interface larger than SLiM-mediated interactions typically do, might be enough for the reliable identification of interprotein evolutionary couplings. No other assumption regarding permanence of the interaction or co-regulation of the corresponding genes needs to be made prior to analysis, although such factors could increase the chance of finding true ECs.

In this first dedicated study of IDP–partner co-evolution, we also show that IDPs are difficult to investigate by methods addressing residue co-variation, due to their fast evolutionary changes, limited sequence representation, increased propensity for moonlighting functions, frequent use of short linear motifs for partner binding and relatively few co-evolving residue pairs both within their chains [39,40] and with their partners. Nevertheless, by demonstrating detectable footprints of IDP–partner co-evolution for interactions with largely different functional readouts, our results are also promising. They imply that the explosion in the number of sequenced genomes, the continuous improvement of techniques of sequence homology detection [58], and advances in sequence alignment approaches optimized for IDPs [40] could soon empower residue covariation analyses of IDPs to provide predictions and new insights into the structures and interactions of IDPs whose experimental investigation proved to be challenging.

4. Materials and Methods

4.1. Dataset Preparation

Bacterial protein complexes were obtained from the DIBS database [45] (Supplementary Table S1). We identified 4 of the 42 as redundant: DIBS: DI2210001–PDB: 3TCJ, DI1200012–3UF7, DI1210005–3C94, and DI1210010–5CW7 were excluded due to redundancy to DI2200001–3HPW, DI1200011–3UF7, DI1210004–3C94, and DI1210007–5CZF, respectively. The DI1200014–5F56 complex was also excluded due to the IDP peptide being <5 residues long. For the remaining 37 complexes, the constituent chains were trimmed or extended to make them optimal for co-evolution analysis. IDP chains shorter than 30 residues were extended based on UniProt [59] to reach the minimum of 30 residues length required for Gremlin analysis. Terminal segments could obviously only be extended to one direction. For the two short IDP chains representing internal segments, more extension was added to the end with residues forming an interface with the partner chain. Partner chains/very long IDP chains were trimmed to interacting domains/subdomains to ensure that the best possible coverage values are obtained (sequence coverage values are highly dependent on the total length of the analyzed sequences). The resulting trimmed UniProt regions used for further analysis are indicated in Table 1. Then for each complex, the IDP counterpart was checked in PFAM 31 [46] to get an idea of their phylogenetic spread. If, for the IDP region in the complex or at least for a neighboring protein domain/region of the protein there were no PFAM families available, or the number of sequences in the full alignments of the relevant PFAM families were <130, then the complex was excluded from correlated mutation analysis (for PFAM families see Supplementary Table S2).

Information on the complexes representing permanent or transient interactions was taken from the literature and UniProt subunit structure annotations. To assign if the genes are encoded within a single operon/gene neighborhood, we used information in the STRING database [60] and Ensembl Bacteria [61]. Assignments on the phylogenetic spread of the interactions were based on literature mining.

4.2. Co-Evolution Analysis

The trimmed regions of the 19 remaining complexes were run with the Gremlin [15] and EVcomplex [14] webservers for interprotein co-variation analysis. When using Gremlin, for regions >60 residues the e-value threshold was set to E-06 and the number of iterations with Jackhmmer to 4, while for regions $\leq$60 residues, we applied a less stringent e-value threshold of E-04 with 8 iterations. The reason for using a less stringent e-value threshold for shorter sequences is explained and supported by the Introduction/Updates section of the Gremlin webserver. Δgene was set to 1–∞ to identify the closest homologs in the analyzed proteomes regardless of their genomic location. In 6 cases, Gremlin stopped due to insufficient alignments for further analysis. It finished the analysis in 13 cases. We accepted interprotein residue pairs with a scaled score $\geq$1.30 and a probability >0.88 as co-varying pairs; evolutionary couplings (ECs). These thresholds were selected based on our observation that among the predicted possible EC pairs with a scaled score >1.3 the probability values are the main determinants of the residue pairs representing true positive predictions with residue–residue distances reflecting direct contacts or false positive predictions with large interresidue distances. Therefore, we have selected a strict probability value threshold (top 12%) to avoid false positive ECs compromising the dataset used for analysis. The 7 complexes with such high-scoring interface ECs identified by Gremlin were analyzed further and compared to the remaining 30 complexes with no high-scoring ECs.

For the EVcomplex analysis, we applied the default e-value (10^{-5}) in searching for homologues and chose the option of selecting the closest homologs of the query sequences from the analyzed proteomes. Here, interprotein ECs with an EVcomplex score >0.9 have been accepted as ECs. EVcomplex identified ECs for 7/19 complexes.

4.3. Interface Properties

For each complex, information on the smallest meaningful biological assembly (chains that can represent the biologically relevant interaction) was taken from DIBS and is indicated in Supplementary Table S1. These contain strictly one IDP chain, which interacts with one or more (but in this case identical) partner chains. An exception is 3O0E, where from the DIBS-indicated L and A chains, L was not present in PDBe PISA, so the equivalent M and B chains have been used. Interface areas, interfacing residues, and interchain physical interactions were derived from PDBe PISA [47] assignments. Total interface areas were defined as the sum of interchain interface areas between the IDP chain and the different interfacing partner chains. The total number of IDP–partner interface H-bonds and salt bridges were also obtained as the sum of such bonds between the IDP chain and the interfacing partner chains, then these were normalized for 1000 Å^2 interface area before statistical comparisons. Interface residues were assigned for each chain based on PISA.

4.4. Amino Acid Compositions

In the residue and residue group composition analyses, we have calculated the compositions of the trimmed protein segments (analyzed sequence ranges), all interfaces (all IFs), interfaces with no ECs (no-EC IFs), EC-carrying interfaces (EC IFs) and EC residues for the IDPs and partners separately. Identical partner chains have been only included once into these composition analyses. The interfaces have been defined based on PDBe PISA, so regions of the trimmed protein segments falling outside the segments with PDB coordinates are not represented by the interface composition values. However, they are represented in the analyzed sequence ranges and among ECs. Residue groups were defined as: hydrophobic (G, A, V, L, M, I), aromatic (F, Y, W), polar (S, T, C, P, N, Q), negative (D, E), and positive (K, R, H). Residue group and residue proportion significances were obtained by the built-in test of equal proportions of R [62].

4.5. Residue–Residue Contact Networks

Interchain residue–residue atomic contacts have been downloaded from the Protein Contacts Atlas [51] for all the X-ray structures (34/37 complexes) using the default 0.5 Å distance cut-off value in PCA. Only the contact lists between the IDP chain and the relevant partner chains have been used for analysis. The number of atomic contacts between each contacting residue pair has been obtained. Also, the total number of atomic contacts (with all contacting residues) has been calculated for all the residues with at least one such interchain atomic contact. If the same partner residue had atomic contacts with the IDP from more than one partner chain, then those different contact numbers were added up to get the total contact number. Also, for IDP residues, all atomic contacts with partner residues were added up regardless of the partner chain. The calculated totals have been compared between EC and non-EC residues for the IDPs and partners separately.

4.6. Secondary Structure Assignments

DSSP secondary structure assignments [63] have been obtained for all the residues present in the complex structures in a way that only one of the identical partner chains has been used. The eight DSSP secondary structure element type assignments have been simplified and grouped in the traditional way into three larger classes: helix (G, H, and I), strand (E and B) and loop (S, T, and C). The distribution of residues among different secondary structure element types have been compared between EC residues and non-EC residues for the IDPs and partners separately using the built-in test of equal proportions of R.

4.7. Residue–Residue Distances

For the ECs being in direct atomic contact according to the Protein Contacts Atlas, the shortest atomic contact distance (between heavy atoms) has been accepted as residue–residue distance. For the rest of the ECs, the shortest distance between the heavy atoms of the two residues has been measured using PyMOL (https://pymol.org/2/; The PyMOL Molecular Graphics System, Version 2.0 by Schrödinger, LLC, New York, USA). Then, for each EC we have randomly selected an interface IDP residue and an interface partner residue from the same interface, and measured their distance similarly. The distances of ECs and the thereby assembled equivalent random reference interface pairs have been compared by Mann–Whitney U test. To check if the observed difference remains consistent, we have obtained 100 additional samples of random residue pairs and compared their distances to those of ECs.

Author Contributions: Conceptualization, R.P.; Data curation, R.P. and F.Z.; Formal analysis, R.P. and F.Z.; Funding acquisition, P.T.; Methodology, R.P.; Resources, P.T.; Supervision, R.P. and P.T.; Visualization, R.P. and F.Z.; Writing—original draft, R.P.; Writing—review & editing, R.P., F.Z. and P.T.

References

1. Stein, R.R.; Marks, D.S.; Sander, C. Inferring Pairwise Interactions from Biological Data Using Maximum-Entropy Probability Models. *PLoS Comput. Biol.* **2015**, *11*, e1004182. [CrossRef] [PubMed]
2. Marks, D.S.; Hopf, T.A.; Sander, C. Protein structure prediction from sequence variation. *Nat. Biotechnol.* **2012**, *30*, 1072–1080. [CrossRef] [PubMed]
3. Taylor, W.R.; Hamilton, R.S.; Sadowski, M.I. Prediction of contacts from correlated sequence substitutions. *Curr. Opin. Struct. Biol.* **2013**, *23*, 473–479. [CrossRef] [PubMed]
4. Adhikari, B.; Bhattacharya, D.; Cao, R.; Cheng, J. CONFOLD: Residue-residue contact-guided ab initio protein folding. *Proteins* **2015**, *83*, 1436–1449. [CrossRef] [PubMed]
5. Jones, D.T.; Singh, T.; Kosciolek, T.; Tetchner, S. MetaPSICOV: Combining coevolution methods for accurate prediction of contacts and long range hydrogen bonding in proteins. *Bioinformatics* **2015**, *31*, 999–1006. [CrossRef] [PubMed]
6. Kosciolek, T.; Jones, D.T. De novo structure prediction of globular proteins aided by sequence variation-derived contacts. *PLoS ONE* **2014**, *9*, e92197. [CrossRef] [PubMed]
7. Kosciolek, T.; Jones, D.T. Accurate contact predictions using covariation techniques and machine learning. *Proteins* **2015**. [CrossRef] [PubMed]
8. Ovchinnikov, S.; Kim, D.E.; Wang, R.Y.; Liu, Y.; DiMaio, F.; Baker, D. Improved de novo structure prediction in CASP11 by incorporating Co-evolution information into rosetta. *Proteins* **2015**. [CrossRef]
9. Ovchinnikov, S.; Kinch, L.; Park, H.; Liao, Y.; Pei, J.; Kim, D.E.; Kamisetty, H.; Grishin, N.V.; Baker, D. Large-scale determination of previously unsolved protein structures using evolutionary information. *eLife* **2015**, *4*, e09248. [CrossRef] [PubMed]
10. Wang, S.; Sun, S.; Li, Z.; Zhang, R.; Xu, J. Accurate De Novo Prediction of Protein Contact Map by Ultra-Deep Learning Model. *PLoS Computat. Biol.* **2017**, *13*, e1005324. [CrossRef] [PubMed]
11. Hayat, S.; Sander, C.; Marks, D.S.; Elofsson, A. All-atom 3D structure prediction of transmembrane beta-barrel proteins from sequences. *Proc. Natl. Acad. Sci. USA* **2015**, *112*, 5413–5418. [CrossRef] [PubMed]

12. Hopf, T.A.; Colwell, L.J.; Sheridan, R.; Rost, B.; Sander, C.; Marks, D.S. Three-dimensional structures of membrane proteins from genomic sequencing. *Cell* **2012**, *149*, 1607–1621. [CrossRef] [PubMed]
13. Halabi, N.; Rivoire, O.; Leibler, S.; Ranganathan, R. Protein sectors: Evolutionary units of three-dimensional structure. *Cell* **2009**, *138*, 774–786. [CrossRef] [PubMed]
14. Hopf, T.A.; Scharfe, C.P.; Rodrigues, J.P.; Green, A.G.; Kohlbacher, O.; Sander, C.; Bonvin, A.M.; Marks, D.S. Sequence co-evolution gives 3D contacts and structures of protein complexes. *eLife* **2014**, *3*. [CrossRef] [PubMed]
15. Ovchinnikov, S.; Kamisetty, H.; Baker, D. Robust and accurate prediction of residue-residue interactions across protein interfaces using evolutionary information. *eLife* **2014**, *3*, e02030. [CrossRef] [PubMed]
16. Kamisetty, H.; Ovchinnikov, S.; Baker, D. Assessing the utility of coevolution-based residue-residue contact predictions in a sequence- and structure-rich era. *Proc. Natl. Acad. Sci. USA* **2013**, *110*, 15674–15679. [CrossRef] [PubMed]
17. Dyson, H.J.; Wright, P.E. Intrinsically unstructured proteins and their functions. *Nat. Rev. Mol. Cell Biol.* **2005**, *6*, 197–208. [CrossRef] [PubMed]
18. Habchi, J.; Tompa, P.; Longhi, S.; Uversky, V.N. Introducing protein intrinsic disorder. *Chem. Rev.* **2014**, *114*, 6561–6588. [CrossRef] [PubMed]
19. Van der Lee, R.; Buljan, M.; Lang, B.; Weatheritt, R.J.; Daughdrill, G.W.; Dunker, A.K.; Fuxreiter, M.; Gough, J.; Gsponer, J.; Jones, D.T.; et al. Classification of intrinsically disordered regions and proteins. *Chem. Rev.* **2014**, *114*, 6589–6631. [CrossRef] [PubMed]
20. Wright, P.E.; Dyson, H.J. Intrinsically unstructured proteins: Re-assessing the protein structure-function paradigm. *J. Mol. Biol.* **1999**, *293*, 321–331. [CrossRef] [PubMed]
21. Pancsa, R.; Fuxreiter, M. Interactions via intrinsically disordered regions: What kind of motifs? *IUBMB Life* **2012**, *64*, 513–520. [CrossRef] [PubMed]
22. Tompa, P.; Davey, N.E.; Gibson, T.J.; Babu, M.M. A Million Peptide Motifs for the Molecular Biologist. *Mol. Cell* **2014**, *55*, 161–169. [CrossRef] [PubMed]
23. Iakoucheva, L.M.; Radivojac, P.; Brown, C.J.; O'Connor, T.R.; Sikes, J.G.; Obradovic, Z.; Dunker, A.K. The importance of intrinsic disorder for protein phosphorylation. *Nucleic Acids Res.* **2004**, *32*, 1037–1049. [CrossRef] [PubMed]
24. Wright, P.E.; Dyson, H.J. Intrinsically disordered proteins in cellular signalling and regulation. *Nat. Rev. Mol. Cell Biol.* **2014**, *16*, 18–29. [CrossRef] [PubMed]
25. Hegyi, H.; Schad, E.; Tompa, P. Structural disorder promotes assembly of protein complexes. *BMC Struct. Biol.* **2007**, *7*, 65. [CrossRef] [PubMed]
26. Balazs, A.; Csizmok, V.; Buday, L.; Rakacs, M.; Kiss, R.; Bokor, M.; Udupa, R.; Tompa, K.; Tompa, P. High levels of structural disorder in scaffold proteins as exemplified by a novel neuronal protein, CASK-interactive protein1. *FEBS J.* **2009**, *276*, 3744–3756. [CrossRef] [PubMed]
27. Mark, W.Y.; Liao, J.C.; Lu, Y.; Ayed, A.; Laister, R.; Szymczyna, B.; Chakrabartty, A.; Arrowsmith, C.H. Characterization of segments from the central region of BRCA1: An intrinsically disordered scaffold for multiple protein-protein and protein-DNA interactions? *J. Mol. Biol.* **2005**, *345*, 275–287. [CrossRef] [PubMed]
28. Dosztanyi, Z.; Chen, J.; Dunker, A.K.; Simon, I.; Tompa, P. Disorder and sequence repeats in hub proteins and their implications for network evolution. *J. Proteome Res.* **2006**, *5*, 2985–2995. [CrossRef] [PubMed]
29. Dunker, A.K.; Cortese, M.S.; Romero, P.; Iakoucheva, L.M.; Uversky, V.N. Flexible nets. The roles of intrinsic disorder in protein interaction networks. *FEBS J.* **2005**, *272*, 5129–5148. [CrossRef] [PubMed]
30. Uversky, V.N.; Dave, V.; Iakoucheva, L.M.; Malaney, P.; Metallo, S.J.; Pathak, R.R.; Joerger, A.C. Pathological unfoldomics of uncontrolled chaos: Intrinsically disordered proteins and human diseases. *Chem. Rev.* **2014**, *114*, 6844–6879. [CrossRef] [PubMed]
31. Davey, N.E.; Van Roey, K.; Weatheritt, R.J.; Toedt, G.; Uyar, B.; Altenberg, B.; Budd, A.; Diella, F.; Dinkel, H.; Gibson, T.J. Attributes of short linear motifs. *Mol. BioSyst.* **2012**, *8*, 268–281. [CrossRef] [PubMed]
32. Van Roey, K.; Uyar, B.; Weatheritt, R.J.; Dinkel, H.; Seiler, M.; Budd, A.; Gibson, T.J.; Davey, N.E. Short linear motifs: Ubiquitous and functionally diverse protein interaction modules directing cell regulation. *Chem. Rev.* **2014**, *114*, 6733–6778. [CrossRef] [PubMed]
33. Fuxreiter, M.; Tompa, P.; Simon, I. Local structural disorder imparts plasticity on linear motifs. *Bioinformatics* **2007**, *23*, 950–956. [CrossRef] [PubMed]

34. Buljan, M.; Chalancon, G.; Eustermann, S.; Wagner, G.P.; Fuxreiter, M.; Bateman, A.; Babu, M.M. Tissue-specific splicing of disordered segments that embed binding motifs rewires protein interaction networks. *Mol. Cell* **2012**, *46*, 871–883. [CrossRef] [PubMed]
35. Weatheritt, R.J.; Davey, N.E.; Gibson, T.J. Linear motifs confer functional diversity onto splice variants. *Nucleic Acids Res.* **2012**, *40*, 7123–7131. [CrossRef] [PubMed]
36. Weatheritt, R.J.; Gibson, T.J. Linear motifs: Lost in (pre)translation. *Trends Biochem. Sci.* **2012**, *37*, 333–341. [CrossRef] [PubMed]
37. Brown, C.J.; Takayama, S.; Campen, A.M.; Vise, P.; Marshall, T.W.; Oldfield, C.J.; Williams, C.J.; Dunker, A.K. Evolutionary rate heterogeneity in proteins with long disordered regions. *J. Mol. Evol.* **2002**, *55*, 104–110. [CrossRef] [PubMed]
38. Brown, C.J.; Johnson, A.K.; Daughdrill, G.W. Comparing models of evolution for ordered and disordered proteins. *Mol. Biol. Evol.* **2010**, *27*, 609–621. [CrossRef] [PubMed]
39. Jeong, C.S.; Kim, D. Coevolved residues and the functional association for intrinsically disordered proteins. In Proceedings of the Pacific Symposium on Biocomputing, Kohala Coast, HI, USA, 3–7 January 2012; pp. 140–151. [CrossRef]
40. Toth-Petroczy, A.; Palmedo, P.; Ingraham, J.; Hopf, T.A.; Berger, B.; Sander, C.; Marks, D.S. Structured States of Disordered Proteins from Genomic Sequences. *Cell* **2016**, *167*, 158–170. [CrossRef] [PubMed]
41. Tompa, P.; Schad, E.; Tantos, A.; Kalmar, L. Intrinsically disordered proteins: Emerging interaction specialists. *Curr. Opin. Struct. Biol.* **2015**, *35*, 49–59. [CrossRef] [PubMed]
42. Dunker, A.K.; Obradovic, Z.; Romero, P.; Garner, E.C.; Brown, C.J. Intrinsic protein disorder in complete genomes. *Genome Inform.* **2000**, *11*, 161–171.
43. Pancsa, R.; Tompa, P. Structural disorder in eukaryotes. *PLoS ONE* **2012**, *7*, e34687. [CrossRef] [PubMed]
44. Ward, J.J.; Sodhi, J.S.; McGuffin, L.J.; Buxton, B.F.; Jones, D.T. Prediction and functional analysis of native disorder in proteins from the three kingdoms of life. *J. Mol. Biol.* **2004**, *337*, 635–645. [CrossRef] [PubMed]
45. Schad, E.; Ficho, E.; Pancsa, R.; Simon, I.; Dosztanyi, Z.; Meszaros, B. DIBS: A repository of disordered binding sites mediating interactions with ordered proteins. *Bioinformatics* **2018**, *34*, 535–537. [CrossRef] [PubMed]
46. Finn, R.D.; Coggill, P.; Eberhardt, R.Y.; Eddy, S.R.; Mistry, J.; Mitchell, A.L.; Potter, S.C.; Punta, M.; Qureshi, M.; Sangrador-Vegas, A.; et al. The Pfam protein families database: Towards a more sustainable future. *Nucleic Acids Res.* **2016**, *44*, D279–D285. [CrossRef] [PubMed]
47. Krissinel, E.; Henrick, K. Inference of macromolecular assemblies from crystalline state. *J. Mol. Biol.* **2007**, *372*, 774–797. [CrossRef] [PubMed]
48. Zor, T.; Mayr, B.M.; Dyson, H.J.; Montminy, M.R.; Wright, P.E. Roles of phosphorylation and helix propensity in the binding of the KIX domain of CREB-binding protein by constitutive (c-Myb) and inducible (CREB) activators. *J. Biol. Chem.* **2002**, *277*, 42241–42248. [CrossRef] [PubMed]
49. Selenko, P.; Gregorovic, G.; Sprangers, R.; Stier, G.; Rhani, Z.; Kramer, A.; Sattler, M. Structural basis for the molecular recognition between human splicing factors U2AF65 and SF1/mBBP. *Mol. Cell* **2003**, *11*, 965–976. [CrossRef]
50. Tompa, P.; Fuxreiter, M. Fuzzy complexes: Polymorphism and structural disorder in protein-protein interactions. *Trends Biochem. Sci.* **2008**, *33*, 2–8. [CrossRef] [PubMed]
51. Kayikci, M.; Venkatakrishnan, A.J.; Scott-Brown, J.; Ravarani, C.N.J.; Flock, T.; Babu, M.M. Visualization and analysis of non-covalent contacts using the Protein Contacts Atlas. *Nat. Struct. Mol. Biol.* **2018**, *25*, 185–194. [CrossRef] [PubMed]
52. Anishchenko, I.; Ovchinnikov, S.; Kamisetty, H.; Baker, D. Origins of coevolution between residues distant in protein 3D structures. *Proc. Natl. Acad. Sci. USA* **2017**, *114*, 9122–9127. [CrossRef] [PubMed]
53. Ait-Bara, S.; Carpousis, A.J.; Quentin, Y. RNase E in the gamma-Proteobacteria: Conservation of intrinsically disordered noncatalytic region and molecular evolution of microdomains. *Mol. Genet. Genom.* **2015**, *290*, 847–862. [CrossRef] [PubMed]
54. Savvides, S.N.; Raghunathan, S.; Futterer, K.; Kozlov, A.G.; Lohman, T.M.; Waksman, G. The C-terminal domain of full-length E. coli SSB is disordered even when bound to DNA. *Protein Sci.* **2004**, *13*, 1942–1947. [CrossRef] [PubMed]
55. Shereda, R.D.; Kozlov, A.G.; Lohman, T.M.; Cox, M.M.; Keck, J.L. SSB as an organizer/mobilizer of genome maintenance complexes. *Crit. Rev. Biochem. Mol. Biol.* **2008**, *43*, 289–318. [CrossRef] [PubMed]

56. Lu, D.; Keck, J.L. Structural basis of Escherichia coli single-stranded DNA-binding protein stimulation of exonuclease I. *Proc. Natl. Acad. Sci. USA* **2008**, *105*, 9169–9174. [CrossRef] [PubMed]
57. Tompa, P.; Szasz, C.; Buday, L. Structural disorder throws new light on moonlighting. *Trends Biochem. Sci.* **2005**, *30*, 484–489. [CrossRef] [PubMed]
58. Finn, R.D.; Clements, J.; Arndt, W.; Miller, B.L.; Wheeler, T.J.; Schreiber, F.; Bateman, A.; Eddy, S.R. HMMER web server: 2015 update. *Nucleic Acids Res.* **2015**, *43*, W30–W38. [CrossRef] [PubMed]
59. Apweiler, R.; Bairoch, A.; Wu, C.H.; Barker, W.C.; Boeckmann, B.; Ferro, S.; Gasteiger, E.; Huang, H.; Lopez, R.; Magrane, M.; et al. UniProt: The Universal Protein knowledgebase. *Nucleic Acids Res.* **2004**, *32*, D115–D119. [CrossRef] [PubMed]
60. Szklarczyk, D.; Franceschini, A.; Wyder, S.; Forslund, K.; Heller, D.; Huerta-Cepas, J.; Simonovic, M.; Roth, A.; Santos, A.; Tsafou, K.P.; et al. STRING v10: Protein-protein interaction networks, integrated over the tree of life. *Nucleic Acids Res.* **2015**, *43*, D447–D452. [CrossRef] [PubMed]
61. Kersey, P.J.; Allen, J.E.; Armean, I.; Boddu, S.; Bolt, B.J.; Carvalho-Silva, D.; Christensen, M.; Davis, P.; Falin, L.J.; Grabmueller, C.; et al. Ensembl Genomes 2016: More genomes, more complexity. *Nucleic Acids Res.* **2016**, *44*, D574–D580. [CrossRef] [PubMed]
62. Newcombe, R.G. Interval estimation for the difference between independent proportions: Comparison of eleven methods. *Stat. Med.* **1998**, *17*, 873–890. [CrossRef]
63. Touw, W.G.; Baakman, C.; Black, J.; te Beek, T.A.; Krieger, E.; Joosten, R.P.; Vriend, G. A series of PDB-related databanks for everyday needs. *Nucleic Acids Res.* **2015**, *43*, D364–D368. [CrossRef] [PubMed]

4

Arabidopsis Heat Stress-Induced Proteins are Enriched in Electrostatically Charged Amino Acids and Intrinsically Disordered Regions

David Alvarez-Ponce [1,2,*], Mario X. Ruiz-González [2,3], Francisco Vera-Sirera [2], Felix Feyertag [1], Miguel A. Perez-Amador [2] and Mario A. Fares [2,3,†]

1 Biology Department, University of Nevada, Reno, NV 89557, USA; ffeyertag@unr.edu
2 Instituto de Biología Molecular y Celular de Plantas, CSIC-UPV, 46022 Valencia, Spain; marioxruizgonzalez@gmail.com (M.X.R.-G.); fravesi@ibmcp.upv.es (F.V.-S.); mpereza@ibmcp.upv.es (M.A.P.-A.)
3 Smurfit Institute of Genetics, University of Dublin, Trinity College, Dublin 2, Ireland
* Correspondence: dap@unr.edu
† Posthumous author.

Abstract: Comparison of the proteins of thermophilic, mesophilic, and psychrophilic prokaryotes has revealed several features characteristic to proteins adapted to high temperatures, which increase their thermostability. These characteristics include a profusion of disulfide bonds, salt bridges, hydrogen bonds, and hydrophobic interactions, and a depletion in intrinsically disordered regions. It is unclear, however, whether such differences can also be observed in eukaryotic proteins or when comparing proteins that are adapted to temperatures that are more subtly different. When an organism is exposed to high temperatures, a subset of its proteins is overexpressed (heat-induced proteins), whereas others are either repressed (heat-repressed proteins) or remain unaffected. Here, we determine the expression levels of all genes in the eukaryotic model system *Arabidopsis thaliana* at 22 and 37 °C, and compare both the amino acid compositions and levels of intrinsic disorder of heat-induced and heat-repressed proteins. We show that, compared to heat-repressed proteins, heat-induced proteins are enriched in electrostatically charged amino acids and depleted in polar amino acids, mirroring thermophile proteins. However, in contrast with thermophile proteins, heat-induced proteins are enriched in intrinsically disordered regions, and depleted in hydrophobic amino acids. Our results indicate that temperature adaptation at the level of amino acid composition and intrinsic disorder can be observed not only in proteins of thermophilic organisms, but also in eukaryotic heat-induced proteins; the underlying adaptation pathways, however, are similar but not the same.

Keywords: temperature response; protein thermostability; salt bridges; intrinsically disordered proteins

---

1. Introduction

Proteins of thermophilic prokaryotes (those adapted to high temperatures) exhibit several distinctive features that increase their thermostability. One of the most consistent observations in thermophile proteins is an enrichment in salt bridges [1,2]. Salt bridges consist of electrostatic interactions among amino acid residues with positive (Lys and Arg) and negative (Glu and Asp) charges, and their contribution to increasing the stability of thermophilic bacteria was first proposed by Perutz and Raidt [3]. In addition, compared with proteins of mesophiles (adapted to intermediate temperatures) and psychrophiles (adapted to low temperatures), thermophile proteins tend to exhibit more disulfide bonds and non-covalent interactions, including hydrogen bonds, and hydrophobic interactions, features that also tend to increase protein stability by linking together distant parts of the amino acid sequence [4,5]. These structural trends have an impact on the amino acid composition of thermophilic proteomes: the proteins of thermophilic bacteria tend to be enriched in charged amino acids and depleted in polar ones such as Ser, Thr, Asn, and Gln [6–12].

A few studies in prokaryotes have also shown that thermophile proteins are depleted in intrinsically disordered regions (IDRs), i.e., regions that lack a defined three-dimensional structure [13–15]. This observation is consistent with the fact that high temperatures induce disorder, but in contrast with the fact that IDRs confer thermoresistance [16–18].

Much less is known about how eukaryotic proteomes adapt to high temperatures. Some studies have suggested that the same biases in amino acid composition observed in thermophilic prokaryotes can be observed in thermophilic fungi (compared to other fungi; ref. [19]) and endothermic vertebrates (compared to ectothermic vertebrates; ref. [20]). In agreement with this notion, comparison of the orthologous proteins of two closely related fish, *Pachycara brachycephalum* (from Antarctica) and *Zoarces viviparous* (from a temperate zone) revealed an excess of Ser and a reduction of Glu and Asn in the cold-adapted species [21]. To our knowledge, the relationship between temperature and intrinsic disorder has not been investigated in eukaryotic proteomes.

Protein adaptation to high temperatures is expected to be observed not only in the proteins of thermophilic organisms, but also in some of the proteins of any mesophilic organism. When an organism is exposed to high temperatures, a subset of its proteins is overexpressed, whereas others are repressed (heat-induced and heat-repressed proteins, respectively, e.g., ref. [22]). As heat-induced function at relatively high temperatures, we hypothesize that they should be similar to those of thermophilic organisms.

Plants represent particularly suitable models to test this hypothesis, as they are sessile organisms that cannot escape from their environment, and they lack the effective thermoregulation mechanisms exhibited by homeotherms. Therefore, plants are expected to have developed adaptations to cope with heat stress [23]. To test our hypothesis, we grew *Arabidopsis thaliana* plants under normal (22 °C) and heat stress conditions (37 °C), and measured gene expression levels. Proteins overexpressed under heat stress were enriched in electrostatically charged amino acids and depleted in polar and hydrophobic amino acids. However, in contrast with our expectations, these proteins were also enriched in IDRs. These results indicate that *Arabidopsis* heat-induced proteins exploit some, but not all the same mechanisms as thermophile proteins to cope with high temperatures.

2. Results

2.1. *Proteins That Are Overexpressed at High Temperatures Are Enriched in Electrostatically Charged Amino Acids and Depleted in Polar and Hydrophobic Amino Acids*

We grew *Arabidopsis* plants at 22 and 37 °C for 24 h, and performed microarray analyses to measure gene expression levels at the beginning of the experiment ($E_{0,22}$ = expression at time 0 and 22 °C) and at the end of the experiment ($E_{24,22}$ and $E_{24,37}$). $E_{0,22}$ strongly correlated with $E_{24,22}$ (Spearman's rank correlation coefficient, $\rho = 0.991$, $p < 10^{-200}$; Figure 1) supporting the robustness of our gene expression measures—the small differences between gene expression at both time points could be due to differences in gene expression during development and to measurement errors. The correlation between $E_{24,22}$ and $E_{24,37}$ was weaker ($\rho = 0.897$, $p = 10^{-200}$; Figure 2), highlighting the effect of heat stress on the expression of many genes.

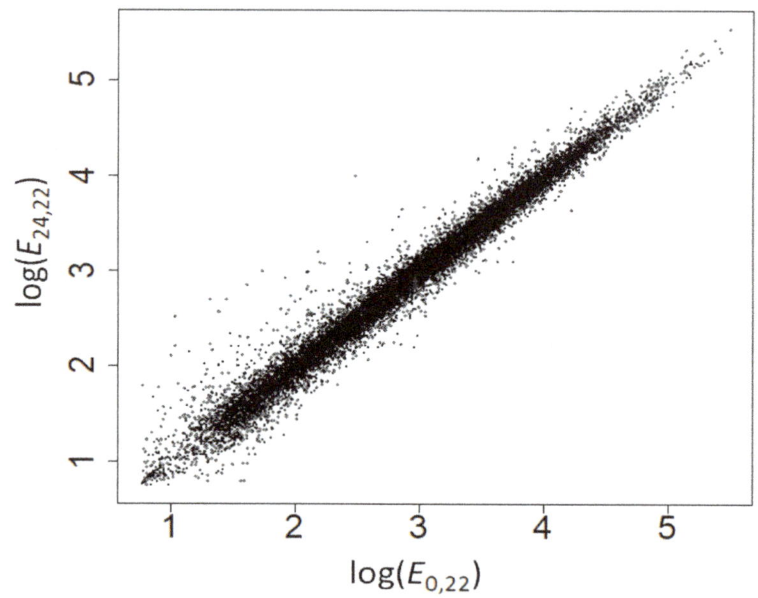

Figure 1. Correlation between gene expression levels at 22 °C at time 0 and at time 24 h.

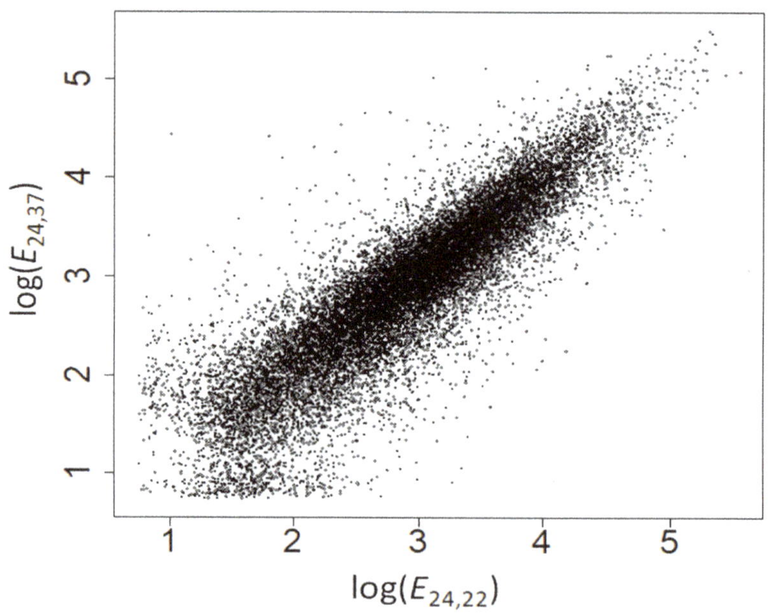

Figure 2. Correlation between gene expression levels at 22 °C at time 24 h and at 37 °C at time 24 h.

For each gene with available probes (n = 20,491), we computed a response to heat stress (R) as the logarithm in base 2 of the ratio of expression levels at 37 and 22 °C (following formula 1). Genes with $R > 0$ are overexpressed at high temperatures, and genes with $R < 0$ are repressed. Genes with $R > 1$ (strongly overexpressed) are enriched in Gene Ontology biological processes "protein refolding", "protein folding", "chaperone cofactor-dependent protein refolding", "chaperone-mediated protein folding", "de novo posttranslational protein folding", "de novo protein folding", "cellular response to heat", "response to heat", "response to temperature stimulus", and "heat acclimation". They are also enriched in molecular functions "misfolded protein binding", "heat shock protein binding", "protein binding involved in protein folding", and "unfolded protein binding" (Tables S1–S3).

We observed a positive correlation between R and the fraction of charged amino acids ($\rho = 0.146$, $p = 2.47 \times 10^{-98}$), and negative correlations between R and both the fraction of polar ($\rho = -0.076$, $p = 1.72 \times 10^{-27}$) and hydrophobic ($\rho = -0.084$, $p = 4.08 \times 10^{-33}$) amino acids (Figure 3). We next computed the correlation between R and the frequency of each amino acid separately. The correlation was significantly positive for all four charged amino acids (Arg, Asp, Glu, and Lys), negative for all hydrophobic amino acids (significant for Gly, Ile, Phe, Pro, and Val), except Met (for which the correlation was non-significantly positive), and negative for all polar amino acids (significant for Asn, Ser, Thr, Trp and Tyr), except for Gln, for which the correlation was significantly positive (Table 1). All these correlations remained significant after controlling for multiple testing (Table 1).

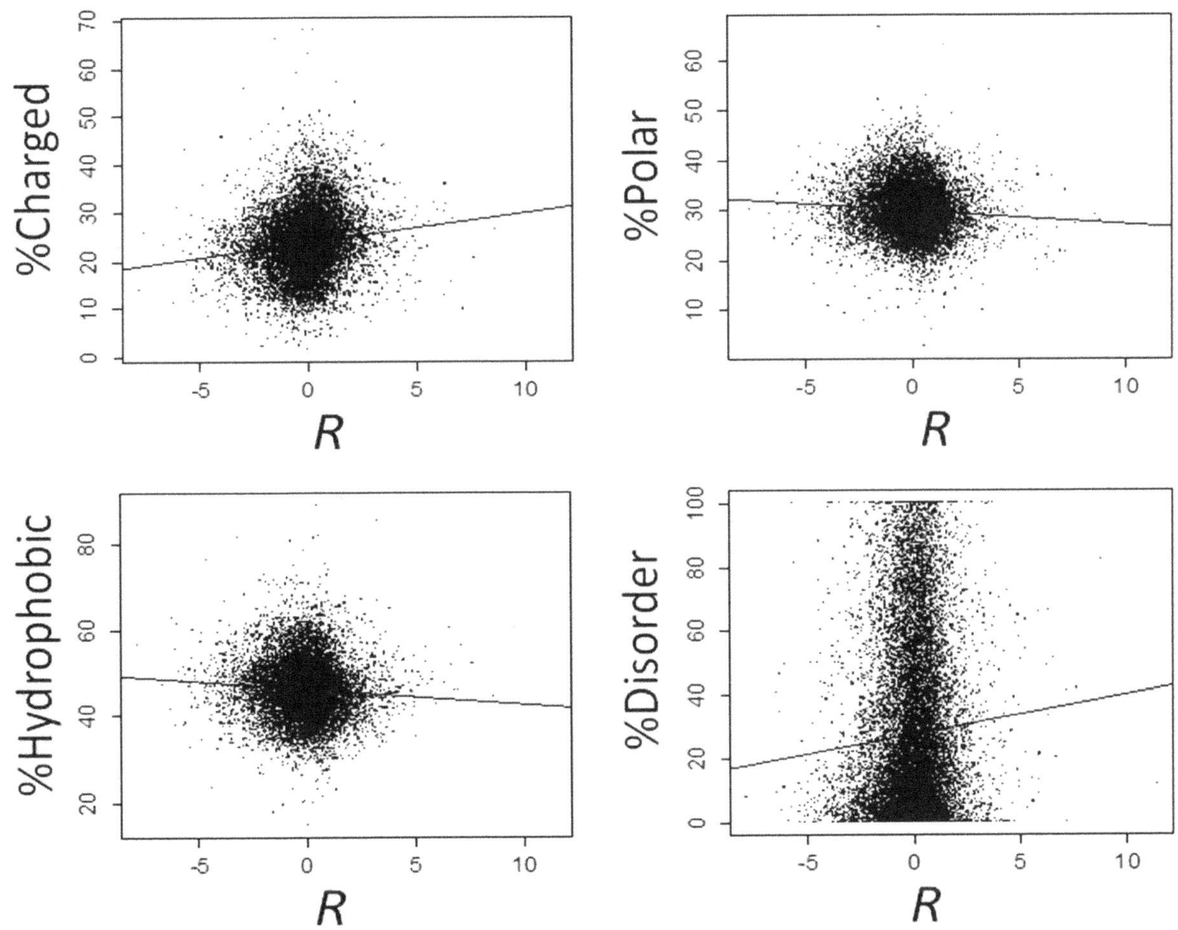

Figure 3. Correlations between response to high temperature (R) and the fraction of charged, polar, hydrophobic and disordered amino acids. Lines represent regression lines.

Table 1. Correlations between amino acid frequencies and response to high temperature.

Type	Amino Acid	No Control			Controlling for GC Content			Controlling for $E_{24,22}$			Controlling for $E_{24,37}$		
		ρ	p-Value	q-Value	ρ	p-Value	q-Value	ρ	p-Value	q-Value	ρ	p-Value	q-Value
Charged	Arg	0.075	**1.31×10^{-26}**	**4.37×10^{-26}**	0.068	**2.05×10^{-22}**	**5.86×10^{-22}**	0.068	**0.013**	**0.015**	0.079	**1.02×10^{-29}**	**3.40×10^{-29}**
	Asp	0.104	**1.62×10^{-50}**	**1.62×10^{-49}**	0.105	**9.95×10^{-52}**	**9.95×10^{-51}**	0.106	**7.16×10^{-53}**	**7.16×10^{-52}**	0.095	**1.84×10^{-42}**	**1.23×10^{-41}**
	Glu	0.118	**5.48×10^{-64}**	**1.10×10^{-62}**	0.122	**5.60×10^{-69}**	**1.12×10^{-67}**	0.115	**2.61×10^{-61}**	**5.22×10^{-60}**	0.115	**2.60×10^{-61}**	**5.20×10^{-60}**
	Lys	0.082	**8.23×10^{-32}**	**4.12×10^{-31}**	0.100	**9.58×10^{-47}**	**4.79×10^{-46}**	0.081	**2.27×10^{-31}**	**9.08×10^{-31}**	0.079	**8.76×10^{-30}**	**3.40×10^{-29}**
	Total	0.146	**2.47×10^{-98}**		0.155	**3.61×10^{-111}**		0.145	**1.04×10^{-97}**		0.140	**1.65×10^{-90}**	
Polar	Asn	−0.025	**3.86×10^{-4}**	**0.001**	−0.018	**0.011**	**0.015**	−0.044	**4.17×10^{-10}**	**8.34×10^{-10}**	0.005	0.433	0.433
	Cys	−0.011	0.127	0.158	−0.009	0.187	0.208	−0.034	**1.54×10^{-6}**	**2.57×10^{-6}**	0.026	**2.07×10^{-4}**	**3.19×10^{-4}**
	Gln	0.046	**3.20×10^{-11}**	**7.11×10^{-11}**	0.053	**2.36×10^{-14}**	**5.24×10^{-14}**	0.046	**6.79×10^{-11}**	**1.51×10^{-10}**	0.044	**3.95×10^{-10}**	**7.90×10^{-10}**
	His	−0.010	0.134	0.158	−0.009	0.210	0.221	−0.024	**0.001**	**0.001**	0.010	0.146	0.154
	Ser	−0.036	**2.25×10^{-7}**	**4.09×10^{-7}**	−0.042	**2.36×10^{-9}**	**4.72×10^{-9}**	−0.052	**1.00×10^{-13}**	**2.86×10^{-13}**	−0.012	0.092	0.102
	Thr	−0.099	**1.10×10^{-45}**	**7.33×10^{-45}**	−0.100	**9.24×10^{-47}**	**4.79×10^{-46}**	−0.098	**1.12×10^{-44}**	**7.47×10^{-44}**	−0.096	**2.75×10^{-43}**	**2.75×10^{-42}**
	Trp	−0.033	**2.26×10^{-6}**	**3.77×10^{-6}**	−0.036	**2.50×10^{-7}**	**4.55×10^{-7}**	−0.039	**2.11×10^{-8}**	**3.84×10^{-8}**	−0.022	**0.002**	**0.002**
	Tyr	−0.024	**0.001**	**0.001**	−0.016	**0.021**	**0.026**	−0.025	**3.72×10^{-4}**	**4.96×10^{-4}**	−0.021	**0.003**	**0.004**
	Total	−0.076	**1.72×10^{-27}**		−0.072	**1.11×10^{-24}**		−0.102	**9.48×10^{-49}**		−0.034	**9.25×10^{-7}**	
Hydro phobic	Ala	−0.008	0.280	0.311	−0.020	**0.004**	**0.006**	0.027	**1.32×10^{-4}**	**1.89×10^{-4}**	−0.060	**1.50×10^{-17}**	**3.75×10^{-17}**
	Gly	−0.054	**1.40×10^{-14}**	**3.50×10^{-14}**	−0.066	**1.99×10^{-21}**	**4.98×10^{-21}**	−0.028	**5.46×10^{-5}**	**8.40×10^{-5}**	−0.092	**1.17×10^{-39}**	**5.85×10^{-39}**
	Ile	−0.045	**1.01×10^{-10}**	**2.02×10^{-10}**	−0.035	**5.63×10^{-7}**	**9.38×10^{-7}**	−0.052	**1.55×10^{-13}**	**3.88×10^{-13}**	−0.033	**2.91×10^{-6}**	**5.29×10^{-6}**
	Leu	−0.004	0.547	0.547	−0.004	0.533	0.533	−0.016	**0.021**	**0.023**	0.015	**0.029**	**0.034**
	Met	0.006	0.387	0.407	0.014	**0.042**	**0.049**	−0.001	0.942	0.942	0.017	**0.017**	**0.021**
	Phe	−0.075	**1.04×10^{-26}**	**4.16×10^{-26}**	−0.070	**9.79×10^{-24}**	**3.26×10^{-23}**	−0.084	**2.59×10^{-33}**	**1.30×10^{-32}**	−0.056	**1.36×10^{-15}**	**3.02×10^{-15}**
	Pro	−0.060	**8.03×10^{-18}**	**2.29×10^{-17}**	−0.074	**1.85×10^{-26}**	**7.40×10^{-26}**	−0.052	**8.41×10^{-14}**	**2.80×10^{-13}**	−0.070	**7.88×10^{-24}**	**2.25×10^{-23}**
	Val	−0.017	**0.012**	**0.017**	−0.024	**0.001**	**0.001**	−0.006	0.370	0.390	−0.033	**3.30×10^{-6}**	**5.50×10^{-6}**
	Total	−0.084	**4.08×10^{-33}**		−0.096	**1.31×10^{-43}**		−0.064	**2.88×10^{-20}**		−0.109	**2.73×10^{-55}**	

p-values and q-values shown in bold face represent significant tests at $\alpha = 0.05$ or $q = 0.05$.

Next, we compared the amino acid composition of proteins encoded by genes that are overexpressed ($R > 0$, n = 10,728) vs. proteins encoded by genes that are repressed ($R < 0$, n = 9763) at 37 °C. Overexpressed proteins were enriched in charged amino acids (median percent in overexpressed proteins: 24.32%; median percent in repressed proteins: 23.20%; Mann-Whitney's U test, $p = 1.90 \times 10^{-66}$) and depleted in both polar (median percent in overexpressed proteins: 29.54%; median percent in repressed proteins: 30.04%; $p = 2.53 \times 10^{-20}$) and hydrophobic (median percent in overexpressed proteins: 45.77%; median percent in repressed proteins: 46.43%; $p = 6.56 \times 10^{-21}$) amino acids. In almost perfect agreement with our correlation analyses, proteins encoded by overexpressed genes were significantly enriched in Arg, Asp, Gln, Glu, and Lys, and significantly depleted in Asn, Gly, Ile, Phe, Pro, Ser, Thr, and Trp (Table 2).

Table 2. Amino acid frequencies in overexpressed ($R > 0$) and repressed ($R < 0$) proteins at high temperatures.

Type	Amino Acid	Median Overexpressed (%)	Median Repressed (%)	p-Value	q-Value
Charged	Arg	5.43	5.19	**8.06×10^{-21}**	**4.61×10^{-20}**
	Asp	5.36	5.10	**1.60×10^{-36}**	**6.40×10^{-35}**
	Glu	6.61	6.15	**8.28×10^{-44}**	**6.62×10^{-42}**
	Lys	6.33	6.06	**1.20×10^{-21}**	**8.00×10^{-21}**
	Total	24.32	23.20	**1.90×10^{-66}**	
Polar	Asn	4.08	4.12	**0.017**	**0.024**
	Cys	1.59	1.60	**0.043**	0.060
	Gln	3.27	3.16	**7.77×10^{-8}**	**1.88×10^{-7}**
	His	2.11	2.10	0.204	0.244
	Ser	8.79	8.96	**2.27×10^{-7}**	**5.19×10^{-7}**
	Thr	4.90	5.13	**5.31×10^{-34}**	**1.42×10^{-32}**
	Trp	1.07	1.11	**4.75×10^{-4}**	**0.001**
	Tyr	2.65	2.68	0.132	0.163
	Total	29.54	30.04	**2.53×10^{-20}**	
Hydrophobic	Ala	6.32	6.30	0.889	0.889
	Gly	6.18	6.41	**2.77×10^{-10}**	**8.21×10^{-10}**
	Ile	5.12	5.23	**1.87×10^{-7}**	**4.40×10^{-7}**
	Leu	9.24	9.27	0.675	0.720
	Met	2.38	2.37	0.399	0.449
	Phe	4.08	4.28	**1.55×10^{-18}**	**7.75×10^{-18}**
	Pro	4.54	4.71	**2.56×10^{-12}**	**8.53×10^{-12}**
	Val	6.67	6.68	0.178	0.215
	Total	45.77	46.43	**6.56×10^{-21}**	

p-values correspond to the Mann-Whitney's U test. p-values and q-values shown in bold face represent significant tests at $\alpha = 0.05$ or $q = 0.05$.

Similar results were obtained when using a more stringent threshold to classify genes as overexpressed ($R > 2$, n = 826) or repressed ($R < -2$, n = 1214) at 37 °C. Overexpressed proteins are enriched in charged amino acids (median percent in overexpressed proteins: 25.30%; median percent in repressed proteins: 22.54%; $p = 1.50 \times 10^{-26}$) and depleted in both polar (median percent in overexpressed proteins: 29.74%; median percent in repressed proteins: 30.17%; $p = 3.20 \times 10^{-8}$) and hydrophobic (median percent in overexpressed proteins: 45.20%; median percent in repressed proteins: 47.24%; $p = 6.04 \times 10^{-11}$) amino acids. More specifically, overexpressed proteins are significantly enriched in Arg, Asp, Gln, Glu, and Lys, and significantly depleted in Asn, Cys, Gly, His, Ile, Phe, Pro, Thr, Trp, and Tyr (Table 3).

Table 3. Amino acid frequencies in highly overexpressed ($R > 2$) and highly repressed ($R < -2$) proteins at high temperatures.

Type	Amino Acid	Median Overexpressed (%)	Median Repressed (%)	p-Value	q-Value
Charged	Arg	5.26	4.80	**7.82×10^{-9}**	**1.04×10^{-7}**
	Asp	5.51	4.95	**1.62×10^{-12}**	**4.32×10^{-11}**
	Glu	6.92	5.92	**1.31×10^{-17}**	**1.05×10^{-15}**
	Lys	6.78	6.17	**1.78×10^{-7}**	**1.78×10^{-6}**
	Total	25.30	22.54	**1.50×10^{-26}**	
Polar	Asn	4.04	4.29	**2.81×10^{-4}**	**0.001**
	Cys	1.66	1.69	**0.031**	**0.045**
	Gln	3.13	2.94	**2.87×10^{-4}**	**6.57×10^{-4}**
	His	2.03	2.12	**0.023**	**0.035**
	Ser	8.47	8.41	0.780	0.810
	Thr	4.95	5.26	**3.52×10^{-6}**	**1.56×10^{-5}**
	Trp	1.05	1.15	**0.035**	**0.050**
	Tyr	2.57	2.86	**6.09×10^{-5}**	**1.87×10^{-4}**
	Total	29.74	30.17	**3.20×10^{-8}**	
Hydrophobic	Ala	6.11	6.12	0.867	0.878
	Gly	6.05	6.50	**5.49×10^{-5}**	**1.76×10^{-4}**
	Ile	5.25	5.48	**0.001**	**0.002**
	Leu	9.01	9.17	0.215	0.292
	Met	2.46	2.52	0.321	0.395
	Phe	4.12	4.65	**9.55×10^{-13}**	**3.82×10^{-11}**
	Pro	4.31	4.62	**9.34×10^{-5}**	**2.58×10^{-4}**
	Val	6.76	6.84	0.294	0.386
	Total	45.20	47.24	**6.04×10^{-11}**	

p-values correspond to the Mann-Whitney's U test. p-values and q-values shown in bold face represent significant tests at $\alpha = 0.05$ or $q = 0.05$.

2.2. The Amino Acid Composition of Heat-Induced Proteins Is Not due to Covariation of Amino Acid Composition with GC Content, Gene Expression Levels, or Subcellular Location

We considered whether our results could be affected by confounding factors. First, GC content is known to affect amino acid composition [24], and R significantly correlates with GC content ($\rho = 0.088$, $p = 9.76 \times 10^{-37}$). Combined, these correlations alone might potentially explain the observed trends. To discard this possibility, we computed partial correlations between R and the frequency of each amino acid, while controlling for GC content, with very similar results. The correlation continued to be significantly positive for charged amino acids and significantly negative for polar and hydrophobic ones (Table 1). More specifically, the correlation was significantly positive for Arg, Asp, Gln, Glu, and Lys and significantly negative for Asn, Gly, Ile, Phe, Pro, Ser, Thr, Trp, Tyr, and Val. Both the negative correlation between R and Ala frequency and the positive correlation between R and Met frequency, which were initially not significant, became significant after controlling for GC content (Table 1).

Second, highly expressed proteins resemble proteins from thermophiles in their amino acid composition [25], and expression levels correlate with R (expression level at 22 °C: $\rho = -0.156$, $p = 4.88 \times 10^{-112}$; expression level at 37 °C: $\rho = 0.241$, $p = 1.18 \times 10^{-268}$). To discard the potential confounding effects of expression levels, we computed partial correlations between R and the frequency of each amino acid, while controlling for expression levels, again with very similar results. When controlling for expression levels at 22 °C, R correlated positively with the frequencies of Ala, Arg, Asp, Gln, Glu, and Lys and negatively with the frequencies of Asn, Cys, Gly, His, Ile, Leu, Phe, Pro, Ser, Thr, Trp, and Tyr. When controlling for expression levels at 37 °C, R correlated positively with the frequencies of Arg, Asp, Cys, Gln, Glu, Leu, Lys, and Met and negatively with the frequencies of Ala, Gly, Ile, Phe, Pro, Thr, Trp, Tyr, and Val. In both cases, the positive correlations between R and the frequency charged amino acids and the negative correlations between R and the frequencies of polar and hydrophobic amino acids remained significant (Table 1).

Proteins locating to different parts of the cell differ in their amino acid compositions and in their response to heat stress ([26,27]; Table 4). To discard subcellular location as a confounding factor, we analyzed the correlation between R and the amino acid composition separately for proteins locating to 10 different subcellular compartments (Table 5). The correlation between R and the fraction of charged amino acids was positive in nine of the compartments, which represents a significant departure from the 50% expected at random (one-tailed binomial test, $p = 0.011$). The correlation was significantly positive for the cytosol, the plastid (the compartments with the higher number of known/inferred proteins), and the mitochondrion. The correlation between R and the fraction of hydrophobic amino acids was negative in eight of the compartments (one-tailed binomial test, $p = 0.055$), significantly negative in the plastid and the mitochondrion, and significantly positive in the nucleus. The correlation between R and the fraction of polar amino acids was negative in half of the compartments, and significantly negative in the cytosol and the nucleus. These results suggest that the enrichment of heat-induced proteins in charged amino acids and their depletion in hydrophobic amino acids are not a byproduct of covariation of both R and amino acid composition with subcellular location. The lack of significance in most of the individual correlations is probably due to the low number of proteins for which location information is available, ranging from 720 for the plastid to 63 in the peroxisome (Table 4), which is expected to greatly reduce the statistical power of our compartment-specific analyses. However, we note an exception: among nuclear proteins R exhibits a significantly positive correlation with the percent of hydrophobic residues (Table 5).

2.3. Proteins That Are Overexpressed at High Temperatures Are Highly Disordered

For each *Arabidopsis* protein, we computed the percentage of amino acids that belong to IDRs using IUPred [28]. This percentage correlates positively with R ($\rho = 0.059$, $p = 4.93 \times 10^{-17}$; Figure 3). Genes that are overexpressed at 37 °C ($R > 0$) encode proteins that are more disordered than those that are repressed ($R < 0$), with median disorder percent of 19.19% and 16.51% for induced and repressed genes, respectively (Mann-Whitney's U test, $p = 2.01 \times 10^{-35}$). The differences are more solid when comparing genes that are strongly overexpressed at 37 °C ($R > 2$) vs. those that are strongly repressed ($R < -2$), with percentages of median disorder of 21.54% and 11.51% for induced and repressed genes, respectively (Mann-Whitney's U test, $P = 2.03 \times 10^{-23}$).

In agreement with previous works [29,30], we found a positive correlation between GC content and the percent of disordered residues ($\rho = 0.044$, $p = 2.84 \times 10^{-10}$). In addition, GC content positively correlates with R ($\rho = 0.088$, $p = 9.76 \times 10^{-37}$), making it possible that the positive correlation between R and disorder might be due to the covariation of both parameters with GC content. The correlation between R and disorder, however, is significant, even after controlling for GC content ($\rho = 0.055$, $p = 3.44 \times 10^{-15}$).

Likewise, intrinsic disorder positively correlates with expression levels (at 22 °C: $\rho = 0.040$, $p = 1.03 \times 10^{-8}$; and at 37 °C: $\rho = 0.072$, $p = 7.75 \times 10^{-25}$), in agreement with previous results in *Escherichia coli* [31], but in contrast with observations in yeasts [32,33]. Disorder, however, significantly correlates with R after controlling for expression levels (at 22 °C: $\rho = 0.066$, $p = 4.64 \times 10^{-21}$; and at 37 °C: $\rho = 0.043$, $p = 1.03 \times 10^{-9}$).

Both intrinsic disorder and R substantially vary among proteins locating to different subcellular compartments (Table 4), thus raising the possibility that covariation of both factors with subcellular location may account for the observed enrichment of stress-induced proteins in IDRs. We analyzed the correlation between intrinsic disorder and R separately for proteins locating to 10 different subcellular compartments. The correlation was positive for eight of the tissues (significantly positive for the cytosol, endoplasmic reticulum, and the vacuole) and significantly negative for the nucleus and the plasma membrane (Table 5). These results indicate that the positive correlation between disorder and R, while generalized, does not apply to proteins locating to all compartments.

Table 4. Amino acid composition, intrinsic disorder and response to heat stress of proteins locating to different subcellular locations.

Subcellular Location	n	Median Charged Amino Acids (%)	Median Polar Amino Acids (%)	Median Hydrophobic Amino Acids (%)	Median Intrinsic Disorder (%)	Median R
Cytosol	633	25.46	26.74	47.31	15.64	0.131
Endoplasmic reticulum	163	24.12	27.22	48.68	10.11	0.147
Extracellular	197	18.94	32.87	48.43	8.61	−0.296
Golgi	375	23.20	29.37	47.38	14.22	0.108
Mitochondrion	286	23.12	27.63	49.26	14.93	0.261
Nucleus	446	26.50	29.16	43.92	42.73	0.406
Peroxisome	63	23.16	26.47	50.00	10.61	−0.207
Plasma membrane	343	22.21	28.63	48.73	14.97	−0.195
Plastid	720	23.33	27.65	48.91	15.28	−0.190
Vacuole	81	21.14	28.24	49.75	8.12	−0.008

Table 5. Correlations between amino acid frequencies and response to high temperature among proteins of different subcellular locations.

Subcellular Location	Correlation R-Charged Amino Acids		Correlation R-Polar Amino Acids		Correlation R-Hydrophobic Amino Acids		Correlation R-Intrinsic Disorder	
	ρ	p-Value	ρ	p-Value	ρ	p-Value	ρ	p-Value
Cytosol	0.171	**1.54×10^{-5}**	−0.142	**3.27×10^{-4}**	−0.069	0.082	0.123	**0.002**
Endoplasmic reticulum	0.061	0.437	−0.015	0.847	−0.112	0.155	0.226	**0.004**
Extracellular	0.054	0.452	0.021	0.765	−0.073	0.309	0.068	0.346
Golgi	−0.046	0.370	0.068	0.191	−0.009	0.866	0.073	0.156
Mitochondrion	0.124	**0.036**	0.060	0.312	−0.125	**0.034**	0.065	0.272
Nucleus	0.020	0.681	−0.119	**0.012**	0.102	**0.031**	−0.207	**1.09×10^{-5}**
Peroxisome	0.016	0.902	−0.104	0.416	0.080	0.535	0.237	0.062
Plasma membrane	0.064	0.234	−0.017	0.750	−0.004	0.947	−0.154	**0.004**
Plastid	0.137	**2.20×10^{-4}**	0.007	0.859	−0.110	**0.003**	0.062	0.095
Vacuole	0.184	0.099	0.082	0.466	−0.189	0.091	0.266	**0.017**

p-values shown in bold face represent significant tests at $\alpha = 0.05$.

3. Discussion

We show that *Arabidopsis* proteins whose expression levels increase at high temperatures (heat-induced proteins) are enriched in charged amino acids, and depleted in polar and hydrophobic amino acids, compared to heat-repressed proteins. The enrichment of heat-induced proteins in charged amino acids and the depletion in polar amino acids are trends that mirror those observed in the proteins of thermophilic prokaryotes. The observed enrichment of heat-induced proteins in electrostatically charged amino acids was expected, as such amino acids can engage in salt bridges, which usually increase protein thermostability [1–3]—it should be noted, nonetheless, that not all charged amino acids participate in salt bridges, and that not all salt bridges increase thermostability [34]. However, the depletion of heat-induced proteins in hydrophobic amino acids was not expected, as the proteins of thermophilic prokaryotes are usually enriched in such amino acids (e.g., ref. [35]).

Despite the overall observed trends (heat-induced proteins being enriched in charged amino acids and depleted in polar and hydrophobic amino acids), not all amino acids vary according to these rules. In particular, the frequencies of Cys (polar), His (polar), Ala (hydrophobic), Leu (hydrophobic), and Met (hydrophobic) do not correlate significantly with R, and Gln (a polar amino acid) is more frequent in heat-induced proteins than in heat-repressed ones (Table 1). The enrichment of heat-induced proteins in Gln is surprising, given its tendency to undergo deamination at high temperatures [36].

We show that the observed overall trends are not due to heat-induced genes/proteins being different in terms of expression levels, GC content or subcellular location. When controlling for these factors, however, the direction of the correlations for certain amino acids change (Table 1). Thus, the observed trends in amino acid composition are likely the result of adaptation of heat-induced and heat-repressed *Arabidopsis* proteins to high and low temperatures, respectively.

Burra et al. [13] predicted that the proteins of thermophilic prokaryotes should be enriched in IDRs, as intrinsically disorder proteins are often resistant to high temperatures [16–18]. However, contradicting their predictions, they observed that thermophiles often are depleted in IDRs, which may compensate for the disorder induced by temperature. Similar observations were made in both another proteome-level analysis [15] and an analysis of FlgM proteins from bacteria adapted to different temperatures [14]. In agreement with Burra et al.'s prediction, we observed that *Arabidopsis* heat-induced proteins are enriched in IDRs. Our results suggest that there are different ways in which ordered/disordered regions can promote thermostability.

The correlations described in the current work are moderate, albeit statistically significant. Several scenarios may account for the weakness of the correlations. First, amino acid composition and protein intrinsic disorder may be affected by factors other than temperature. Second, the difference between the temperatures used in this study (22 vs. 37 °C) is small compared to the differences between the optimal temperatures of psychrophiles, mesophiles, and thermophiles. Third, certain plant genes may have changed their patterns of response to heat stress during the recent evolutionary history of *Arabidopsis*. i.e., certain genes that are currently heat-induced may have been heat-repressed in the past, and certain genes that are currently heat-repressed may have been heat-induced in the past. As amino acid and disorder adjustment to temperature is expected to take a relatively long amount of time, such switches in expression profiles may have limited the adaptation of proteomes to temperatures. Fourth, the adaptability of plant proteomes to temperatures may be more limited than that of prokaryotic proteomes, e.g., due to the higher complexity of protein-protein interaction networks and the smaller effective population size of plants [37].

In summary, the amino acid composition of heat-induced proteins in *Arabidopsis* mirrors to some extent, but not completely, that of the proteomes of thermophilic prokaryotes. This indicates that protein adaptation to high temperatures takes place partly through similar molecular mechanisms in prokaryotes and eukaryotes. Our observations also indicate that adaptation of proteins at the level of amino acid composition and protein intrinsic disorder can be detected not only when comparing the proteomes of species adapted to very different temperatures, but also among the proteins of the same species with different temperature response profiles. These observations expand our view of how eukaryotic proteomes adapt to different temperatures.

4. Materials and Methods

4.1. Plant Material, Growth Conditions, and Experimental Design

Arabidopsis thaliana Columbia ecotype seeds were sterilized with 70% ethanol for 20 min, 2.5% sodium hypochlorite (commercial bleach) with 0.05% Triton X-100 for 10 min, and finally, four washes with sterile dH_2O. Seeds were placed onto Whatmann paper in Murashige and Skoog (MS) medium plates (Duchefa, Haarlem, The Netherlands). Plates were kept in the dark at 4 °C for 96 h for stratification, and incubated during 8 h in light at 22 °C to promote germination. Plates were transferred to darkness at 22 °C for 72 h. At this moment plates were either kept at 22 °C or transferred to 37 °C. Seedlings were harvested at 0 and 24 h with four biological replicates. Samples were frozen in liquid nitrogen and stored at −80 °C.

4.2. Microarray Analysis

Total RNA was extracted using the RNeasy Plant Mini Kit (Qiagen, Hilden, Germany), and RNA integrity was tested with the 2100 Bioanalyzer (Agilent, Santa Clara, CA, USA). Transcriptome analyses were carried out according to Minimum Information About a Microarray Experiment (MIAME) guidelines. We used the Agilent *Arabidopsis* (V4) Gene Expression 4 × 44K Microarray in a one-color experimental design. The microarray contained 43,803 probes (60-mer oligonucleotides). Four biological replicates were analyzed for each treatment (time points 0 and 24 h at 22 °C and 24 h at 37 °C).

Half a µg of RNA was amplified and labeled with the Agilent Low Input Quick Amp Labeling Kit. To assess the labeling and hybridization efficiencies we used an Agilent Spike-In Kit. Hybridization and slide washing were performed with the Gene Expression Hybridization Kit (Agilent) and Gene Expression Wash Buffers (Agilent), respectively. Then, slides were scanned at 5 µm resolution in an Agilent G2565AA microarray scanner, and image files were analyzed with the Feature Extraction software 9.5.1. We used the GeneSpring 12.1 software (Agilent) to perform the interarray analyses. To ensure a high-quality data set we removed control features, and selected only features for which the 'IsWellAboveBG' parameter was one in at least three out of four biological replicates (31,921 features from 43,803). Our microarray data sets have been submitted to the Gene Expression Omnibus database (accession number: GSE116592).

A new gene annotation of probes in the microarray was carried out using BLASTN searches (https://blast.ncbi.nlm.nih.gov/Blast.cgi), using the sequences of each probe as query against the Arabidopsis genome annotation in The Arabidopsis Information Resource (TAIR; www.arabidopsis.org), version 10. BLAST results for each probe were filtered with a minimum *E*-value of 9.9×10^{-6}, a minimum sequence identity of 98% between probe and transcript, and a minimum overlap of the 75% of the probe sequence length. Probes matching multiple genes were not considered. Results for this gene annotation are quite similar to those obtained in similar analyses performed by TAIR (ftp://ftp.arabidopsis.org/Microarrays/Agilent/).

4.3. Gene Overexpression/Repression Analysis

For each probe and experimental condition (three conditions: 0 h at 22 °C, 24 h at 22 °C, and 24 h at 37 °C), expression levels were averaged across the four biological replicates. For those genes that mapped to more than one probe, expression levels were averaged across all probes. As a result, a single expression level was obtained for each gene and experimental condition.

For each gene with available probes (n = 20,491), the response (R) of its expression to heat stress was computed as:

$$R = \log_2 \frac{E_{24,37}}{E_{24,22}} \quad (1)$$

where $E_{24,37}$ is expression level at 37 °C at 24 h, and $E_{24,22}$ is expression level at 22 °C at 24 h. R takes positive values for genes that are overexpressed at 37 °C compared to 22 °C, and negative values for those that are repressed.

4.4. Protein and Gene Sequence Analysis

All *Arabidopsis* protein sequences were obtained from Ensembl Plants [38] (assembly: TAIR10). For each gene encoding multiple proteins (alternative splicing isoforms), the longest protein was selected for analysis. For each protein, the frequency of each amino acid was computed by dividing the number of occurrences of the amino acid by the length of the protein. GC content of each gene was retrieved from Ensembl Plants' Biomart [38,39]. For each protein, the most likely subcellular location was retrieved from the SUBA4 database [40]. The consensus location was used. Only proteins located to a single compartment were used in compartment-specific analyses.

4.5. Prediction of Protein Intrinsic Disorder

Protein intrinsic disorder prediction was carried out using IUPred [28] for regions of disorder of at least 30 amino acids ("long" option). IUPred predicts tendency for polypeptide chains to be intrinsically disordered or ordered by analyzing the composition of amino acids within a window of 30 consecutive amino acids. It does so by utilizing an energy predictor matrix to estimate the tendency for pairs of amino acids to form strong stabilizing connections, the underlying assumption being that globular proteins form strong stabilizing contacts whereas structurally disordered proteins lack this capacity. IUPred reports a disorder score for each residue ranging from 0 to 1, conferring complete order to disorder, respectively. In this study, we used a threshold of >0.4 to calculate the proportion of amino acids within each protein that were likely to be in disordered regions.

4.6. Statistical Analyses

Statistical analyses were conducted using R [41]. Partial correlation analyses were conducted using the R function pcor.test [42]. Tests repeated on all 20 amino acids were corrected for multiple testing using the Benjamini-Hochberg approach [43].

Author Contributions: M.X.R.-G., F.V.-S., M.A.P.-A. and M.A.F. conceived, designed, and conducted the experiments and microarray analyses. D.A.-P. conceived and designed the bioinformatics analyses. D.A.-P. and F.F. conducted the bioinformatics analyses. D.A.-P. and M.X.R.-G. wrote the paper. All authors contributed to editing the manuscript, read, and approved the manuscript, except M.A.F., who is a posthumous author.

Acknowledgments: D.A.-P., M.X.R.-G., F.V.-S., F.F. and M.A.P.-A. dedicate this work to the memory of M.A.F. Current address of F.F.: Structural Genomics Consortium; Target Discovery Institute, Nuffield Department of Medicine, University of Oxford, Oxford OX3 7FZ, United Kingdom.

Abbreviations

IDR	Intrinsically Disordered Region
MS	Murashige and Skoog
TAIR	The Arabidopsis Information Resource
BLAST	Basic Local Alignment Search Tool

References

1. Karshikoff, A.; Ladenstein, R. Ion pairs and the thermotolerance of proteins from hyperthermophiles: A 'traffic rule' for hot roads. *Trends Biochem. Sci.* **2001**, *26*, 550–557. [CrossRef]
2. Strop, P.; Mayo, S.L. Contribution of surface salt bridges to protein stability. *Biochemistry* **2000**, *39*, 1251–1255. [CrossRef] [PubMed]
3. Perutz, M.; Raidt, H. Stereochemical basis of heat stability in bacterial ferredoxins and in haemoglobin A2. *Nature* **1975**, *255*, 256–259. [CrossRef] [PubMed]
4. Argos, P.; Rossmann, M.G.; Grau, U.M.; Zuber, H.; Frank, G.; Tratschin, J.D. Thermal stability and protein structure. *Biochemistry* **1979**, *18*, 5698–5703. [CrossRef] [PubMed]
5. Beeby, M.; D O'Connor, B.; Ryttersgaard, C.; Boutz, D.R.; Perry, L.J.; Yeates, T.O. The genomics of disulfide bonding and protein stabilization in thermophiles. *PLoS Biol.* **2005**, *3*, e309. [CrossRef] [PubMed]
6. Farias, S.T.; Bonato, M. Preferred amino acids and thermostability. *Genet. Mol. Res.* **2003**, *2*, 383–393. [PubMed]
7. Haney, P.J.; Badger, J.H.; Buldak, G.L.; Reich, C.I.; Woese, C.R.; Olsen, G.J. Thermal adaptation analyzed by comparison of protein sequences from mesophilic and extremely thermophilic *Methanococcus* species. *Proc. Natl. Acad. Sci. USA* **1999**, *96*, 3578–3583. [CrossRef] [PubMed]
8. Kreil, D.P.; Ouzounis, C.A. Identification of thermophilic species by the amino acid compositions deduced from their genomes. *Nucleic Acids Res.* **2001**, *29*, 1608–1615. [CrossRef] [PubMed]
9. Tekaia, F.; Yeramian, E.; Dujon, B. Amino acid composition of genomes, lifestyles of organisms, and evolutionary trends: A global picture with correspondence analysis. *Gene* **2002**, *297*, 51–60. [CrossRef]
10. Zeldovich, K.B.; Berezovsky, I.N.; Shakhnovich, E.I. Protein and DNA sequence determinants of thermophilic adaptation. *PLoS Comput. Biol.* **2007**, *3*, e5. [CrossRef] [PubMed]
11. Chakravarty, S.; Varadarajan, R. Elucidation of determinants of protein stability through genome sequence analysis. *FEBS Lett.* **2000**, *470*, 65–69. [CrossRef]
12. Cambillau, C.; Claverie, J.-M. Structural and genomic correlates of hyperthermostability. *J. Biol. Chem.* **2000**, *275*, 32383–32386. [CrossRef] [PubMed]
13. Burra, P.V.; Kalmar, L.; Tompa, P. Reduction in structural disorder and functional complexity in the thermal adaptation of prokaryotes. *PLoS ONE* **2010**, *5*, e12069. [CrossRef] [PubMed]
14. Wang, J.; Yang, Y.; Cao, Z.; Li, Z.; Zhao, H.; Zhou, Y. The role of semidisorder in temperature adaptation of bacterial FlgM proteins. *Biophys. J.* **2013**, *105*, 2598–2605. [CrossRef] [PubMed]
15. Vicedo, E.; Schlessinger, A.; Rost, B. Environmental pressure may change the composition protein disorder in prokaryotes. *PLoS ONE* **2015**, *10*, e0133990. [CrossRef] [PubMed]
16. Galea, C.A.; High, A.A.; Obenauer, J.C.; Mishra, A.; Park, C.-G.; Punta, M.; Schlessinger, A.; Ma, J.; Rost, B.; Slaughter, C.A. Large-scale analysis of thermostable, mammalian proteins provides insights into the intrinsically disordered proteome. *J. Proteome Res.* **2008**, *8*, 211–226. [CrossRef] [PubMed]
17. Tsvetkov, P.; Myers, N.; Moscovitz, O.; Sharon, M.; Prilusky, J.; Shaul, Y. Thermo-resistant intrinsically disordered proteins are efficient 20S proteasome substrates. *Mol. Biosyst.* **2012**, *8*, 368–373. [CrossRef] [PubMed]
18. Galea, C.A.; Nourse, A.; Wang, Y.; Sivakolundu, S.G.; Heller, W.T.; Kriwacki, R.W. Role of intrinsic flexibility in signal transduction mediated by the cell cycle regulator, $p27^{Kip1}$. *J. Mol. Biol.* **2008**, *376*, 827–838. [CrossRef] [PubMed]
19. Van Noort, V.; Bradatsch, B.; Arumugam, M.; Amlacher, S.; Bange, G.; Creevey, C.; Falk, S.; Mende, D.R.; Sinning, I.; Hurt, E. Consistent mutational paths predict eukaryotic thermostability. *BMC Evol. Biol.* **2013**, *13*, 7. [CrossRef] [PubMed]
20. Wang, G.-Z.; Lercher, M.J. Amino acid composition in endothermic vertebrates is biased in the same direction as in thermophilic prokaryotes. *BMC Evol. Biol.* **2010**, *10*, 263. [CrossRef] [PubMed]
21. Windisch, H.S.; Lucassen, M.; Frickenhaus, S. Evolutionary force in confamiliar marine vertebrates of different temperature realms: Adaptive trends in zoarcid fish transcriptomes. *BMC Genom.* **2012**, *13*, 549. [CrossRef] [PubMed]
22. Albanese, V.; Yam, A.Y.-W.; Baughman, J.; Parnot, C.; Frydman, J. Systems analyses reveal two chaperone networks with distinct functions in eukaryotic cells. *Cell* **2006**, *124*, 75–88. [CrossRef] [PubMed]

23. Berry, J.; Bjorkman, O. Photosynthetic response and adaptation to temperature in higher plants. *Annu. Rev. Plant Physiol.* **1980**, *31*, 491–543. [CrossRef]
24. Sueoka, N. Correlation between base composition of deoxyribonucleic acid and amino acid composition of protein. *Proc. Natl. Acad. Sci. USA* **1961**, *47*, 1141–1149. [CrossRef] [PubMed]
25. Cherry, J.L. Highly expressed and slowly evolving proteins share compositional properties with thermophilic proteins. *Mol. Biol. Evol.* **2010**, *27*, 735–741. [CrossRef] [PubMed]
26. Nakashima, H.; Nishikawa, K. The amino acid composition is different between the cytoplasmic and extracellular sides in membrane proteins. *FEBS Lett.* **1992**, *303*, 141–146. [PubMed]
27. Nakashima, H.; Nishikawa, K. Discrimination of intracellular and extracellular proteins using amino acid composition and residue-pair frequencies. *J. Mol. Biol.* **1994**, *238*, 54–61. [CrossRef] [PubMed]
28. Dosztanyi, Z.; Csizmok, V.; Tompa, P.; Simon, I. IUPred: Web server for the prediction of intrinsically unstructured regions of proteins based on estimated energy content. *Bioinformatics* **2005**, *21*, 3433–3434. [CrossRef] [PubMed]
29. Peng, Z.; Uversky, V.N.; Kurgan, L. Genes encoding intrinsic disorder in eukaryota have high GC content. *Intrinsically Disord. Proteins* **2016**, *4*, e1262225. [CrossRef] [PubMed]
30. Yruela, I.; Contreras-Moreira, B. Genetic recombination is associated with intrinsic disorder in plant proteomes. *BMC Genom.* **2013**, *14*, 772. [CrossRef] [PubMed]
31. Paliy, O.; Gargac, S.M.; Cheng, Y.; Uversky, V.N.; Dunker, A.K. Protein disorder is positively correlated with gene expression in *Escherichia coli*. *J. Proteome Res.* **2008**, *7*, 2234–2245. [CrossRef] [PubMed]
32. Singh, G.P.; Dash, D. How expression level influences the disorderness of proteins. *Biochem. Biophys. Res. Commun.* **2008**, *371*, 401–404. [CrossRef] [PubMed]
33. Yang, J.R.; Liao, B.Y.; Zhuang, S.M.; Zhang, J. Protein misinteraction avoidance causes highly expressed proteins to evolve slowly. *Proc. Natl. Acad. Sci. USA* **2012**, *109*, E831–840. [CrossRef] [PubMed]
34. Hendsch, Z.S.; Tidor, B. Do salt bridges stabilize proteins? A continuum electrostatic analysis. *Protein Sci.* **1994**, *3*, 211–226. [CrossRef] [PubMed]
35. Zhou, X.-X.; Wang, Y.-B.; Pan, Y.-J.; Li, W.-F. Differences in amino acids composition and coupling patterns between mesophilic and thermophilic proteins. *Amino Acids* **2008**, *34*, 25–33. [CrossRef] [PubMed]
36. Catanzano, F.; Barone, G.; Graziano, G.; Capasso, S. Thermodynamic analysis of the effect of selective monodeamidation at asparagine 67 in ribonuclease a. *Protein Sci.* **1997**, *6*, 1682–1693. [CrossRef] [PubMed]
37. Charlesworth, B. Fundamental concepts in genetics: Effective population size and patterns of molecular evolution and variation. *Nat. Rev. Genet.* **2009**, *10*, 195–205. [CrossRef] [PubMed]
38. Bolser, D.; Staines, D.M.; Pritchard, E.; Kersey, P. Ensembl Plants: Integrating tools for visualizing, mining, and analyzing plant genomics data. *Methods Mol. Biol.* **2016**, *1374*, 115–140. [PubMed]
39. Kasprzyk, A.; Keefe, D.; Smedley, D.; London, D.; Spooner, W.; Melsopp, C.; Hammond, M.; Rocca-Serra, P.; Cox, T.; Birney, E. EnsMart: A generic system for fast and flexible access to biological data. *Genome Res.* **2004**, *14*, 160–169. [CrossRef] [PubMed]
40. Hooper, C.M.; Castleden, I.R.; Tanz, S.K.; Aryamanesh, N.; Millar, A.H. SUBA4: The interactive data analysis centre for arabidopsis subcellular protein locations. *Nucleic Acids Res.* **2016**, *45*, D1064–D1074. [CrossRef] [PubMed]
41. R Core Team. R: A language and environment for statistical computing. R Foundation for Statistical Computing. Available online: http://www.R-project.org/ (accessed on 31 October 2014).
42. Kim, S. Ppcor: An R package for a fast calculation to semi-partial correlation coefficients. *Commun. Stat. Appl. Methods* **2015**, *22*, 665. [CrossRef] [PubMed]
43. Benjamini, Y.; Hochberg, Y. Controlling the false discovery rate: A practical and powerful approach to multiple testing. *J. R. Stat. Soc. B* **1995**, *57*, 289–300.

5

Direct Single-Molecule Observation of Sequential DNA Bending Transitions by the Sox2 HMG Box

Mahdi Muhammad Moosa, Phoebe S. Tsoi, Kyoung-Jae Choi, Allan Chris M. Ferreon * and Josephine C. Ferreon *

Department of Pharmacology and Chemical Biology, Baylor College of Medicine, Houston, TX 77030, USA; Mahdi.Moosa@bcm.edu (M.M.M.); Phoebe.Tsoi@bcm.edu (P.S.T.); Kyoungjae.Choi@bcm.edu (K.-J.C.)
* Correspondence: Allan.Ferreon@bcm.edu (A.C.M.F.); Josephine.Ferreon@bcm.edu (J.C.F.)

Abstract: Sox2 is a pioneer transcription factor that initiates cell fate reprogramming through locus-specific differential regulation. Mechanistically, it was assumed that Sox2 achieves its regulatory diversity via heterodimerization with partner transcription factors. Here, utilizing single-molecule fluorescence spectroscopy, we show that Sox2 alone can modulate DNA structural landscape in a dosage-dependent manner. We propose that such stoichiometric tuning of regulatory DNAs is crucial to the diverse biological functions of Sox2, and represents a generic mechanism of conferring functional plasticity and multiplicity to transcription factors.

Keywords: transcription factors; DNA-protein interactions; Sox2 sequential DNA loading; smFRET; DNA conformational landscape; sequential DNA bending; transcription factor dosage

---

1. Introduction

Sox2 regulates a remarkable variety of genes differentially; it activates some and represses others [1–3]. This functional diversity is assumed to be mediated by Sox2 heterodimerization with other transcription factors (TFs) such as Oct4, Oct1, Pax6, and Nanog [4,5]. Recent reports, however, suggest that these canonical partners often remain spatiotemporally separated from Sox2 during genome engagement [6–10]. This raises an important question regarding the TF's mechanism of action as to how Sox2 alone can exert differential loci-specific regulatory effects.

Sox2 is a sequence-specific high-mobility group transcription factor (HMG-TF) [11]. These TFs have conserved DNA binding domains [12,13], also known as HMG box. These DNA binding domains are partly disordered and are assumed to undergo binding-induced functional disorder-to-order transitions [14]. HMG-TFs are known to cooperatively form heterodimers on DNA regulatory elements [13,15–17]; each heteromeric TF pair induces characteristic DNA bend and differentially regulates target gene transcription [18–20]. Interestingly, a number of recent studies suggested that Sox2 can also function as homodimers [21–23]. Whether and how such Sox2 assemblies alter DNA conformations remain largely unknown. Here, we utilize the strengths of single-molecule Förster/fluorescence resonance energy transfer (smFRET) measurements along with ensemble methods to understand the effects of Sox2 binding on regulatory DNA structural landscape in the context of the HMG box (Sox2^{HMG}). Our results suggest that Sox2^{HMG} induces stoichiometry-dependent alternate DNA bends and we propose that the resulting alternate DNA conformations may drive different transcriptional outcomes.

2. Results

2.1. Multiple Sox2^{HMG} Domains Cooperatively Interact with dsDNA^{NANOG}

In our initial ensemble experiments, we observe that Sox2^{HMG} cooperatively binds to the *NANOG* composite promoter (DNA^{NANOG}; Figure 1). We utilized fluorescence anisotropy to detect Sox2^{HMG} binding to dsDNA^{NANOG} (Supplementary Methods). Anisotropy reports on fluorophore rotational properties, dependent on both probe local and global environment perturbations; fluorescence anisotropy of labeled macromolecule usually increases upon ligand binding. To characterize Sox2^{HMG}-DNA binding, we singly-labeled dsDNA^{NANOG} with Alexa Fluor 647 (Supplementary Methods) and monitored changes in DNA fluorescence anisotropy with increasing Sox2^{HMG} concentrations (Figure 1a). Nonlinear least squares (NLS) fitting of the anisotropy data to a Hill equation yields an apparent dissociation constant (K_D) of 15.1 ($\pm$2.0) nM and Hill coefficient of 1.5 ($\pm$0.3). The estimated K_D is similar to that previously reported for specific DNA-Sox2 interactions [18]. A Hill coefficient greater than 1 indicates that multiple Sox2 HMG boxes bind to the DNA in a TF concentration-dependent fashion [24]. Anisotropy measurements also indicate that Sox2^{HMG} alone (i.e., the DNA-binding domain in the absence of dsDNA^{NANOG}) fails to dimerize/oligomerize (Figure S1). To verify the binding of multiple Sox2 molecules to DNA^{NANOG}, we carried out fluorescence electrophoretic mobility shift assay (fEMSA) of DNA with increasing [Sox2^{HMG}]. The fEMSA micrograph shows concentration-dependent appearance of multiple electrophoretic species (Figure 1b). This suggests a multistep Sox2^{HMG} interaction with the *NANOG* proximal promoter. The non-equilibrium nature of mobility shift assays, however, precludes precise estimation of binding affinities of individual Sox2-DNA assemblies on the basis the fEMSA micrograph [25].

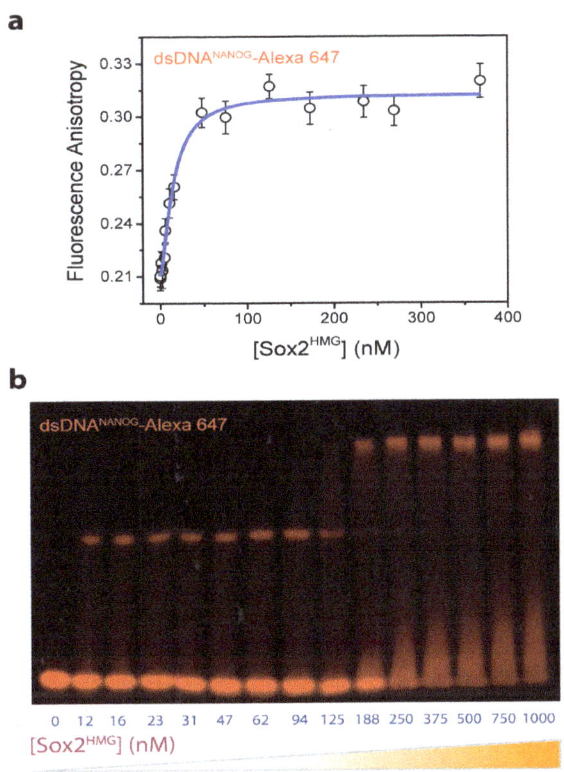

Figure 1. Sox2 cooperatively binds to the *NANOG* upstream promoter (DNA^{NANOG}). (**a**) DNA binding of Sox2^{HMG} was probed by monitoring changes in fluorescence anisotropy of Alexa Fluor 647-labeled dsDNA with increasing [Sox2^{HMG}]. The solid line represents nonlinear least squares (NLS) fit of the data to a Hill equation. NLS-derived parameters: K_D = 15.1 ($\pm$2.0) nM, Hill coefficient = 1.5 ($\pm$0.3). (**b**) Fluorescence electrophoretic mobility assay (fEMSA) of Sox2^{HMG}-DNA^{NANOG} binding suggests a multistep Sox2^{HMG} complex formation with dsDNA^{NANOG} involving multiple protein molecules that are able to bind the DNA partner. (See also Figure S2.)

2.2. Sox2^{HMG} Induces Sequential dsDNA^{NANOG} Bending Transitions

Next, we focused on understanding the mechanism of the TF-DNA complex formation. Although the mobility shift assay clearly demonstrates a multistep higher-order Sox2^{HMG} complex formation with the dsDNA (Figure 1b), our ensemble experiments (i.e., fluorescence anisotropy and fEMSA) were not sensitive enough to determine the stoichiometries of respective TF-DNA complexes. To directly observe Sox2^{HMG}-DNA^{NANOG} binding steps, we performed single-molecule fluorescence microscopy experiments that provide key advantages over conventional ensemble methods: (1) individual conformational sub-populations that are averaged out in ensemble measurements can be directly detected; and (2) experiments can be performed with extremely low concentrations of the labeled molecule (typically 50–100 pM). The ability to carry out experiments at low biomolecule concentration provides access and resolution for characterizing individual interaction steps in tightly interacting systems.

We utilized the distance-dependence of FRET to characterize Sox2^{HMG}-DNA^{NANOG} interaction at single-molecule resolution. smFRET is sensitive to distance changes in the 20–70 Å range [26], and provides the necessary spatial resolution to probe changes in dsDNA^{NANOG} conformations as induced by TF binding (estimated end-to-end distance of DNA^{NANOG} is 57.4 Å, assuming inter-base axial rise of 3.4 Å [27]). For the smFRET experiments, we labeled DNA^{NANOG} with Alexa Fluor 488 and 594 donor-accepter dye-pair (Supplementary Methods). Bursts of fluorescence from donor and acceptor dyes were recorded as dual-labeled *NANOG* promoter DNA passed through the sub-fL observation volume of our custom-built ISS Alba confocal laser microscopy system (described previously [28]). These fluorescence intensities were converted to FRET efficiency (E_{FRET}) histograms, providing a scheme for direct visualization of DNA conformational distributions. Without Sox2^{HMG}, the dual-labeled DNA showed a single-peak in its E_{FRET} histogram with histogram width typical of smFRET studies of freely diffusing dsDNA molecules [29,30] (Figure 2a; top panel). An NLS fit of the histogram to a Gaussian function yielded E_{FRET} value of 0.39 ($\pm$0.04). On the basis of this E_{FRET} value, we estimate the apparent distance between the two dyes to be approximately 64.6 Å (assuming a Förster distance of 60 Å between Alexa 488/594 dyes [31]). This is consistent with the estimated end-to-end distance of dsDNA^{NANOG}, where the slight increase in the apparent distance (compared to the estimated distance) can be attributed to the linkers present in Alexa dyes.

Often, histograms of data collected in diffusion-based smFRET experiments show an additional peak at zero E_{FRET} that arise from molecules with active donor(s) and either inactive or absent acceptor [29,30,32–35]. These zero E_{FRET} peaks tend to significantly overlap with low E_{FRET} peak populations and hamper direct estimation of the position of the non-zero peak(s) [36–39]. Interestingly, our smFRET histograms lack zero E_{FRET} peaks (Figure 2a). We attribute this to the absence of dual donor-labeled dsDNA molecules as ensured by sequential labeling of individual DNA strands (Supplementary Methods). Therefore, sequential labeling and purification of individual fluorophore-conjugated oligos prior to duplex formation can be utilized to minimize zero peaks.

DNA bending (also known as DNA looping) is critical for many eukaryotic TF function [40–44]. Accordingly, Sox2 was shown to induce binding-mediated *FGF* (fibroblast growth factor) enhancer bending [18]. We postulate that similar spatially precise bending is induced in Sox2-DNA^{NANOG} complexes during gene regulation. To characterize Sox2^{HMG} binding-induced *NANOG* promoter DNA bending, we carried out isothermal smFRET Sox2^{HMG} titration against approximately 100 pM dual-labeled DNA (Figure 2). Our smFRET experiments provide a direct way to distinguish between subtle conformational changes of DNA^{NANOG} induced upon Sox2 binding. In our smFRET experiments, we observed a multistep bending transition in the DNA structural landscape (Figure 2a). Initially, DNA^{NANOG} undergoes a cooperative bending to a 0.45 ($\pm$0.01) E_{FRET} state that corresponds to 32.1° ($\pm$1.4°) apparent bend angle at low Sox2^{HMG} concentrations ($\leq$4 nM) (see Supplementary Methods for the details of FRET-to-apparent-angle conversion). NLS fit of the data yields an estimated K_D of 305 ($\pm$39) pM (Figure 2c). Such a tight interaction is unlikely to be driven by higher order

$Sox2^{HMG}$ assemblies and we therefore postulate that this dsDNA conformation (henceforth referred as B_I) is induced by binding to single $Sox2^{HMG}$ molecules.

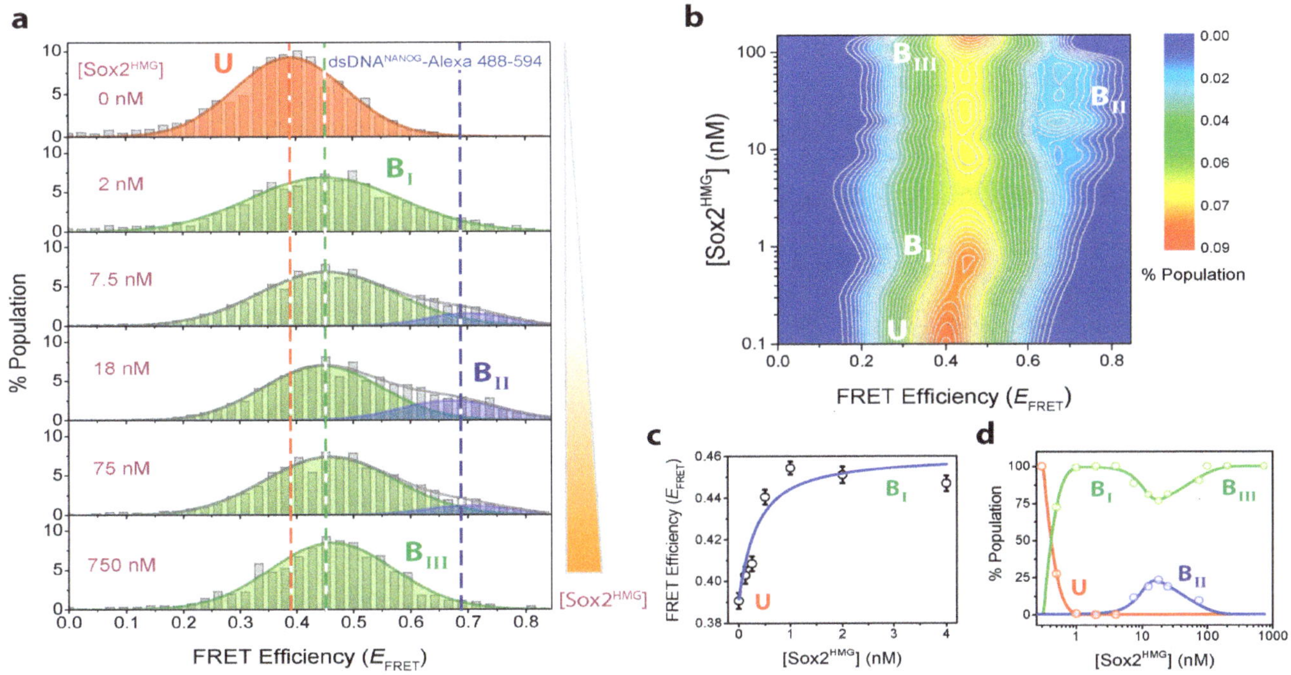

Figure 2. smFRET reveals $Sox2^{HMG}$ concentration-dependent multistep bending of DNA^{NANOG}. (**a**) E_{FRET} histograms of DNA^{NANOG} with increasing [$Sox2^{HMG}$]. (**b**) [$Sox2^{HMG}$]-E_{FRET} contour map color coded based on fractional occupancy of individual DNA conformations. Corresponding DNA conformations are marked on the contour map. (**c**) Sox2 binding isotherm of the U $\rightleftharpoons$ B_I transition as probed by detecting changes in E_{FRET}, linked to dsDNA bending transition. The NLS-derived apparent K_D for this binding step is 0.30 (±0.04) nM (binding equation with fixed Hill coefficient of 1). (**d**) $dsDNA^{NANOG}$ conformational distributions as modulated by $Sox2^{HMG}$ concentration, determined from NLS fitting of individual smFRET histograms to Gaussian functions.

Our ensemble results suggested that multiple $Sox2^{HMG}$ can form higher order TF-DNA assemblies (Figure 1). To characterize the complex formation, we probed for changes in DNA^{NANOG} conformations upon further addition of Sox2 on preformed monomeric $Sox2^{HMG}$-DNA^{NANOG} complexes. With increasing [$Sox2^{HMG}$], we observe a progressive reduction of the B_I population and the emergence of a new population exhibiting higher E_{FRET} (~0.68). This higher E_{FRET} population corresponds to a DNA^{NANOG} apparent bend angle of 70° (±2.4°; henceforth referred to as B_{II} DNA conformation). We infer that this DNA conformation is induced by sequential binding of two individual Sox2 TFs on the dsDNA, where binding of each monomer induces an approximate 32° bend at respective binding sites. Our observed apparent bend angle in the ternary complex (two Sox2 monomers and DNA) is similar to the DNA bend angle previously resolved for heterodimeric HMG box TF-DNA complexes [11,17].

Interestingly, an additional transition is visible in our isothermal smFRET titration when additional $Sox2^{HMG}$ is added (i.e., >75 nM [$Sox2^{HMG}$]). We observe progressive depopulation of the B_{II} bent DNA conformation and coupled emergence of a population at E_{FRET} ~0.44 as [$Sox2^{HMG}$] increases further (henceforth referred as B_{III}; Figure 2a). We estimate the apparent bend angle for the B_{III} population to be 30.4° (±4.5°) from the E_{FRET} data (Supplementary Methods). A longer fEMSA run also indicates higher-order oligomer formation that is consistent with the formation of B_{III} population (Figure S2). Mechanistically, Sox family TFs induce DNA bends via FM dipeptide intercalation between two Thymine (T) bases at the minor groove interface [45,46]. Within the *NANOG* composite promoter, three TT pairs are present: two within the two HMG-TF binding sites (Oct/Sox motifs) identified by

Rodda et al. [47] and one in between. We hypothesize that the initial two DNA bends are induced by sequential Sox2 binding to the two high-affinity HMG-TF binding motifs, where each binding induces an apparent 32° bend at the sites of interactions (a net 70° DNA apparent bend angle in the ternary complex). As Sox2^{HMG} concentration further increases (>75 nM), an additional TF molecule interacts with the DNA at the remaining TT site and induces similar bend albeit at the opposite DNA face. This results in effective reversal of the second bend as evidenced by the increased inter-dye distance (i.e., reduced E_{FRET}) at higher [Sox2^{HMG}]. The final bend remains relatively unchanged upon further increase in Sox2 (up to 1 μM; Figure 2d). Overall, our smFRET data directly demonstrates multistep sequential DNA bending transitions dependent on Sox2 concentration.

3. Discussion

Sox2 is a tightly regulated transcription factor; both significant increases and decreases in Sox2 dosage can be detrimental to its biological function [48,49]. Alterations in Sox2 dosage result in multiple developmental and acquired disorders [50–54]. We show that the Sox2 HMG box can induce concentration-dependent alternate DNA bends (Figure 3). Alternate promoter bends are likely to regulate genes differentially and initiate downstream cascades crucial for Sox2's diverse functions. Our results provide a mechanism for Sox2's strict dosage dependence in its function-dysfunction dichotomy [50,55–58].

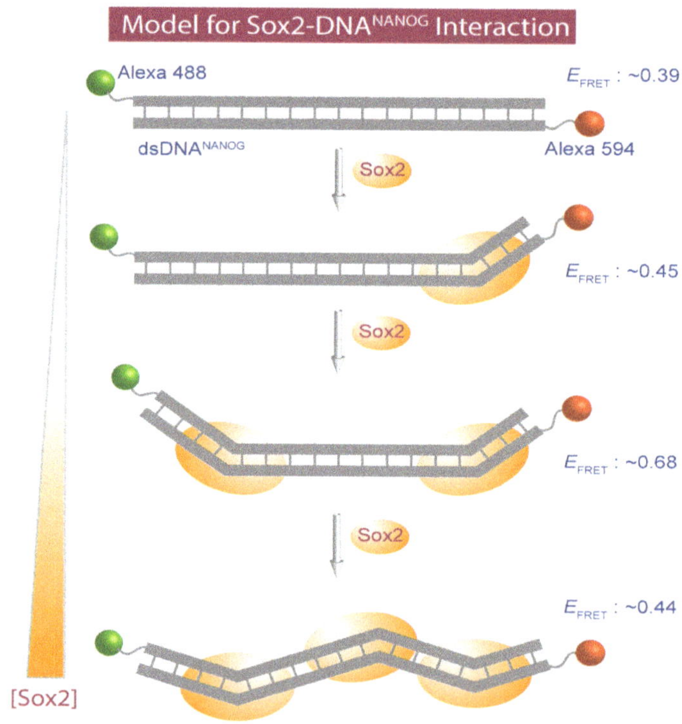

Figure 3. Schematic representation of the Sox2 stoichiometry-dependent dsDNA bending transitions.

In summary, our smFRET experiments clearly demonstrate the role of Sox2 dosage in modulating the conformational landscape of HMG box-binding DNA motifs. Previous studies on Sox family members suggested that heterodimeric homeodomain TFs can induce sequential bending as they interact with their DNA partners [59–62]. Here, we utilize the strengths of smFRET to demonstrate that a representative sequence-specific HMG-TF alone induces concentration-dependent multistep DNA bending transitions. We envision additional layers of tunability for heteromeric HMG-TFs in respective regulatory complexes where affinities of individual transcription factors for DNAs as well as inter-TF interactions can vary dramatically.

4. Materials and Methods

Experimental details are provided in the Supplementary Materials. Briefly, ensemble fluorescence anisotropy and fluorescence electrophoretic mobility assay (fEMSA) experiments were performed in Buffer E (20 mM Tris, 50 mM NaCl, 0.10 mg/mL BSA, 5% glycerol, 0.1 mM DTT/0.05 mM TCEP, pH 8) with Alexa Fluor-647 labeled $dsDNA^{NANOG}$ (Forward: ACTTTTGCATTACAATG; 17 bp). smFRET experiments were performed in the same buffer using a custom-built confocal fluorescence microscopy set up as described previously [28].

Author Contributions: A.C.M.F. and J.C.F. conceived and designed the experiments; M.M.M., P.S.T., K.-J.C., A.C.M.F. and J.C.F. performed the experiments; M.M.M., P.S.T., A.C.M.F. and J.C.F. analyzed the data; M.M.M., A.C.M.F. and J.C.F. wrote the paper.

Abbreviations

fEMSA	fluorescence electrophoretic mobility shift assay
FGF	fibroblast growth factor
HMG	high mobility group
NLS	nonlinear least squares
smFRET	single-molecule Förster resonance energy transfer
TF	transcription factor

References

1. Chew, L.J.; Gallo, V. The Yin and Yang of Sox proteins: Activation and repression in development and disease. *J. Neurosci. Res.* **2009**, *87*, 3277–3287. [CrossRef] [PubMed]
2. Zhang, S.; Cui, W. Sox2, a key factor in the regulation of pluripotency and neural differentiation. *World J. Stem Cells* **2014**, *6*, 305–311. [CrossRef] [PubMed]
3. Liu, Y.R.; Laghari, Z.A.; Novoa, C.A.; Hughes, J.; Webster, J.R.; Goodwin, P.E.; Wheatley, S.P.; Scotting, P.J. Sox2 acts as a transcriptional repressor in neural stem cells. *BMC Neurosci.* **2014**, *15*, 95. [CrossRef] [PubMed]
4. Kondoh, H.; Kamachi, Y. SOX-partner code for cell specification: Regulatory target selection and underlying molecular mechanisms. *Int. J. Biochem. Cell Biol.* **2010**, *42*, 391–399. [CrossRef] [PubMed]
5. Kondoh, H.; Kamachi, Y. Chapter 8—SOX2–Partner Factor Interactions and Enhancer Regulation. In *Sox2*; Academic Press: Boston, MA, USA, 2016; pp. 131–144.
6. Thomson, M.; Liu, S.J.; Zou, L.N.; Smith, Z.; Meissner, A.; Ramanathan, S. Pluripotency factors in embryonic stem cells regulate differentiation into germ layers. *Cell* **2011**, *145*, 875–889. [CrossRef] [PubMed]
7. Soufi, A.; Donahue, G.; Zaret, K.S. Facilitators and impediments of the pluripotency reprogramming factors' initial engagement with the genome. *Cell* **2012**, *151*, 994–1004. [CrossRef] [PubMed]
8. Chen, J.; Zhang, Z.; Li, L.; Chen, B.C.; Revyakin, A.; Hajj, B.; Legant, W.; Dahan, M.; Lionnet, T.; Betzig, E.; Tjian, R. Single-molecule dynamics of enhanceosome assembly in embryonic stem cells. *Cell* **2014**, *156*, 1274–1285. [CrossRef]
9. Soufi, A.; Garcia, M.F.; Jaroszewicz, A.; Osman, N.; Pellegrini, M.; Zaret, K.S. Pioneer transcription factors target partial DNA motifs on nucleosomes to initiate reprogramming. *Cell* **2015**, *161*, 555–568. [CrossRef]
10. White, M.D.; Angiolini, J.F.; Alvarez, Y.D.; Kaur, G.; Zhao, Z.W.; Mocskos, E.; Bruno, L.; Bissiere, S.; Levi, V.; Plachta, N. Long-Lived Binding of Sox2 to DNA Predicts Cell Fate in the Four-Cell Mouse Embryo. *Cell* **2016**, *165*, 75–87. [CrossRef]
11. Hou, L.; Srivastava, Y.; Jauch, R. Molecular basis for the genome engagement by Sox proteins. *Semin. Cell Dev. Biol.* **2017**, *63*, 2–12. [CrossRef]
12. Soullier, S.; Jay, P.; Poulat, F.; Vanacker, J.M.; Berta, P.; Laudet, V. Diversification pattern of the HMG and SOX family members during evolution. *J. Mol. Evol.* **1999**, *48*, 517–527. [CrossRef] [PubMed]
13. Malarkey, C.S.; Churchill, M.E. The high mobility group box: The ultimate utility player of a cell. *Trends Biochem. Sci.* **2012**, *37*, 553–562. [CrossRef] [PubMed]

14. Weiss, M.A. Floppy SOX: Mutual induced fit in hmg (high-mobility group) box-DNA recognition. *Mol. Endocrinol.* **2001**, *15*, 353–362. [CrossRef] [PubMed]
15. Schlierf, B.; Ludwig, A.; Klenovsek, K.; Wegner, M. Cooperative binding of Sox10 to DNA: Requirements and consequences. *Nucleic Acids Res.* **2002**, *30*, 5509–5516. [CrossRef] [PubMed]
16. Ng, C.K.; Li, N.X.; Chee, S.; Prabhakar, S.; Kolatkar, P.R.; Jauch, R. Deciphering the Sox-Oct partner code by quantitative cooperativity measurements. *Nucleic Acids Res.* **2012**, *40*, 4933–4941. [CrossRef] [PubMed]
17. Clore, G.M. Chapter 3—Dynamics of SOX2 Interactions with DNA A2—Kondoh, Hisato. In *Sox2*; Lovell-Badge, R., Ed.; Academic Press: Boston, MA, USA, 2016; pp. 25–41.
18. Scaffidi, P.; Bianchi, M.E. Spatially precise DNA bending is an essential activity of the sox2 transcription factor. *J. Biol. Chem.* **2001**, *276*, 47296–47302. [CrossRef] [PubMed]
19. Dragan, A.I.; Read, C.M.; Makeyeva, E.N.; Milgotina, E.I.; Churchill, M.E.; Crane-Robinson, C.; Privalov, P.L. DNA binding and bending by HMG boxes: Energetic determinants of specificity. *J. Mol. Biol.* **2004**, *343*, 371–393. [CrossRef]
20. Slattery, M.; Riley, T.; Liu, P.; Abe, N.; Gomez-Alcala, P.; Dror, I.; Zhou, T.; Rohs, R.; Honig, B.; Bussemaker, H.J.; Mann, R.S. Cofactor binding evokes latent differences in DNA binding specificity between Hox proteins. *Cell* **2011**, *147*, 1270–1282. [CrossRef] [PubMed]
21. Li, J.; Pan, G.; Cui, K.; Liu, Y.; Xu, S.; Pei, D. A dominant-negative form of mouse SOX2 induces trophectoderm differentiation and progressive polyploidy in mouse embryonic stem cells. *J. Biol. Chem.* **2007**, *282*, 19481–19492. [CrossRef]
22. Cox, J.L.; Mallanna, S.K.; Luo, X.; Rizzino, A. Sox2 uses multiple domains to associate with proteins present in Sox2-protein complexes. *PLoS ONE* **2010**, *5*, e15486. [CrossRef]
23. Xia, P.; Wang, S.; Ye, B.; Du, Y.; Huang, G.; Zhu, P.; Fan, Z. Sox2 functions as a sequence-specific DNA sensor in neutrophils to initiate innate immunity against microbial infection. *Nat. Immunol.* **2015**, *16*, 366–375. [CrossRef] [PubMed]
24. Weiss, J.N. The Hill equation revisited: Uses and misuses. *FASEB J.* **1997**, *11*, 835–841. [CrossRef] [PubMed]
25. Hellman, L.M.; Fried, M.G. Electrophoretic mobility shift assay (EMSA) for detecting protein-nucleic acid interactions. *Nat. Protoc.* **2007**, *2*, 1849–1861. [CrossRef] [PubMed]
26. Ferreon, A.C.; Deniz, A.A. Protein folding at single-molecule resolution. *Biochim. Biophys. Acta* **2011**, *1814*, 1021–1029. [CrossRef]
27. Watson, J.D.; Crick, F.H.C. Molecular Structure of Nucleic Acids: A Structure for Deoxyribose Nucleic Acid. *Nature* **1953**, *171*, 737. [CrossRef] [PubMed]
28. Tsoi, P.S.; Choi, K.J.; Leonard, P.G.; Sizovs, A.; Moosa, M.M.; MacKenzie, K.R.; Ferreon, J.C.; Ferreon, A.C. The N-Terminal Domain of ALS-Linked TDP-43 Assembles without Misfolding. *Angew. Chem. Int. Ed. Engl.* **2017**, *56*, 12590–12593. [CrossRef] [PubMed]
29. Deniz, A.A.; Dahan, M.; Grunwell, J.R.; Ha, T.; Faulhaber, A.E.; Chemla, D.S.; Weiss, S.; Schultz, P.G. Single-pair fluorescence resonance energy transfer on freely diffusing molecules: Observation of Forster distance dependence and subpopulations. *Proc. Natl. Acad. Sci. USA* **1999**, *96*, 3670–3675. [CrossRef]
30. Dey, S.K.; Pettersson, J.R.; Topacio, A.Z.; Das, S.R.; Peteanu, L.A. Eliminating Spurious Zero-Efficiency FRET States in Diffusion-Based Single-Molecule Confocal Microscopy. *J. Phys. Chem. Lett.* **2018**, *9*, 2259–2265. [CrossRef]
31. Johnson, I.D.; Spence, M.T.Z. *The Molecular Probes Handbook: A Guide to Fluorescent Probes and Labeling Technologies*; Molecular Probes: Eugene, OR, USA, 2010.
32. Pljevaljcic, G.; Millar, D.P.; Deniz, A.A. Freely diffusing single hairpin ribozymes provide insights into the role of secondary structure and partially folded states in RNA folding. *Biophys. J.* **2004**, *87*, 457–467. [CrossRef]
33. Morgan, M.A.; Okamoto, K.; Kahn, J.D.; English, D.S. Single-molecule spectroscopic determination of lac repressor-DNA loop conformation. *Biophys. J.* **2005**, *89*, 2588–2596. [CrossRef]
34. Schuler, B. Single-molecule FRET of protein structure and dynamics—A primer. *J. Nanobiotechnol.* **2013**, *11* (Suppl. 1), S2. [CrossRef]
35. Tyagi, S.; VanDelinder, V.; Banterle, N.; Fuertes, G.; Milles, S.; Agez, M.; Lemke, E.A. Continuous throughput and long-term observation of single-molecule FRET without immobilization. *Nat. Methods* **2014**, *11*, 297–300. [CrossRef] [PubMed]
36. Ferreon, A.C.; Gambin, Y.; Lemke, E.A.; Deniz, A.A. Interplay of α-synuclein binding and conformational

switching probed by single-molecule fluorescence. *Proc. Natl. Acad. Sci. USA* **2009**, *106*, 5645–5650. [CrossRef] [PubMed]

37. Ferreon, A.C.; Moran, C.R.; Ferreon, J.C.; Deniz, A.A. Alteration of the α-synuclein folding landscape by a mutation related to Parkinson's disease. *Angew. Chem. Int. Ed. Engl.* **2010**, *49*, 3469–3472. [CrossRef]
38. Gambin, Y.; VanDelinder, V.; Ferreon, A.C.; Lemke, E.A.; Groisman, A.; Deniz, A.A. Visualizing a one-way protein encounter complex by ultrafast single-molecule mixing. *Nat. Methods* **2011**, *8*, 239–241. [CrossRef] [PubMed]
39. Moosa, M.M.; Ferreon, A.C.; Deniz, A.A. Forced folding of a disordered protein accesses an alternative folding landscape. *Chemphyschem* **2015**, *16*, 90–94. [CrossRef]
40. Su, W.; Jackson, S.; Tjian, R.; Echols, H. DNA looping between sites for transcriptional activation: Self-association of DNA-bound Sp1. *Genes Dev.* **1991**, *5*, 820–826. [CrossRef]
41. Lim, F.L.; Hayes, A.; West, A.G.; Pic-Taylor, A.; Darieva, Z.; Morgan, B.A.; Oliver, S.G.; Sharrocks, A.D. Mcm1p-induced DNA bending regulates the formation of ternary transcription factor complexes. *Mol. Cell Biol.* **2003**, *23*, 450–461. [CrossRef]
42. Petrascheck, M.; Escher, D.; Mahmoudi, T.; Verrijzer, C.P.; Schaffner, W.; Barberis, A. DNA looping induced by a transcriptional enhancer in vivo. *Nucleic Acids Res.* **2005**, *33*, 3743–3750. [CrossRef]
43. Whittington, J.E.; Delgadillo, R.F.; Attebury, T.J.; Parkhurst, L.K.; Daugherty, M.A.; Parkhurst, L.J. TATA-binding protein recognition and bending of a consensus promoter are protein species dependent. *Biochemistry* **2008**, *47*, 7264–7273. [CrossRef]
44. Gietl, A.; Grohmann, D. Modern biophysical approaches probe transcription-factor-induced DNA bending and looping. *Biochem. Soc. Trans.* **2013**, *41*, 368–373. [CrossRef] [PubMed]
45. Williams, D.C.; Cai, M., Jr.; Clore, G.M. Molecular basis for synergistic transcriptional activation by Oct1 and Sox2 revealed from the solution structure of the 42-kDa Oct1.Sox2.Hoxb1-DNA ternary transcription factor complex. *J. Biol. Chem.* **2004**, *279*, 1449–1457. [CrossRef]
46. Palasingam, P.; Jauch, R.; Ng, C.K.; Kolatkar, P.R. The structure of Sox17 bound to DNA reveals a conserved bending topology but selective protein interaction platforms. *J. Mol. Biol.* **2009**, *388*, 619–630. [CrossRef] [PubMed]
47. Rodda, D.J.; Chew, J.L.; Lim, L.H.; Loh, Y.H.; Wang, B.; Ng, H.H.; Robson, P. Transcriptional regulation of nanog by OCT4 and SOX2. *J. Biol. Chem.* **2005**, *280*, 24731–24737. [CrossRef] [PubMed]
48. Yamaguchi, S.; Hirano, K.; Nagata, S.; Tada, T. Sox2 expression effects on direct reprogramming efficiency as determined by alternative somatic cell fate. *Stem Cell Res.* **2011**, *6*, 177–186. [CrossRef] [PubMed]
49. Prakash, N. Chapter 4—Posttranscriptional Modulation of Sox2 Activity by miRNAs A2—Kondoh, Hisato. In *Sox2*; Lovell-Badge, R., Ed.; Academic Press: Boston, MA, USA, 2016; pp. 43–71.
50. Bertolini, J.; Mercurio, S.; Favaro, R.; Mariani, J.; Ottolenghi, S.; Nicolis, S.K. Chapter 11—Sox2-Dependent Regulation of Neural Stem Cells and CNS Development A2—Kondoh, Hisato. In *Sox2*; Lovell-Badge, R., Ed.; Academic Press: Boston, MA, USA, 2016; pp. 187–216.
51. Van Heyningen, V. Chapter 13—Congenital Abnormalities and SOX2 Mutations A2—Kondoh, Hisato. In *Sox2*; Lovell-Badge, R., Ed.; Academic Press: Boston, MA, USA, 2016; pp. 235–242.
52. Rizzoti, K.; Lovell-Badge, R. Chapter 14—Role of SOX2 in the Hypothalamo–Pituitary Axis. In *Sox2*; Academic Press: Boston, MA, USA, 2016; pp. 243–262.
53. Iwafuchi-Doi, M.; Zaret, K.S. Cell fate control by pioneer transcription factors. *Development* **2016**, *143*, 1833–1837. [CrossRef] [PubMed]
54. Wuebben, E.L.; Rizzino, A. The dark side of SOX2: Cancer—A comprehensive overview. *Oncotarget* **2017**, *8*, 44917–44943. [CrossRef] [PubMed]
55. Sarkar, A.; Hochedlinger, K. The sox family of transcription factors: Versatile regulators of stem and progenitor cell fate. *Cell Stem Cell* **2013**, *12*, 15–30. [CrossRef] [PubMed]
56. Liu, K.; Lin, B.; Zhao, M.; Yang, X.; Chen, M.; Gao, A.; Liu, F.; Que, J.; Lan, X. The multiple roles for Sox2 in stem cell maintenance and tumorigenesis. *Cell Signal.* **2013**, *25*, 1264–1271. [CrossRef] [PubMed]
57. Kamachi, Y.; Kondoh, H. Sox proteins: Regulators of cell fate specification and differentiation. *Development* **2013**, *140*, 4129–4144. [CrossRef]
58. Hagey, D.W.; Muhr, J. Sox2 acts in a dose-dependent fashion to regulate proliferation of cortical progenitors. *Cell Rep.* **2014**, *9*, 1908–1920. [CrossRef] [PubMed]

59. Peirano, R.I.; Wegner, M. The glial transcription factor Sox10 binds to DNA both as monomer and dimer with different functional consequences. *Nucleic Acids Res.* **2000**, *28*, 3047–3055. [CrossRef] [PubMed]
60. Takayama, Y.; Clore, G.M. Impact of protein/protein interactions on global intermolecular translocation rates of the transcription factors Sox2 and Oct1 between DNA cognate sites analyzed by z-exchange NMR spectroscopy. *J. Biol. Chem.* **2012**, *287*, 26962–26970. [CrossRef] [PubMed]
61. Morimura, H.; Tanaka, S.I.; Ishitobi, H.; Mikami, T.; Kamachi, Y.; Kondoh, H.; Inouye, Y. Nano-analysis of DNA conformation changes induced by transcription factor complex binding using plasmonic nanodimers. *ACS Nano* **2013**, *7*, 10733–10740. [CrossRef] [PubMed]
62. Yamamoto, S.; De, D.; Hidaka, K.; Kim, K.K.; Endo, M.; Sugiyama, H. Single molecule visualization and characterization of Sox2-Pax6 complex formation on a regulatory DNA element using a DNA origami frame. *Nano Lett.* **2014**, *14*, 2286–2292. [CrossRef] [PubMed]

6

Both Intrinsically Disordered Regions and Structural Domains Evolve Rapidly in Immune-Related Mammalian Proteins

Keiichi Homma [1,*], Hiroto Anbo [1], Tamotsu Noguchi [2] and Satoshi Fukuchi [1]

[1] Department of Life Science and Informatics, Maebashi Institute of Technology, 460-1 Kamisadori-machi, Maebashi-shi 371-0816, Japan; koume8@icloud.com (H.A.); sfukuchi@maebashi-it.ac.jp (S.F.)
[2] Pharmaceutical Education Research Center, Meiji Pharmaceutical University, 2-522-1 Noshio, Kiyose-shi, Tokyo 204-8588, Japan; noguchit@my-pharm.ac.jp
* Correspondence: khomma@maebashi-it.ac.jp

Abstract: Eukaryotic proteins consist of structural domains (SDs) and intrinsically disordered regions (IDRs), i.e., regions that by themselves do not assume unique three-dimensional structures. IDRs are generally subject to less constraint and evolve more rapidly than SDs. Proteins with a lower number of protein-to-protein interactions (PPIs) are also less constrained and tend to evolve fast. Extracellular proteins of mammals, especially immune-related extracellular proteins, on average have relatively high evolution rates. This article aims to examine if a high evolution rate in IDRs or that in SDs accounts for the rapid evolution of extracellular proteins. To this end, we classified eukaryotic proteins based on their cellular localizations and analyzed them. Moreover, we divided proteins into SDs and IDRs and calculated the respective evolution rate. Fractional IDR content is positively correlated with evolution rate. For their fractional IDR content, immune-related extracellular proteins show an aberrantly high evolution rate. IDRs evolve more rapidly than SDs in most subcellular localizations. In extracellular proteins, however, the difference is diminished. For immune-related proteins in mammals in particular, the evolution rates in SDs come close to those in IDRs. Thus high evolution rates in both IDRs and SDs account for the rapid evolution of immune-related proteins.

Keywords: secretion; immune; extracellular; protein-protein interaction; intrinsically disordered region; structural domain; evolution

---

1. Introduction

Mature eukaryotic proteins consist not only of structural domains (SDs), but also of intrinsically disordered regions (IDRs), i.e., regions that by themselves do not fold into unique three-dimensional structures [1]. Although some IDRs interact with proteins or other macromolecules, they are generally under less constraint than SDs and thus have higher evolution rates [2]. A positive correlation between fractional IDR contents of proteins and evolution rates is thus expected.

Proteins with more protein-to-protein interactions (PPIs) tend to be more evolutionarily constrained and have lower evolution rates [3,4]. Highly expressed proteins are also more constrained and evolve slowly [4–6]. These two factors partially account for the evolution rate of proteins.

Eukaryotic proteins have specific subcellular localizations in general, with different average fractional IDR contents in different cellular localizations [7]. For instance, IDR contents are generally high in nuclear proteins [7,8], while they tend to be low in mitochondrial proteins [9,10]. It is plausible that different fractional IDR contents in different subcellular localizations result in varied evolution rates.

Interestingly, extracellular proteins (synonymously called secreted proteins) in mammalian species were often found to evolve faster than intracellular proteins [11,12]. This finding is partly explainable by rapid evolution of immune-related extracellular proteins as many of the coding genes are subject to positive selection [13,14]. That is, the evolution rate, ω, defined by the nonsynonymous to synonymous substitution rate ratio, exceeds unity at sites under positive selection and the existence of many such sites result in high evolution rates of many immune-related genes. For instance, antimicrobial peptides, α- and β-defensins and cathelicidins, are reportedly subject to positive selection and evolve rapidly in mammals [15–17]. We consider it worthwhile to carry out research on evolutionary characteristics of immune-related secreted proteins, as they are involved in host defences [18], pathogen–host interactions [19,20], production of antibodies [21], colony-stimulating factors [22], haematopoiesis [23], and triggering proteolytic cascades [24,25], as well as enzyme replacement therapies [26]. The generally high evolution of immune-related proteins evinces their importance in evolution of mammalian species [27]. Further research may reveal how immune-related proteins function and may lead to pharmaceutical applications.

However, the difference in evolution rate with intracellular proteins remained significant even if analyses were limited to non-immune-related extracellular proteins. The generally low expression levels in secreted proteins partially explain the rapid evolution. Whether the substitution frequency in IDRs or SDs or both contributes to the increased evolution rate of extracellular proteins, however, has not been explored.

We examined the correlation of fractional IDR content and evolution rate and found it positive. We then analyzed the evolution rates of SDs and IDRs of proteins in different localizations. In most localizations, IDRs were found to evolve faster than SDs, as expected. Immune-related secreted proteins in mammals, however, exhibited extremely high evolution rates in SDs that approach those in IDRs. This surprising finding indicates that positive selection that is said to function on a number of immune-related genes operates strongly both on IDRs and SDs of the coded proteins.

2. Results and Discussion

2.1. Classification of Eukaryotic Proteins by Subcellular Localizations

For accurate analyses of evolution rates in different subcellular localizations, reliable localization annotations of most proteins are necessary. At present, only four species satisfy this criterion in UniProt: *Homo sapiens*, *Mus musculus* (mouse), *Arabidopsis thaliana* (thale cress), and *Saccharomyces cerevisiae* (budding yeast). We thus selected the human, mouse, thale cress, and budding yeast proteins with orthologs and classified the selected proteins by subcellular localization (Table 1). Proteins that are localized to both the nucleus and the cytosol were specifically grouped (abbreviated as NC), as the group reportedly contains many proteins with multiple PPIs [28]. We combined proteins residing in the endoplasmic reticulum and the Golgi apparatus (termed EG), since many proteins cycle between the two organells. Secreted proteins were divided into immune-related (SI) and non-immune-related (SN), because immune-related proteins generally evolve rapidly [11]. Thale cress had a limited number of immune-related proteins, while unsurprisingly budding yeast had none. Multiply localized proteins except for the aforementioned NC proteins were classified as one group (ML). Note that many proteins with orthologs were not classifiable due to the unavailability of pertinent information.

Table 1. Number of proteins in each subcellular localization.

Species	All	NU	NC	CY	MT	EG	PM	SN	SI	ML
H. sapiens	10,348	1639	632	455	377	400	1116	584	139	3023
M. musculus	10,068	1719	546	224	426	514	998	796	125	2787
A. thaliana	8910	1032	163	331	356	348	534	431	6	594
S. cerevisiae	5304	1532	232	241	639	458	281	69	0	416

NU: Nucleus; NC: Nucleus and cytosol; CY: Cytosol; MT: Mitochondria; EG: Endoplasmic reticulum or Golgi apparatus; PM: Plasma membrane; SN: Secreted, non-immune-related; SI: Secreted, immune-related; ML: Multiple localizations except NC (ML).

2.2. Evolution Rates and Other Properties of Proteins in Different Subcellular Localizations

For each pair of orthologs, we determined the evolution rate, ω, defined by the ratio of nonsynonymous to synonymous substitution rate, i.e., dN/dS. The median ω at each localization is shown (Figures 1 and 2). Note that for this and other data presented in the figure, different scales were used in different species. As the number of immune-related proteins (SI) in *A. thaliana* is small, no corresponding data were plotted in this species. Proteins of the four species showed similar patterns. For instance proteins in the cytosol (CY) and those that reside both in the nucleus and the cytosol (NC) had the median evolution rates lower than the overall median in all four species. In general the median evolution rates in intracellular proteins (NU, NC, CY, MT, and EG; shown in blue) were lower than those of secreted proteins (SN and SI; shown in red). Among the secreted proteins, immune-related proteins (SI) exhibited particularly high evolution rates, in agreement with the literature [13,14].

The fractional IDR content of each protein was predicted by DISOPRED [29], DICHOT [30], and POODLE-L [31] and the median in each localization was calculated (Figures 1 and 2). Although the medians of most localizations (Figures 1 and 2) were nearly always the lowest by DISOPRED, higher by DICHOT, and the highest by POODLE, we note that the overall averages by the three methods generally do not differ much. For instance, the fractions of IDRs in human proteins by DISOPRED, DICHOT, POODLE are 30.2%, 26.4%, and 30.1%, respectively. The differences in the medians are thus mostly attributable to differences in the distributions of fractional IDRs. Nevertheless the corresponding medians by the three prediction methods showed similar patterns. For instance, by all three methods in the four species, we got high fractional IDR contents in the nuclear proteins (NU) and low values in the mitochondrial proteins (MT), consistent with previous reports [7–10]. Intriguingly, the secreted non-immune proteins (SN) in budding yeast were revealed to have a high median IDR content, unlike the counterparts of the three multicellular eukaryotes. The difference may reflect the difference between unicellular and multicellular organisms. This issue needs to be addressed later with analyses of more eukaryotes.

We also calculated and graphed the median numbers of PPIs of proteins in the localizations (Figures 1 and 2). PPIs have been less studied in mouse and thale cress proteins than in human and budding yeast counterparts, as evidenced by the reduced numbers of PPIs in mouse and thale cress (Figures 1 and 2). The mouse and thale cress PPI data are therefore less reliable as those of the other two species. As reported [28], multiply localized proteins (NC and ML) generally showed more interactions with other proteins. Immune-related secreted proteins (SI), however, had fewer interacting partners on average.

Additionally, the median expression level in ppm of the proteins at each localization was determined and graphed as logarithms to the base of ten (Figures 1 and 2). Yeast proteins were generally expressed much more than mammalian proteins. The expression levels of the human immune-related proteins (SI) were generally high, but those of the mouse counterparts were indistinguishable from the average.

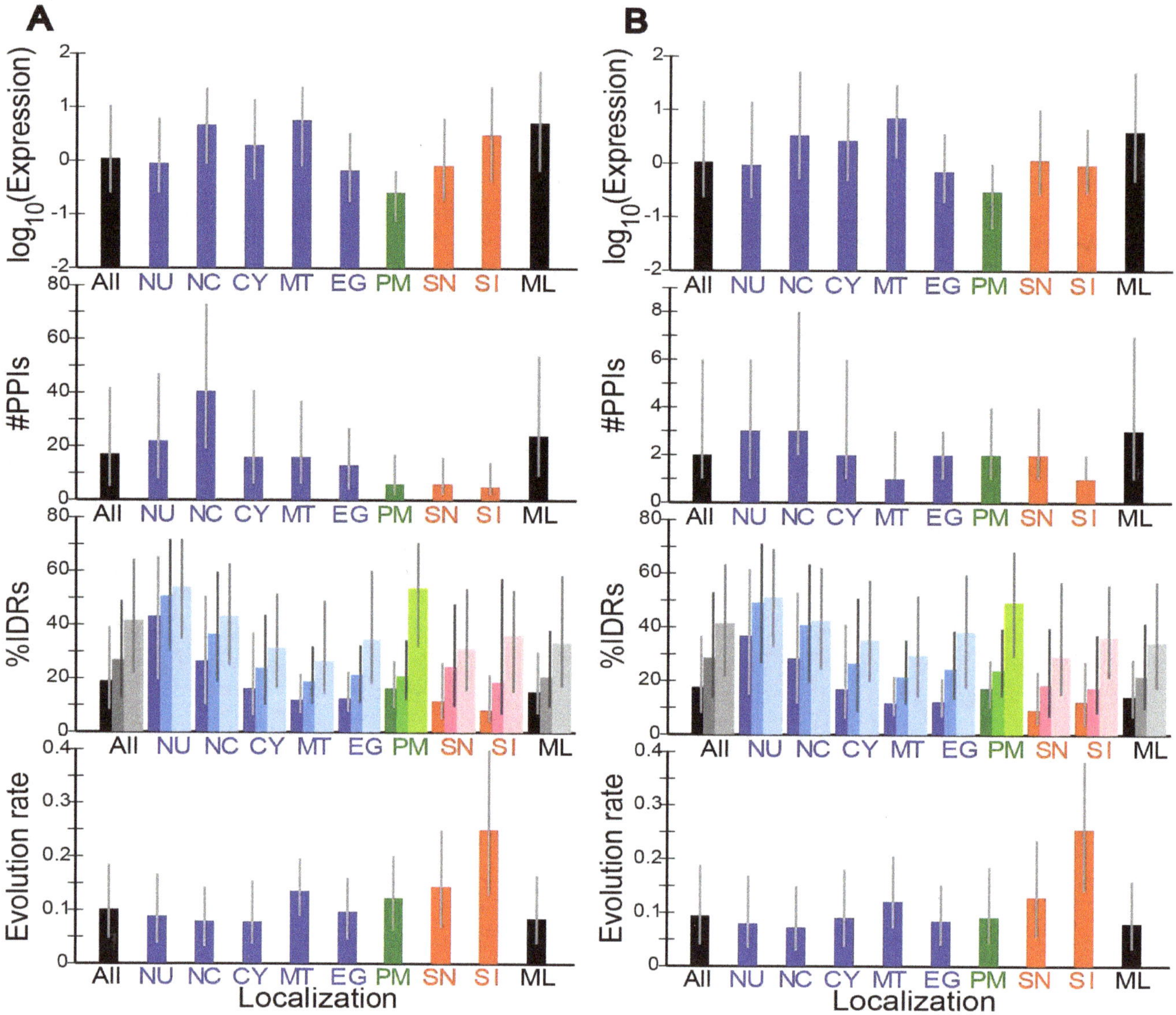

Figure 1. Medians and ranges of four quantities in different localizations in two mammals. (**A**) *H. sapiens*; (**B**) *M. musculus*; Rectangles in each panel from the bottom to the top represent the medians in evolution rate, fractional IDR content by DISOPRED (left), DICHOT (middle) and POODLE (right), the number of PPIs, and expression level. Grey vertical bars represent interquartile ranges, with their bottom and top corresponding to the 25th to the 75th percentile, respectively. The abbreviations for localizations are as in Table 1.

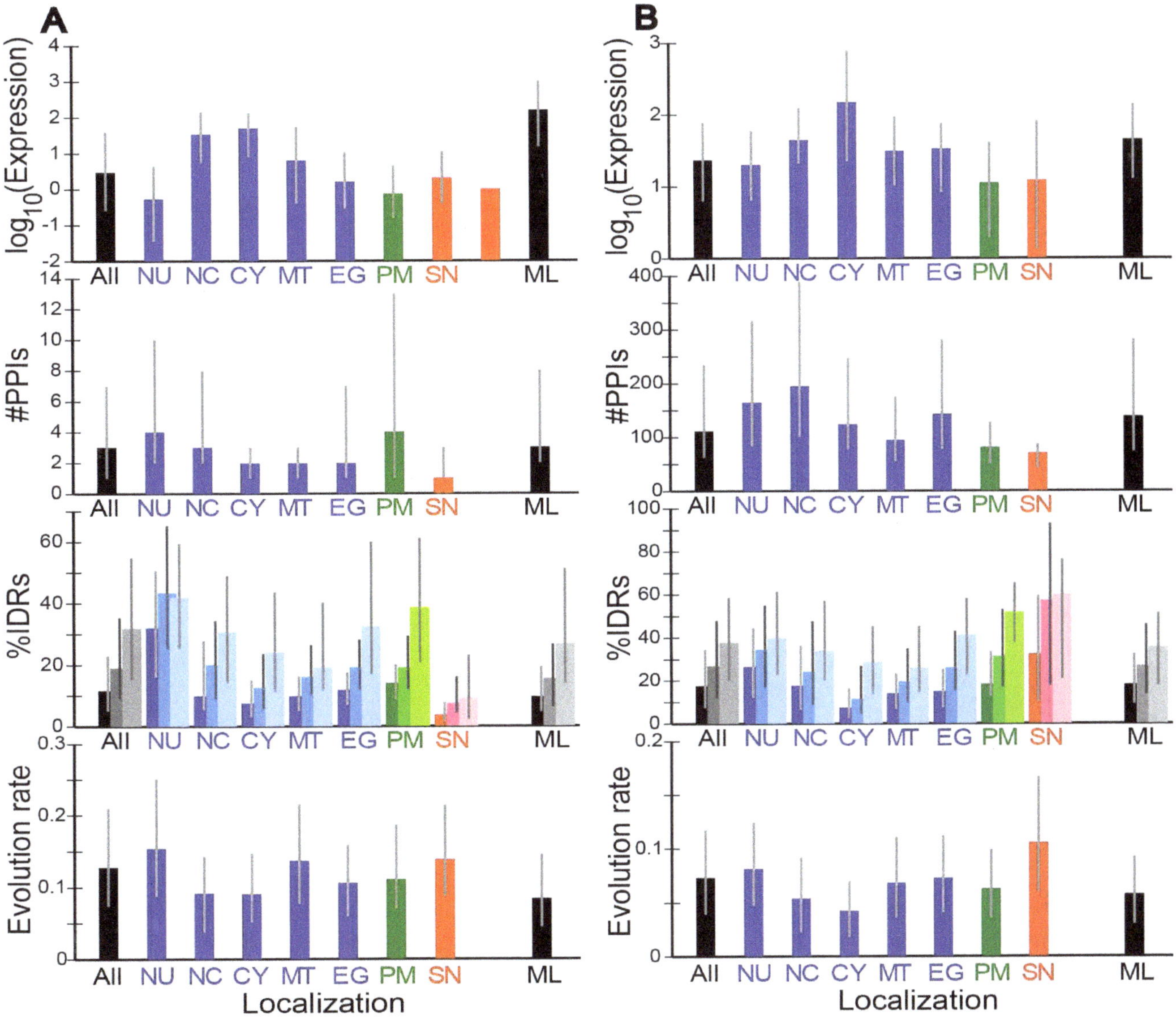

Figure 2. Medians of four quantities in different localizations in two non-mammalian eukaryotes. (**A**) *A. thaliana*; (**B**) *S. cerevisiae*; the data are presented as in Figure 1.

2.3. Correlation of Evolution Rates with Protein Properties

We computed Spearman's correlation coefficients (Rhos) of number of PPIs with evolution rate (ω) and found them to be weakly negative but significantly different from zero (all at $p < 0.01$) (Table 2). The negative correlation is consistent with previous results [3,4]. As the number of PPIs was generally low in extracellular proteins (SN and SI, Figures 1 and 2), the negative correlation partially explains their high evolution rates.

We also found small but significant (all at $p < 1 \times 10^{-113}$) negative correlations between expression level and ω (Table 2), corroborating previous findings [5,6]. The negative correlation was stronger in budding yeast. Since the expression levels of non-immune-related secreted proteins (SN) were not high (Figures 1 and 2), the negative correlation at least in part explains the high evolution rates of these proteins. By contrast the expression levels of immune-related secreted proteins (SI) were not significantly low (Figure 1) and do not contribute to the extremely high evolution rates.

As IDRs have a propensity to evolve faster than SDs, the more IDRs a protein has, the faster it is expected to evolve. To test this possibility, correlation coefficients of %IDR with ω were calculated. Fractional IDR content was positively correlated with evolution rate in all the four species (Table 2).

Although the correlation coefficients were generally small, they all significantly differed from zero (at $p < 1 \times 10^{-4}$). As the median fractional IDR contents in immune-related secreted proteins (SI) were lower than average, this factor does not make positive contribution to the evolution rates.

Table 2. Correlations between three properties and evolution rate ω.

Correlation with	*H. sapiens*	*M. musculus*	*A. thaliana*	*S. cerevisiae*
#PPI with ω	−0.293	−0.194	−0.054	−0.195
Expression level with ω	−0.264	−0.231	−0.337	−0.459
%IDR (DISOPRED) with ω	0.093	0.094	0.168	0.264
%IDR (DICHOT) with ω	0.113	0.146	0.052	0.303
%IDR (POODLE) with ω	0.096	0.097	0.113	0.179

Spearman's correlation coefficient (Rho) of each pair is shown.

2.4. Evolution Rates in SDs and IDRs in Different Subcellular Localizations

In order to see whether IDRs or SDs in immune-related proteins mostly account for the high evolution rates, we calculated the evolution rates in IDRs and SDs separately and compared the two. The median evolution rate in IDRs in all proteins was significantly higher than that in SDs, irrespective of species (Figures 3 and S1). We detected the same disparity at most localizations.

Upon closer examination of the mammalian rates, we noticed that the IDR/SD evolution rate ratio tended to be higher in intracellular localizations (NU, NC, CY, MT, and EG) than in extracellular ones (SN and SI). In the plant *A. thaliana* the inside–outside difference in evolution rate was detectable but was less pronounced (Figure S1A). In contrast, budding yeast failed to show this tendency (Figure S1B). In immune-relate secreted proteins (SI), the rates in IDRs and SDs were both higher than average, with the difference between them statistically insignificant in a majority of cases (Figure 3). SDs apparently evolve quite rapidly in immune-related proteins to approach the rates of IDRs to give rise to the anomalously high evolution rates. So far as we are aware, the phenomenon of the evolution rate in SDs that comes close that in IDRs in immune-related proteins is the first to be reported. The non-immune related extracellular proteins (SN) also tended to have higher than average evolution rate in SDs in *H. sapiens* and *M. musculus*, and *A. thaliana*, although the difference from the average was more conspicuous in the two mammals (Figure 3) than in the plant species (Figure S1A). In contrast SDs in non-immune related extracellular proteins (SN) did not show an above-average evolution rate in *S. cerevisiaie* (Figure S1B). In mammalian mitochondrial (MT) and plasma membrane (PM) proteins, the evolution rates of SDs and IDRs were close to each other (Figure 3), although the former was significantly higher than the latter in all cases. By contrast the counterparts in the two non-mammalian species failed to show the tendency (Figure S1). We need to investigate other species before attaching any significance to this possibly mammalian-specific phenomenon.

We recognize the need to analyze more animal species to check the generality of our finding on immune-related extracellular proteins. For accurate analyses by the same methodology, however, two closely related and entirely sequenced species must be available and at least one of them must have a majority of proteins annotated by UniProt to provide reliable subcellular localizations. Unfortunately no animal species other than *H. sapiens* and *M. musculus* currently meet the latter criterion. Since 3463 (~22% of the total) *Drosophila melanogaster* proteins have been annotated, however, we carried out preliminary analyses of this fly. Thirty-eight annotated immune-related extracellular proteins were identified in 13,957 orthologs. The results showed that the evolution rates in IDRs and SDs were both high in immune-related proteins but the former was much higher than the latter. The ratio of the median evolution rate in IDRs to that in SDs was 2.37, 1.60, and 2.99 by DISOPRED, DICHOT, and POODLE, respectively. As the corresponding ratios of all *Drosophila* proteins were 2.10, 2.45, and 1.79, the ratio was not necessarily diminished in immune-related proteins in fruit fly. Thus, the preliminary results indicate that the phenomenon of rapid evolution in both SD and IDRs in immune-related secreted proteins is possibly limited to vertebrates.

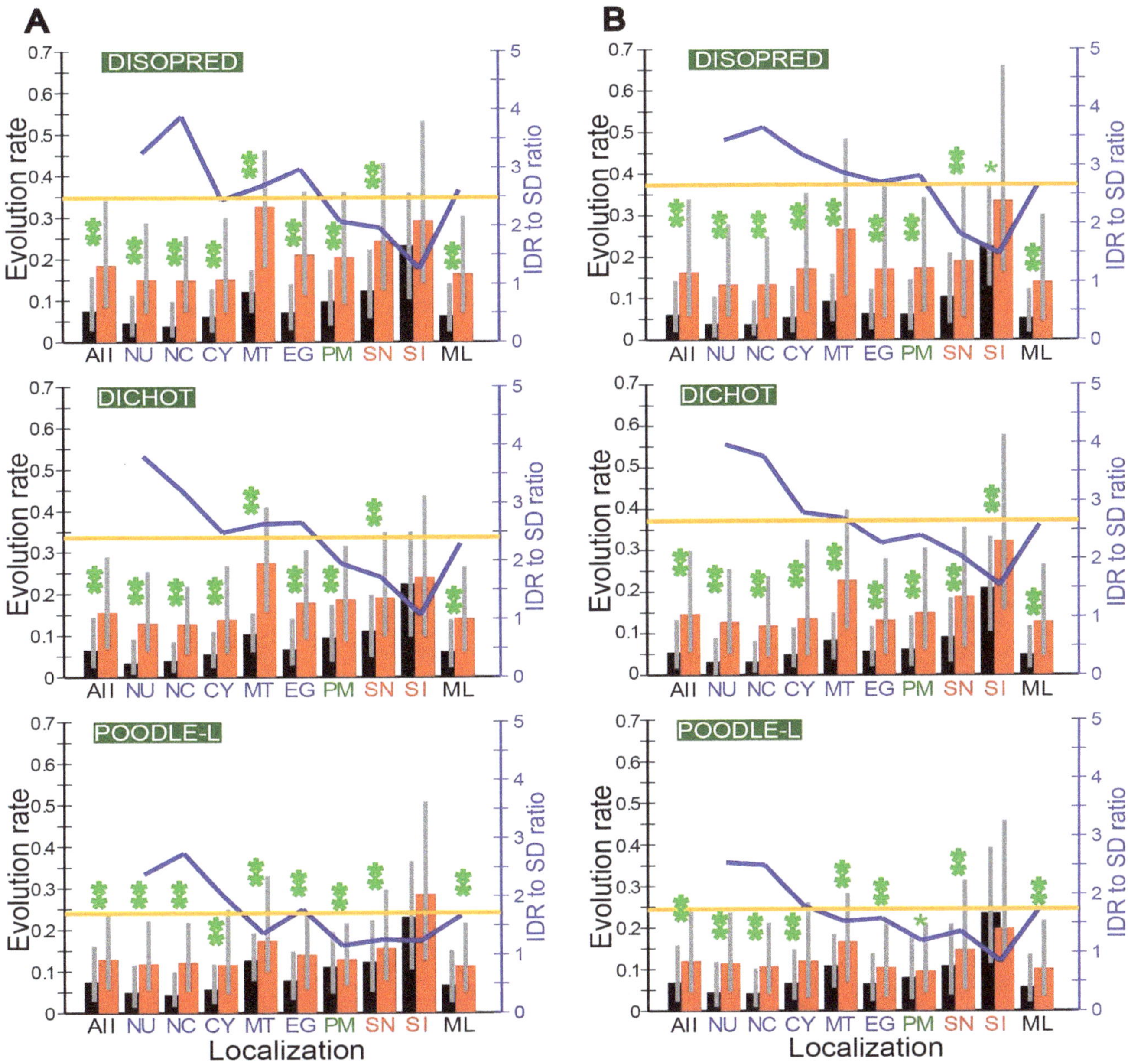

Figure 3. Evolution rates are higher in IDRs than in SDs except possibly for immune-related secreted proteins in mammals. (**A**) *H. sapiens*; (**B**) *M. musculus*; the diagrams (top to bottom) in each panel are based on DISOPRED, DICHOT and POODLE-L predictions. The median evolution rates in SDs are shown in black rectangles, while those in IDRs are depicted in red (left scale). Grey vertical lines show ranges from the 25th to the 75th percentile. Blue lines represent the median evolution rate ratios of IDRs to SDs at respective localizations, while horizontal orange lines show the ratio of all proteins (right scale). One asterisk signifies a statistically significant difference between the evolution rate distributions of IDRs and SDs at $p < 0.01$, while two asterisks denote a statistically significant difference at $p < 0.001$ (U-test). The same abbreviations for localizations as those in Table 1 are used.

In the cytosolic proteins (CY) of budding yeast, the median evolution rate in IDRs was only a little higher than that in SDs (Figure S1B). As noted before, budding yeast proteins generally interact with much more proteins than human proteins and did not exhibit intracellular-extracellular disparity in the IDR to SD evolution ratio.

2.5. Examles of Proteins with Nonsynonymous and Synonymous Substitutions

To give specific examples, we diagramed some human and mouse proteins with locations of nonsynonymous and synonymous substitutions (Figure 4). As we selected the proteins as they exhibit close-to-median ratios of nonsynonymous to synonymous substitution rates in SDs and IDRs, the frequencies of nonsynonymous to synonymous substitutions do not necessarily show median values. Although the three prediction methods gave different results, the major disparities were found in the boundaries of IDRs and did not affect main results. In immune-related secreted proteins (Figure 4A–D), nonsynonymous mutations (red bars) were almost as frequent as synonymous ones (black bars) both in IDRs (pink rectangles) and SDs (gray regions). In comparison, in proteins of other subcellular localizations, nonsynonymous substitutions occurred much less frequently than synonymous substitutions in SDs, while the difference was less pronounced in IDRs (Figure 4E–H).

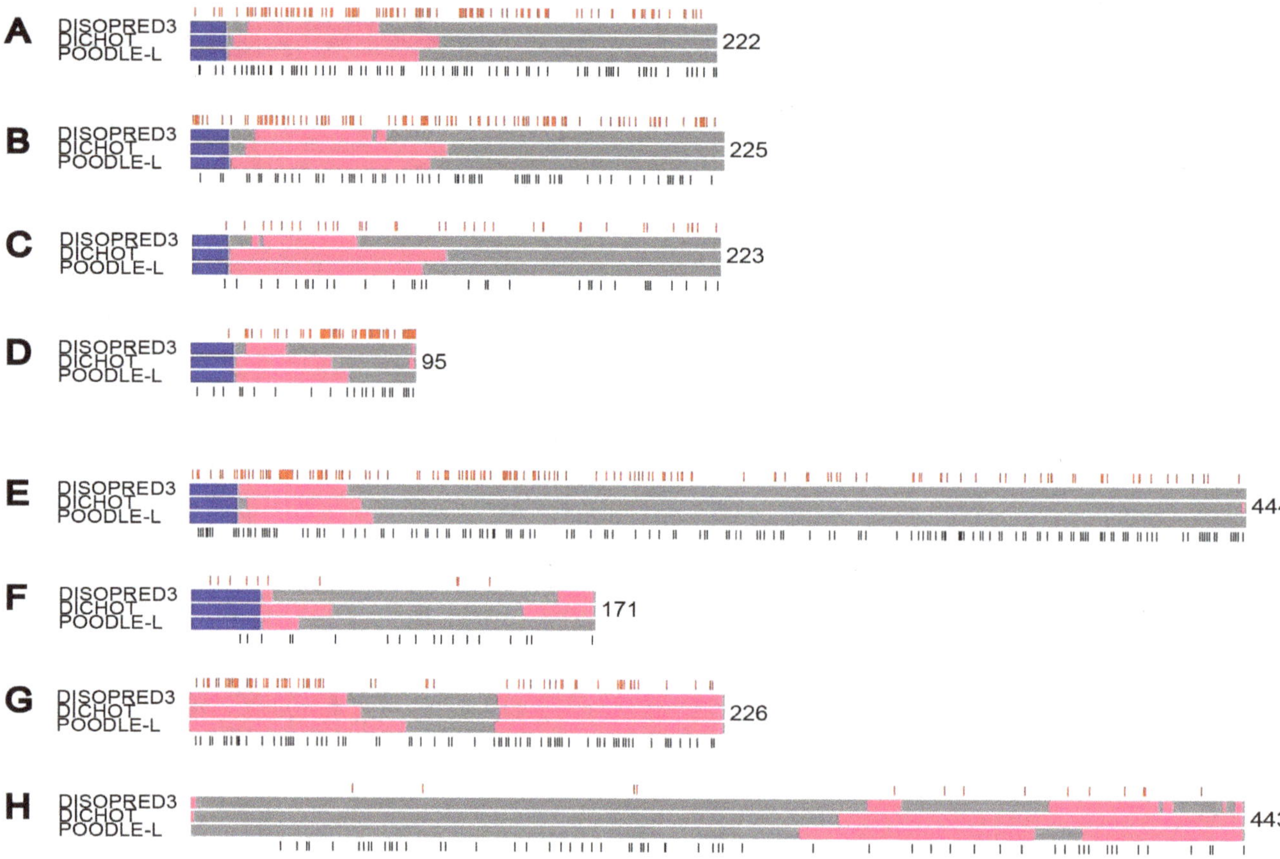

Figure 4. Examples of proteins with locations of nonsynonymous and synonymous substitutions. Each protein is represented by three rectangles with DISOPRED, DICHOT, and POODLE predictions (top to bottom) of IDRs (pink) and SDs (gray) as well as signal sequences (blue), if any, and the length shown on the right. The locations of nonsynonymous and synonymous substitutions are shown above (red lines) and below (black lines), respectively. (**A–D**): Immune-related secreted proteins, (**E**,**F**): non-immune-related secreted proteins, (**G**,**H**): nuclear proteins. (**A**) The human PRG2, (**B**) human PRG3, (**C**) mouse PRG2, (**D**) mouse DEFA20, (**E**) human SERPINA10, (**F**) mouse NENF, (**G**) human PROP1, (**H**) mouse NEK2 proteins.

2.6. Significance and Remaining Issues

The generally much lower frequency of nonsynonymous substitutions than synonymous substitutions in SDs reflects the fact that nonsynonymous changes very often destabilize the structures. By contrast, the difference between nonsynonymous and synonymous substitution rates is diminished in IDRs as nonsynonymous changes are frequently accommodated in IDRs. Consequently the ratio of nonsynonymous to synonymous substitution rate (ω) is expected to be much smaller in SDs than in IDRs. Although the actual results obtained in this research were mostly consistent with this expectation, those of immune-related secreted proteins of the two mammalian species were not; ω in SDs approaches that in IDRs.

As ω is larger than 1 at positively selected sites, the existence of numerous such sites in a region increases the average ω. Since many sites in immune-related secreted proteins were reported to be under positive selection [13–17], the heightened ω in SDs of such proteins indicates that many positively selected sites fall in SDs. The observation that IDRs of immune-related proteins exhibit higher ω than those of other proteins also implies that IDRs contain positively selected sites, too. The classification of positively selected sites in immune-related proteins into SDs and IDRs will probably lead to a better understanding of mechanisms of immunity. It is plausible that many nonsynonymous changes occur at the surface of SDs that interacts with other proteins.

It is also of interest to investigate known genes under positive selection that are associated with gamete recognition [32,33] and male reproduction [34,35] to find if SDs as well as IDRs of the encoded proteins evolve rapidly. We note that extracellular domains receive a number of posttranslational modifications such as phosphorylations, glycosylation, and lipidation. Investigation of evolution rates at posttranslational modification sites of immune-related proteins is another prospective area.

3. Materials and Methods

The nucleotide sequences of *H. sapience*, *M. musculus*, and *Rattus norvegicus* genes were downloaded from Ensembl (Release 91) [36]. The nucleotide sequences of *A. thaliana* (TAIR10), *Arabidopsis lyrata*, *Drosophila melanogaster* (BDGP6) genes were obtained from Ensembl, too. Ensembl also provided the orthologous relationships between *H. sapience* and *M. musculus* as well as those of *M. musculus* and *R. norvegicus*. The sequences of *S. cerevisiae* and *Saccharomyces paradoxus* were obtained from the Saccharomyces Genome Database [37], while those of *Drosophila pseudoobscura* genes were downloaded from FlyBase [38]. The orthologs of the two *Arabisopsis* species, the two yeast species, and the two *Drosophila* species were selected by bidirectional best hit analysis. The proteins were classified by subcellular localizations based on the Gene-Ontology (GO) annotations in UniProt (Release 2017_05) [39]. Specifically, the following GO IDs were used for subcellular classifications: nucleus: GO:0005634; cytoplasm: GO:0005829; mitochondria: GO:0005739; endoplasmic reticulum/Golgi apparatus: GO:0005783, GO:0005794, and GO:0005793; plasma membrane: GO:0005886; secreted: GO:0005576 and GO:0005615; immune-related: GO:0002376.

From the coding sequences, the signal peptides were removed based on UniProt annotations because they are unclassifiable as SDs or IDRs due to their absence in mature proteins. The remaining amino acid sequences of orthologs were aligned by MAFFT [40] and the corresponding nucleotide sequences were aligned according to the MAFFT results. dn/ds values were then computed using the codeml program (model M0) in PAML (version 4.9d) [41]. Statistical differences between two quantities were tested by Mann-Whitney's *U*-test, while statistical significance of correlations was evaluated by Spearman's rank correlation by means of in-house programs.

Number of PPIs and expression levels were taken from the BioGRID (version 3.4.158) [42] and the PaxDb (version 4.1) [43] databases, respectively. BioGRID is a curated database of interactions including protein-protein interactions obtained by two-hybrid studies, affinity purification coupled to mass spectrometry, and other methods, while PaxDB contains whole genome protein abundance information obtained by integrating numerous datasets using scores and ranks. Each protein was divided into SDs and IDRs by three methods: DISOPRED3 [29], DICHOT [30] and POODLE-L [31]. Briefly, DISOPRED3 assigns IDRs based on sequence profiles and other sequence-derived features, DICHOT classifies proteins into SDs and IDRs using sequence characteristics, alignments to existing protein structures, and sequence divergence, while POODLE-L is a prediction method for long IDRs that makes use of support-vector machine with 10 kinds of simple physico-chemical properties of amino acids. Based on the overall MAFFT alignments, the alignments of the corresponding sections were made. The evolution rate of each section was then determined as above.

4. Conclusions

In human and mouse, the SDs of immune-related proteins evolve at a high rate which comes close to that of the IDRs. This observation indicates that positive selection operates on both SDs and IDRs of the encoded proteins in many immune-related genes. Comparatively high evolution in SDs is also observed in non-immune-related secreted proteins in human and mouse, and to a lesser extent in thale cress, but not in budding yeast. Thus accelerated evolution in SDs as well as in IDRs contributes to rapid evolution of extracellular proteins in mammals.

Author Contributions: Conceptualization, K.H.; Methodology, K.H.; Software, H.A., T.N., S.F.; Validation, K.H., H.A., T.N., S.F.; Formal Analysis, K.H.; Investigation, K.H., H.A.; Resources, K.H.; Data Curation, K.H.; Writing-Original Draft Preparation, K.H.; Writing-Review & Editing, K.H.; Visualization, K.H.; Supervision, K.H.; Project Administration, K.H.; Funding Acquisition, K.H.

Abbreviations

SD	Structural domain
IDR	Intrinsically disordered region
PPI	Protein-to-protein interaction
NU	Nucleus
NC	Nucleus and cytosol
CY	Cytosol
MT	Mitochondria
EG	Endoplasmic reticulum or Golgi apparatus
PM	Plasma membrane
SN	Secreted, non-immune-related
SI	Secreted, immune-related
ML	Multiple localizations except NC
dN	Nonsynonymous substitution rate
dS	Synonymous substitution rate
GO	Gene Ontology

References

1. Wright, P.E.; Dyson, H.J. Intrinsically unstructured proteins: Re-assessing the protein structure-function paradigm. *J. Mol. Biol.* **1999**, *293*, 321–331. [CrossRef] [PubMed]
2. Brown, C.J.; Takayama, S.; Campen, A.M.; Vise, P.; Marshall, T.W.; Oldfield, C.J.; Williams, C.J.; Dunker, A.K. Evolutionary rate heterogeneity in proteins with long disordered regions. *J. Mol. Evol.* **2002**, *55*, 104–110. [CrossRef] [PubMed]
3. Fraser, H.B.; Hirsh, A.E.; Steinmetz, L.M.; Scharfe, C.; Feldman, M.W. Evolutionary rate in the protein interaction network. *Science* **2002**, *296*, 750–752. [CrossRef] [PubMed]
4. Krylov, D.M.; Wolf, Y.I.; Rogozin, I.B.; Koonin, E.V. Gene loss, protein sequence divergence, gene dispensability, expression level, and interactivity are correlated in eukaryotic evolution. *Genome Res.* **2003**, *13*, 2229–2235. [CrossRef] [PubMed]
5. Pál, C.; Papp, B.; Hurst, L.D. Highly expressed genes in yeast evolve slowly. *Genetics* **2001**, *158*, 927–931. [PubMed]
6. Subramanian, S.; Kumar, S. Gene expression intensity shapes evolutionary rates of the proteins encoded by the vertebrate genome. *Genetics* **2004**, *168*, 373–381. [CrossRef]
7. Liu, J.; Perumal, N.E.; Oldfield, C.J.; Su, E.W.; Uversky, V.N.; Dunker, A.K. Intrinsic disorder in transcription factors. *Biochemistry* **2006**, *45*, 6873–6888. [CrossRef]

8. Minezaki, Y.; Homma, K.; Kinjo, A.R.; Nishikawa, K. Human transcription factors contain a high fraction of intrinsically disordered regions essential for transcriptional regulation. *J. Mol. Biol.* **2006**, *359*, 1137–1149. [CrossRef]
9. Homma, K.; Fukuchi, S.; Nishikawa, K.; Sakamoto, S.; Sugawara, H. Intrinsically disordered regions have specific functions in mitochondrial and nuclear proteins. *Mol. Biosyst.* **2012**, *8*, 247–255. [CrossRef]
10. Ito, M.; Tohsato, Y.; Sugisawa, H.; Kohara, S.; Fukuchi, S.; Nishikawa, I.; Nishikawa, K. Intrinsically disordered proteins in human mitochondria. *Genes Cells* **2012**, *17*, 817–825. [CrossRef]
11. Julenius, K.; Pedersen, A.G. Protein evolution is faster outside the cell. *Mol. Biol. Evol.* **2006**, *23*, 2039–2048. [CrossRef]
12. Liao, B.Y.; Weng, M.P.; Zhang, J. Impact of extracellularity on the evolutionary rate of mammalian proteins. *Genome Biol. Evol.* **2010**, *2*, 39–43. [CrossRef]
13. Hughes, A.L.; Nei, M. Pattern of nucleotide substitution at major histocompatibility complex class I loci reveals overdominant selection. *Nature* **1988**, *335*, 167–170. [CrossRef]
14. Hughes, A.L.; Nei, M. Nucleotide substitution at major histocompatibility complex class II loci: Evidence for overdominant selection. *Proc. Natl. Acad. Sci. USA* **1989**, *86*, 958–962. [CrossRef]
15. Patil, A.; Hughes, A.L.; Zhang, G. Rapid evolution and diversification of mammalian α-defensins as revealed by comparative analysis of rodent and primate genes. *Physiol. Genom.* **2004**, *20*, 1–11. [CrossRef]
16. Morrison, G.M.; Semple, C.A.M.; Kilanowski, F.M.; Hill, R.E.; Dorin, J.R. Signal sequence conservation and mature peptide divergence within subgroups of the murine β-defensin gene family. *Mol. Biol. Evol.* **2003**, *20*, 460–470. [CrossRef]
17. Zelezetsky, I.; Pontillo, A.; Puzzi, L.; Antcheva, N.; Segat, L.; Pacor, S.; Crovella, S.; Tossi, A. Evolution of the primate cathelicidin. Correlation between structural variations and antimicrobial activity. *J. Biol. Chem.* **2006**, *281*, 19861–19871. [CrossRef]
18. Baxt, L.A.; Garza-Mayers, A.C.; Goldberg, M.B. Bacterial subversion of host innate immune pathways. *Science* **2013**, *340*, 697–701. [CrossRef]
19. Sánchez, B.; Urdaci, M.C.; Margolles, A. Extracellular proteins secreted by probiotic bacteria as mediators of effects that promote mucosa-bacteria interactions. *Microbiology* **2010**, *156*, 3232–3242. [CrossRef]
20. Nobre, T.M.; Martynowicz, M.W.; Andreev, K.; Kuzmenko, I.; Nikaido, H.; Gidalevitz, D. Modification of Salmonella lipopolysaccharides prevents the outer membrane penetration of novobiocin. *Biophys. J.* **2015**, *109*, 2537–2545. [CrossRef]
21. Horlick, R.A.; Macomber, J.L.; Bowers, P.M.; Nebern, T.Y.; Tomlinson, G.L.; Krapf, I.P.; Dalton, J.L.; Verdino, P.; King, D.J. Simultaneous surface display and secretion of proteins from mammalian cells facilitate efficient in vitro selection and maturation of antibodies. *J. Biol. Chem.* **2013**, *288*, 19861–19869. [CrossRef]
22. Lieschke, G.J.; Burgess, A.W. Granulocyte colony-stimulating factor and granulocyte-macrophage colony-stimulating factor (1). *N. Engl. J. Med.* **1992**, *327*, 28–35. [CrossRef]
23. Jelkmann, W. Regulation of erythropoietin production. *J. Physiol.* **2011**, *589*, 1251–1258. [CrossRef]
24. Lucas, A.; McFadden, G. Secreted immunomodulatory viral proteins as novel biotherapeutics. *J. Immunol.* **2004**, *173*, 4765–4774. [CrossRef]
25. Lubbers, R.; van Essen, M.F.; van Kooten, C.; Trouw, L.A. Production of complement components by cells of the immune system. *Clin. Exp. Immunol.* **2017**, *188*, 183–194. [CrossRef]
26. Bonin-Debs, A.L.; Boche, I.; Gille, H.; Brinkmann, U. Development of secreted proteins as biotherapeutic agents. *Expert Opin. Biol. Ther.* **2004**, *4*, 551–558. [CrossRef]
27. Castillo-Davis, C.I.; Kondrashov, F.A.; Hartl, D.L.; Kulathinal, R.J. The functional genomic distribution of protein divergence in two animal phyla: Coevolution, genomic conflict, and constraint. *Genome Res.* **2004**, *14*, 802–811. [CrossRef]
28. Ota, M.; Gonja, H.; Koike, R.; Fukuchi, S. Multiple-localization and hub proteins. *PLoS ONE* **2016**, *11*, e0156455. [CrossRef]
29. Jones, D.T.; Cozetto, D. DISOPRED3: Precise disordered region predictions with annotated protein-binding activity. *Bioinformatics* **2015**, *31*, 857–863. [CrossRef]
30. Fukuchi, S.; Homma, K.; Minezaki, Y.; Gojobori, T.; Nishikawa, K. Development of an accurate classification system of proteins into structured and unstructured regions that uncovers novel structural domains: Its application to human transcription factors. *BMC Struct. Biol.* **2009**, *9*, 26. [CrossRef]

31. Hirose, S.; Shimizu, K.; Kanai, S.; Kuroda, Y.; Noguchi, T. POODLE-L: A two-level SVM prediction system for reliably predicting long disordered regions. *Bioinformatics* **2007**, *23*, 2046–2053. [CrossRef]
32. Lee, Y.H.; Ota, T.; Vacquier, V.D. Positive selection is a general phenomenon in the evolution of abalone sperm lysin. *Mol. Biol. Evol.* **1995**, *12*, 231–238. [CrossRef]
33. Swanson, W.J.; Vacquier, V.D. Extraordinary divergence and positive Darwinian selection in a fusagenic protein coating the acrosomal process of abalone spermatozoa. *Proc. Natl. Acad. Sci. USA* **1995**, *92*, 4957–4961. [CrossRef]
34. Tsauer, S.-C.; Wu, C.-I. Positive selection and the molecular evolution of a gene of male reproduction, *Acp26Aa*, of *Drosophila*. *Mol. Biol. Evol.* **1997**, *14*, 544–549. [CrossRef]
35. Wyckoff, G.J.; Wang, W.; Wu, C.-I. Rapid evolution of male reproductive genes in the descent of man. *Nature* **2000**, *403*, 304–309. [CrossRef]
36. Zerbino, D.R.; Achuthan, P.; Akanni, W.; Amode, M.R.; Barrell, D.; Bhai, J.; Billis, K.; Cummins, C.; Gall, A.; Girón, C.G.; et al. Ensembl 2018. *Nucleic Acids Res.* **2018**, *46*, D754–D761. [CrossRef]
37. Skrzypek, M.S.; Nash, R.S.; Wong, E.D.; MacPherson, K.A.; Hellerstedt, S.T.; Engel, S.R.; Karra, K.; Weng, S.; Sheppard, T.K.; Binkley, G.; et al. Saccharomyces genome database informs human biology. *Nucleic Acids Res.* **2018**, *46*, D736–D742. [CrossRef]
38. Gramates, L.S.; Marygold, S.J.; dos Santos, G.; Urbano, J.M.; Antonazzo, G.; Matthews, B.B.; Rey, A.J.; Tabone, C.J.; Crosby, M.A.; Emmert, D.B.; et al. FlyBase at 25: Looking to the future. *Nucleic Acids Res.* **2017**, *45*, D663–D671. [CrossRef]
39. The UniProt Consortium. UniProt: The universal protein knowledgebase. *Nucleic Acids Res.* **2017**, *45*, D158–D169. [CrossRef]
40. Kuraku, S.; Zmasek, C.M.; Nishimura, O.; Katoh, K. aLeaves facilitates on-demand exploration of metazoan gene family trees on MAFFT sequence alignment server with enhanced interactivity. *Nucleic Acids Res.* **2013**, *41*, W22–W28. [CrossRef]
41. Yang, Z. PAML4: Phylogenetic analysis by maximum likelihood. *Mol. Biol. Evol.* **2007**, *24*, 1586–1591. [CrossRef]
42. Chatr-Aryamontri, A.; Oughtred, R.; Boucher, L.; Rust, J.; Chang, C.; Kolas, N.K.; O'Donnell, L.; Oster, S.; Theesfeld, C.; Sellam, A.; et al. The BioGRID interaction database: 2017 update. *Nucleic Acids Res.* **2017**, *45*, D369–D379. [CrossRef]
43. Wang, M.; Herrmann, C.J.; Simonovic, M.; Szklarczyk, D.; von Mering, C. Version 4.0 of PaxDB: Protein abundance data, integrated across model organisms, tissues, and cell-lines. *Proteomics* **2015**, *15*, 3163–3168. [CrossRef]

7

The Cyanobacterial Ribosomal-Associated Protein LrtA from *Synechocystis* sp. PCC 6803 is an Oligomeric Protein in Solution with Chameleonic Sequence Properties

Lellys M. Contreras [1,†], Paz Sevilla [2,3], Ana Cámara-Artigas [4], José G. Hernández-Cifre [5], Bruno Rizzuti [6,*], Francisco J. Florencio [7], María Isabel Muro-Pastor [7], José García de la Torre [5] and José L. Neira [8,9,*,†]

1 Center for Environmental Biology and Chemistry Research, Facultad Experimental de Ciencias y Tecnología, Universidad de Carabobo, 2001 Valencia, Venezuela; lellyscontreras@gmail.com
2 Facultad de Farmacia, Departamento de Química Física II, Universidad Complutense de Madrid, 28040 Madrid, Spain; paz@farm.ucm.es
3 Instituto de Estructura de la Materia, IEM-CSIC, Serrano 121, 28006 Madrid, Spain
4 Department of Chemistry and Physics, Research Centre CIAIMBITAL, University of Almería- ceiA3, 04120 Almería, Spain; acamara@ual.es
5 Department of Physical Chemistry, University of Murcia, 30003 Murcia, Spain; jghc@um.es (J.G.H.-C.); jgt@um.es (J.G.d.l.T.)
6 CNR-NANOTEC, Licryl-UOS Cosenza and CEMIF.Cal, Department of Physics, University of Calabria, 87036 Rende, Italy
7 Instituto de Bioquímica Vegetal y Fotosíntesis, CSIC-Universidad de Sevilla, 41092 Seville, Spain; floren@us.es (F.J.F.); imuro@ibvf.csic.es (M.I.M.-P.)
8 Instituto de Biología Molecular y Celular, Universidad Miguel Hernández, 03202 Elche (Alicante), Spain
9 Instituto de Biocomputación y Física de Sistemas Complejos, Joint Units IQFR-CSIC-BIFI, and GBsC-CSIC-BIFI, Universidad de Zaragoza, 50009 Zaragoza, Spain
* Correspondence: bruno.rizzuti@cnr.it (B.R.); jlneira@umh.es (J.L.N.)

† These authors contributed equally to this work.

Abstract: The LrtA protein of *Synechocystis* sp. PCC 6803 intervenes in cyanobacterial post-stress survival and in stabilizing 70S ribosomal particles. It belongs to the hibernating promoting factor (HPF) family of proteins, involved in protein synthesis. In this work, we studied the conformational preferences and stability of isolated LrtA in solution. At physiological conditions, as shown by hydrodynamic techniques, LrtA was involved in a self-association equilibrium. As indicated by Nuclear Magnetic Resonance (NMR), circular dichroism (CD) and fluorescence, the protein acquired a folded, native-like conformation between pH 6.0 and 9.0. However, that conformation was not very stable, as suggested by thermal and chemical denaturations followed by CD and fluorescence. Theoretical studies of its highly-charged sequence suggest that LrtA had a Janus sequence, with a context-dependent fold. Our modelling and molecular dynamics (MD) simulations indicate that the protein adopted the same fold observed in other members of the HPF family (β-α-β-β-β-α) at its N-terminal region (residues 1–100), whereas the C terminus (residues 100–197) appeared disordered and collapsed, supporting the overall percentage of overall secondary structure obtained by CD deconvolution. Then, LrtA has a chameleonic sequence and it is the first member of the HPF family involved in a self-association equilibrium, when isolated in solution.

Keywords: conformational plasticity; disordered protein; folding; ribosomal protein; spectroscopy; protein stability

1. Introduction

The *lrtA* gene was first identified in *Synechococcus* sp. PCC 7002 as a sequence encoding a light-repressed protein [1], with a larger half-life in dark conditions than in the presence of light [2]. Although the exact functions of the LrtA protein are unknown, recent studies have shown that it is involved in post-stress survival in *Synechocystis* sp. PCC 6803, stabilizing the 70S ribosomal particles [3].

LrtA is related to other proteins, which are highly present among bacteria and associated with ribosomes. These proteins modulate ribosome activity to preserve their integrity and aid in cell survival during stress circumstances. Under these conditions, stalling of the protein synthesis, a major energy-consuming process in living cells, is downregulated, usually by proteins involved in ribosome inhibition. Reduction of translation activity is associated with: (i) dimerization of 70S particles to form the translationally inactive 100S disome (also known as hibernating ribosomes [4]), mediated by intermolecular interactions among proteins; or alternatively; (ii) interaction of canonical ribosomal proteins with the ribosome [5,6]. Among the most studied members of this protein family are two *Escherichia coli* proteins: YfiA (also known as PY or RaiA, ribosome associated inhibitor A); and YhbH (also known as HPF, hibernation promoting factor). YfiA is thought to inhibit translation indirectly, by modulating a more stringent proofreading mechanism involving 70S particles [7,8]. On the other hand, HPF stops translation by stabilizing 100S dimers [8–10]. Formation of 100S disomes is also mediated by other proteins, known as ribosome modulation factors (RMFs), in γ-proteobacteria species or double YfiA- and YhbH- knocked-out cells [10]. Phylogenetic analyses have shown that most bacteria have at least one of those HPF or YfiA homologues [10]. These homologues have been classified in three classes, based on the presence of a conserved domain and, in some cases, additional sequence extensions: long HPF, short HPF, and YFiA. The conserved domain has a β-α-β-β-β-α fold, with the two α-helices packed against one side of the four-stranded β-sheet [5,11]. According to its sequence, the LrtA from *Synechocystis* sp. PCC 6803 could be classified within the long HPF sub-family; in addition, it also bears similarity to the spinach plastid-specific ribosomal protein, which is present in the chloroplast stroma, either associated or unbound to the 30S ribosomal unit [3]. Although we have shown that LrtA stabilizes 70S particles [3], nothing is known about the conformation or stability of the isolated protein in solution.

In this work, we embarked in the characterization of the conformational stability and structure of LrtA from *Synechocystis* sp. PCC 6803 by using experimental and *in silico* approaches. Our results are the first characterization of the conformation and stability of a member of the long HPF subfamily. At physiological pH, LrtA was involved in a self-association equilibrium, as shown by hydrodynamic techniques. The protein acquired a native-like conformation around pH 6.0, as judged from intrinsic fluorescence (monitoring tertiary structure), ANS (8-anilino-1-naphthalene sulfonic acid) fluorescence (revealing hydrophobic solvent-exposed patches), CD (reporting secondary structure) and 1D ^{1}H NMR experiments. The MD simulations and analyses of sequence suggested that the protein had a Janus sequence, and its conformation was solvent-dependent, with an N-terminal region acquiring the fold of other members of the HPF family and the C-terminal region appearing disordered and collapsed. Therefore, LrtA is the first member of the HPF family with chameleonic features encoded in its sequence and shown to be involved in a self-association equilibrium when isolated in solution.

2. Results

2.1. Isolated LrtA Was Involved in a Self-Association Equilibrium in Solution

We first tried to elucidate the oligomerization state of the protein to identify the protein-concentration range where we must characterize the conformational stability of the protein.

To map the hydrodynamic properties of LrtA we used three hydrodynamic techniques: DOSY-NMR (diffusion ordered NMR spectroscopy), DLS (dynamic light scattering), size exclusion chromatography (SEC), glutaraldehyde cross-linking, and lifetime fluorescence measurements. Furthermore, we tried to measure the self-association of LrtA by using isothermal titration calorimetry

(ITC), but in all attempts, protein precipitated at the concentrations required to carry out the experiments. It is important to pinpoint the differences among the different hydrodynamic techniques used in this work. With NMR, we shall obtain information about the low-molecular weight species, whose overall rotational tumbling is very fast. By using DLS, we shall obtain information about the hydrodynamic parameters, assuming a spherical shape, for all the species (high or low molecular weight) present in solution, and we shall be able to see whether those hydrodynamic parameters are protein-concentration-dependent. By using SEC, we shall be able to monitor the elution volume of LrtA, which will depend on the molecular weight and the shape of the molecule, but that volume could be also affected by possible interactions with the column. Finally, by lifetime fluorescence measurements, we shall determine how the decay of the electronic excited states can be affected by: (i) the presence of conformational isomers; (ii) energy transfer among the eight Tyr residues in LrtA; or (iii) even transient electronic effects in collisional quenching [12,13]. To elucidate the oligomerization state of the protein, we also tried to carry out T_2 echo measurements estimating the averaged correlation time of the species present in solution from the amide region. However, at the times used (2.9 ms and 400 μs) in our echo experiments, most of the amide peaks disappeared, and only the proton resonances of the His ring (around 8.5 ppm) could be clearly measured, yielding a very long value for the T_2, unreliable to estimate the mobility of the backbone of the polypeptide chain.

The DOSY-NMR measurements at pH 8.0 yielded a translational diffusion coefficient (D) with a value of $(7.2 \pm 0.2) \times 10^{-7}$ cm^2 s^{-1} (Figure S1A). By taking into account the hydrodynamic radius, R_S, of dioxane (2.12 Å), and its D under our conditions ($(6.8 \pm 0.2) \times 10^{-6}$ cm^2 s^{-1}), the R_S (Stokes radius) estimated for LrtA was 20 ± 2 Å. We can compare this value with that theoretically determined for a polypeptide with the length of LrtA. The R value for an unsolvated, ideal, spherical molecule can be estimated from [14]: $R = \sqrt[3]{3M\overline{V}/4N_A\pi}$, where N_A is Avogadro's number, M is the molecular weight (22.717 kDa) and $\overline{V}$ the specific volume of LrtA (0.729 mL/g). The calculated R for LrtA is 18.7 Å, but since the hydration shell is 3.2 Å wide [15], the hydration radius would be 21.9 Å, which is similar to the R_S from DOSY-NMR. On the other hand, it has been shown that the R_S of a folded spherical protein can be approximated by [16]: $R_S = (4.75 \pm 1.11)N^{0.29}$, where N is the number of residues; in a 197-residue-long protein such as LrtA, this expression yields 21 ± 6 Å, similar to that determined by DOSY-NMR experiments. Therefore, by the DOSY-NMR measurements, we are only detecting a monomeric globular species of LrtA. In fact, the 1D ^{1}H-NMR spectrum of LrtA at pH 8.0 (with 500 mM NaCl) (Figure S2) corresponds to that of a well-folded protein with dispersed peaks in the methyl and amide regions; interestingly enough, the spectrum had down-field shifted H$_\alpha$ protons (between 5.0 and 6.0 ppm), suggesting the presence of residues involved in β-strands. It is interesting to note that, as possible higher-order molecular species could not be observed in the NMR spectrum due to their molecular weight (and therefore signal broadening), the presence of the majority of the amide protons belonging to possible disordered regions in the protein (see below, 2.4.) would not be observed in the amide region due to: (i) solvent hydrogen-exchange at pH 8.0; (ii) conformational exchange broadening; or (iii) overlapping with the signals from the well-folded region (as indicated by the largest increase of intensity around 8.3 ppm, Figure S2B). In addition, the alkyl resonances belonging to possible polypeptide disordered regions would be hindered by the rest of the methyl groups of the protein in the up-field shifted region. On the other hand, the 1D ^{1}H-NMR spectrum at pH 4.5 showed a smaller intensity and a poorer signal-to-noise ratio (due probably to the precipitation during sample preparation) than the spectrum at pH 8.0. In addition, the spectrum also showed broader peaks (and less intense, as it is evident by comparing Figure S3B with Figure S2B), and the absence of well-dispersed signals in the amide and methyl regions (Figure S3), as shown, for instance, by the lack of peaks around 0.3 and 0.5 ppm, which appeared at pH 8.0. The broader peaks at pH 4.5 (when compared to pH 8.0) could be due to the presence of uni-molecular conformational-exchange equilibria or, alternatively, to the presence of self-associated species. Thus, to elucidate whether at pH 4.5 there were concentration-dependent equilibria, we carried out far UV CD experiments at protein concentrations of 4.5 and 9.8 μM (in protomer units); these experiments (Figure S4) showed that the molar ellipticity and the shape of the spectra were protein-concentration-dependent. Therefore, these results suggest that the conformation of the protein and its self-associated features were different at the two pH values of 4.5 and 8.0.

Second, we measured the hydrodynamic features of LrtA by using DLS at several protein concentrations. Two peaks were identified in the size distribution analysis for the concentration of 68 μM (in protomer units): the first peak with $R_S = 39 \pm 5$ Å that accounts for the 97.5% of the protein in the solution and a second peak with $R_S = 409 \pm 200$ Å corresponding to a small amount of aggregates (Figure 1A). Taking into account that particle scattering intensity is proportional to the square of the molecular weight, a small percentage (in this case 2.5%) of protein aggregates dominates the intensity distribution, which can be misleading, and therefore, the results in Figure 1A are shown as size distribution by volume instead of intensity. That R_S, obtained from the first peak, corresponds to a molecular weight of 81 kDa, by using an empirical mass vs. size calibration curve in the instrument software. The experiments at different LrtA concentrations indicate that the R_S (obtained from the volume peak measurements of the first observed peak) varied with the protein concentration. These findings suggest the presence of a self-associated equilibrium at pH 8.0 (Figure 1B) involving the protein, as also described in other oligomeric proteins [17]. It is interesting to note that these oligomeric species should not be observed in NMR due to their large molecular weights [15].

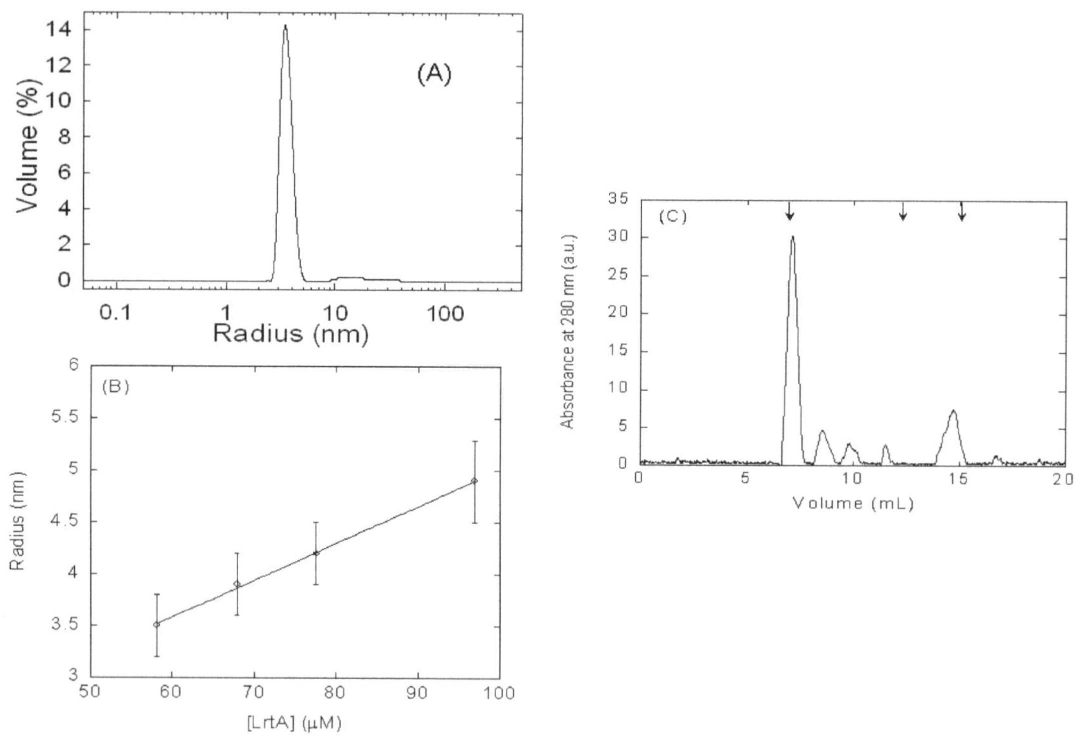

Figure 1. Hydrodynamic measurements of LrtA by dynamic light scattering (DLS) and size exclusion chromatography (SEC): (**A**) DLS measurements of the hydrodynamic radius RS of LrtA as a function of the percentage of the volume peak at 68 μM concentration (in protomer units). (**B**) Variation of the calculated R_S with LrtA concentration (in protomer units). Error bars are standard deviations from the fitting to a spherical shape. (**C**) SEC chromatogram of LrtA at 97 μM (in protomer units) at pH 8.0 (50 mM Tris) in 0.7 M NaCl; the arrows at the top indicate (from left to right) the elution volumes of blue dextran (7.1 ± 0.1 mL), albumin (12.1 ± 0.1 mL; 63.7 kDa), and bovine RNase A (15.1 ± 0.1 mL; 15.7 kDa) (the errors are standard deviations of three independent measurements). Chromatogram was baseline-corrected by UNICORN 5.01 software (GE Healthcare), and therefore, the origin of the sharpening observed in the peaks. Experiments were carried out at 25 °C.

Third, we also detected the presence of oligomeric species using glutaraldehyde cross-linking, which can react as monomer, but also as a heterogeneous polymer, involving accessible lysine residues. Our results (Figure S5) indicate that in the presence of a final concentration of 1% gutaraldehyde, there were dimers (which appear close to the band of the protein marker at 48 kDa) and other high-molecular-weight species with molecular weights larger than 210 kDa, at the top of the SDS-PAGE (sodium dodecyl sulfate polyacrylamide gel electrophoresis) gel lanes. These high molecular weight species could be due to the presence of cross-linked dimeric species.

Next, we used SEC in a Superose 12 10/300 GL, in buffer pH 8.0 (50 mM Tris) with 0.7 M NaCl to elucidate whether the protein behaved as an oligomer. We used such high concentrations of NaCl to avoid, as much as possible, any kind of protein-column interactions (as those we have observed to occur with other kind of matrix columns, see below Section 4). The protein markers used to calibrate the column were also loaded in the same buffer. At loading concentrations of 97 μM (in protomer units) of LrtA, the protein eluted as several peaks (errors are standard deviations from three independent measurements) at 7.3 ± 0.2, 8.5 ± 0.1, 9.8 ± 0.2, 11.5 ± 0.1, and 14.7 ± 0.2 mL (Figure 1C). These peaks, especially that at 7.3 mL, indicate that the protein behaved as an oligomer, with molecular weights larger than those of albumin (63.7 kDa) and bovine RNase A (15.7 kDa), although other higher-order molecular species of LrtA were present in solution. The peak at 14.7 mL could be due to the monomeric species, which was observed under these conditions. The other oligomeric species could be assigned to hexamers (11.5 mL) and dodecamers (9.8 mL), whereas the other two could be due to the presence of aggregates (as those species detected in DLS, see above); however, it is important to indicate that some of the peaks could be also due to protein-column interactions even in the presence of high NaCl concentration.

Finally, we measured the fluorescence lifetimes of LrtA at different protein concentrations. The experimental decay of the total protein fluorescence was best fit to bi-exponential functions (Table 1, Figure S6), and thus, two lifetimes were observed; attempts to fit the experimental data to more than two exponentials led to an increase in the χ^2. At any of the protein concentrations, the shortest lifetime corresponded to the largest amplitude (a_1), and it did not change with the protein concentration. Interestingly enough, the longest lifetime (as well as its amplitude, a_2) was concentration-dependent (Table 1). Furthermore, the $<\tau>$ showed also a concentration-dependence: going from a value of 6 ns (at the smallest concentration) to 1 ns at 98 μM. It is well-known that the intrinsic fluorescence lifetime of the first excited electronic singlet state does not change, but due to various quenching processes, changes in the environment around the fluorophores or even conformational changes in the molecular species, the measured lifetime is different from the intrinsic one [12,13]. Thus, even ruling out a possible fitting of the data to a monomer ↔ oligomer equilibria because we are observing the lifetimes of eight Tyr residues, each of them with a different environment, we can conclude that the concentration-dependence observed in the $<\tau>$ (Table 1) should have its origin in association-dissociation events.

Table 1. Fluorescence lifetimes of LrtA (100 mM, phosphate buffer (pH 8.0), with 500 mM NaCl) at 25 °C [a].

Concentration (μM)	τ_1 (ns)	a_1	τ_2 (ns)	a_2	$<\tau>$ (ns)	χ^2
98	0.49 ± 0.08	1.9 ± 0.2	3.4 ± 0.1	0.072 ± 0.005	1.101	1.384
9.8	0.46 ± 0.07	2.0 ± 0.2	3.6 ± 0.2	0.048 ± 0.006	0.9622	1.528
7.8	0.54 ± 0.08	1.8 ± 0.2	2.7 ± 0. 2	0.07 ± 0.01	0.9037	1.35
1.9	0.69 ± 0.09	0.71 ± 0.09	14.9 ± 0. 8	0.0156 ± 0.0008	5.233	1.22
0.98	0.48 ± 0.07	1.0 ± 0.1	19.5 ± 0.4	0.0104 ± 0.0005	5.996	1.238

[a] Errors are from fitting to a bi-exponential function.

It could be thought that the detected self-associated species could be random-oligomers; although we cannot rule out the presence of aggregates (from the DLS, SEC, and glutaraldehyde cross-linking results), however, there are at least three pieces of evidence suggesting that the self-associated species do not oligomerize un-specifically: (i) the DLS results show a linear dependence with the concentration (Figure 1B); (ii) the protein-dependent, and almost exponential, variation of the life-times; and, (iii) the presence of bands at particular molecular weights in the glutaraldehyde experiments. Therefore, those results, together with experiments from (GdmCl) chemical denaturations (see below, Section 2.3.), must be due to the presence of self-associated equilibria, involving the regions around some of the eight Tyr residues.

2.2. LrtA Acquired a Native-Like Conformation Between pH 6.0 and 9.0

We analyzed the structure of LrtA at varying pH to find out in which interval the protein acquired a native-like conformation. To this end, we used several biophysical techniques, namely, intrinsic and ANS fluorescence, CD and NMR. We used intrinsic fluorescence to monitor changes in the tertiary structure around its eight Tyr residues. Furthermore, ANS fluorescence was used to monitor the burial of solvent-exposed hydrophobic patches. We acquired far-UV CD spectra to monitor the changes in secondary structure. Finally, we acquired 1D ^{1}H NMR spectra that show the presence of secondary and tertiary structure at physiological pH (see above, Section 2.1.). These spectra indicate (Figures S2 and S3) that the secondary and tertiary structures of the protein at pH 4.5 and 8.0 were completely different.

2.2.1. Fluorescence

Intrinsic Steady-State Fluorescence and Thermal Denaturations—The fluorescence spectrum of LrtA at physiological pH showed a maximum at 308 nm, as expected for a polypeptide chain with fluorescent Tyr residues. The pH-dependence of the intrinsic $<1/\lambda>$ showed two transitions (Figure 2A, left axis, filled circles). The first transition finished at pH 6.0, but we could not determine its pK_a due to the absence of an acidic baseline. This transition was probably due to the titration of some of the seventeen Glu and/or twelve Asp residues of the LrtA sequence [18,19], which can alter the environment around some of the Tyr residues. However, we cannot rule out that it could be also due to the titration of some of the eight naturally-occurring His residues, taking place at an unusually low pK_a. The second transition occurred at basic pH, starting at pH > 9.0, but, in this case, we could not determine the pK_a due to the absence of a baseline at the highest pH values. This transition was probably due to the titration of at least some of the eight Tyr residues in the sequence. Therefore, the changes observed in fluorescence as the pH was changed could be due to titrations of specific residues around the Tyr residues, or alternatively, to conformational changes involving those fluorescent amino acids.

Thermal denaturations of LrtA were carried out at several pH values with a protein concentration of 9.8 μM, in protomer units. At pH values larger than 6.0, we observed an irreversible broad transition (Figure 2C, left axis, blank circles). Below pH 6.0, we did not observe any sigmoidal behavior, and we did not observe any sigmoidal transition at pH 13.0 either (Figure S7A). It could be argued that as fluorescence is intrinsically temperature-sensitive [12,13], we are not monitoring the denaturation of the protein. However, it must be kept in mind that fluorescence temperature sensitivity is linear (as observed at low pH values, Figure S7A; or in the native and unfolded baselines of the curve shown in Figure 2C), but it is not sigmoidal as observed in the denaturations at pH 7.0, with a midpoint around 45 °C (Figure 2C), or at pH 8.4 (Figure S7A). We also carried out experiments at LrtA concentrations of 5 μM, in protomer units (Figure 2C), and denaturation was also irreversible. Therefore, irreversibility was not associated with the amount of protein used during thermal denaturations.

ANS-Binding—At low pH, the ANS fluorescence intensity at 480 nm was large and decreased as the pH was raised (Figure 2A, right axis, blank squares), suggesting that LrtA had solvent-exposed hydrophobic regions. We could not determine the pK_a of this titration due to the absence of an acidic baseline. The burial of solvent-exposed hydrophobic residues was complete at pH 6.0, as it happens with the transition observed by following the intrinsic fluorescence (see above). Since ANS reports on burial of hydrophobic surface, and therefore it monitors conformational changes, we must conclude that the protein had structural changes at acidic pH values; then, the variations monitored by intrinsic fluorescence (see above) at acidic pH must be associated with conformational changes due to the protonation of Asp and Glu residues (or the other amino acids described above).

In conclusion, our results indicate that, at low pH values, LrtA had solvent-exposed hydrophobic regions.

Solvent-Exposure of Tyr Residues Monitored by Iodide and Acrylamide Quenching—We carried out quenching experiments at pH 3.0, 7.0, and 11.0, because these are the three regions where we observed a different intrinsic fluorescence behavior of LrtA (Figure 2A). We used two quenching agents because of the charge effects probably occurring at extreme pH value with I^-. We have assumed that the fluorescence lifetimes of the self-associated protein (for a fixed protein concentration) did not change in the whole pH interval. The K_{SV} values for KI and acrylamide in the absence of denaturant were smaller than those measured in other proteins containing only Tyr residues [20,21] (Table 2). As a general trend, the K_{SV} values of LrtA in the presence of acrylamide were smaller at acidic pH values than at physiological or basic ones; these differences could be due to the presence of higher-order self-associated species at the acidic pH values (as suggested by the ANS results (Figure 2A) and the CD data at low pH (Figure S4)). Furthermore, these results indicate that the structure of LrtA underwent some conformational changes at acidic pH (in agreement with results from intrinsic and ANS fluorescence, Figure 2A, and the NMR results, Figures S2 and S3). In the presence of GdmCl (guanidine hydrochloride), the K_{SV} values were larger (either in KI or acrylamide) than those in the absence of denaturant (Table 2), suggesting that Tyr residues were more solvent-exposed.

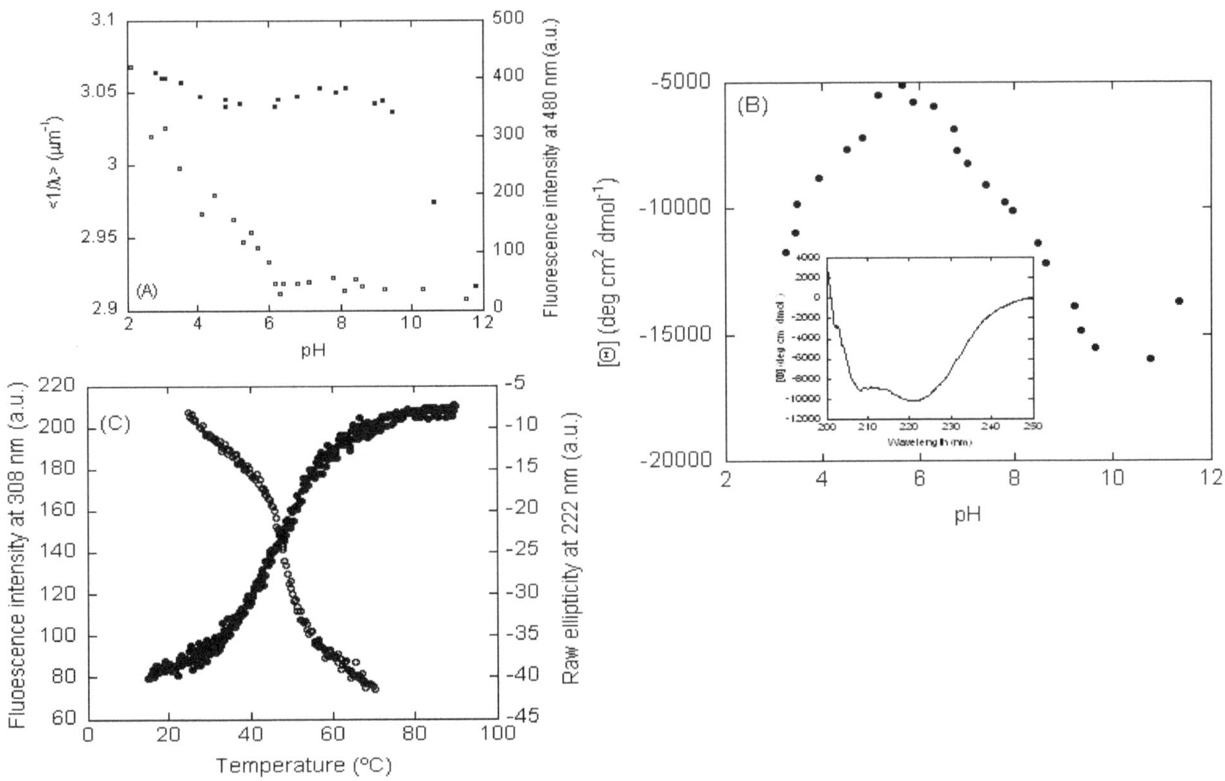

Figure 2. pH-denaturation of LrtA: (**A**) Intrinsic (left axis, filled circles) and ANS (right axis, blank squares) fluorescence of LrtA, as the pH was modified. (**B**) Changes in the [Θ] at 222 nm as the pH was varied (filled circles). Inset: far-UV CD spectrum of LrtA at 100 mM phosphate buffer (pH 8.0), with 500 μM NaCl at 25 °C. (**C**) Thermal denaturations followed by intrinsic fluorescence (left axis, blank circles) at pH 7.0 and 5 μM (in protomer units) of LrtA, and raw ellipticity at 222 nm at pH 7.0 and 9.8 μM, in protomer units (right axis, filled circles).

Table 2. Quenching parameters for LrtA under several conditions at 25 °C.

Solution Conditions	KI	Acrylamide	
	K_{sv} (M^{-1})	K_{sv} (M^{-1})	υ (M^{-1})
pH 3.0	- [a]	7.7 ± 0.5	2.1 ± 0.1
pH 7.0	1.63 ± 0.03	11 ± 1	2.2 ± 0.3
pH 11	0.78 ± 0.02	9.7 ± 0.5	0.5 ± 0.1
GdmCl (pH 7.0)	3.1 ± 0.1	22 ± 4	2.0 ± 0.3

[a] Not determined due to protein precipitation.

2.2.2. CD

The far-UV (ultraviolet) CD spectrum of LrtA at pH 8.0 had minima at 222 nm and 210 nm (Figure 2B inset), suggesting the presence of helix- or turn-like conformations. Decomposition of the far-UV CD spectrum at pH 8.0, by using the k2d algorithm, available online at the DICHROWEB site [22,23], yields a 30% of helical structure, 17% of β-sheet and 52% of random-coil. However, as LrtA has 8 Tyr, 4 Phe, and 14 His residues (eight naturally-occurring residues and six in the purification tail, see Section 4), we cannot rule out the absorbance of aromatic residues at this wavelength [24,25].

The molar ellipticity, [⊖], at 222 nm showed a dumb-bell shape, with a maximum value at pH 6.0 (Figure 2B). These results suggest that there were changes in the secondary structure (or alternatively in the environment around aromatic residues [24,25]) above and below pH 6.0; interestingly enough, the changes at low pH mirrored those observed by intrinsic and ANS fluorescence (Figure 2A). Since the fluorescence results indicated that the environment around Tyr residues remained essentially unaltered until pH 9.0 (Figure 2A, filled circles), and taking into account the ANS-fluorescence (Figure 2A, blank circles), we can conclude that between pH 6.0 and 9.0, although the protein had a folded conformation (Figure S2A,B), either the secondary structure of the protein changed or, alternatively, the environment around some of the 4 Phe and 14 His residues in LrtA. We could not determine the pK_a values corresponding to the titrations at the two sides of the curve due to the absence of acidic and basic baselines, respectively.

As it happened with the thermal denaturations followed by fluorescence, the transitions followed by the ellipticity at 222 nm did not show any sigmoidal behavior below pH 6.0 (Figure S7B), but above that pH there was an irreversible broad transition (Figure 2C, right axis, filled circles).

To sum up, the spectroscopic probes (intrinsic and ANS fluorescence, CD, and NMR) indicate that LrtA acquired a native, with well-folded regions (Figure S2) from pH 6.0 to 9.0.

2.3. *LrtA Showed an Irreversible Complex Unfolding Equilibrium*

As the thermal denaturations were irreversible (either followed by fluorescence or CD, Figure 2C and Figure S7), we tried to determine the conformational stability of LrtA by using GdmCl-denaturations followed by fluorescence and CD (urea-denaturations followed by fluorescence did not show any sigmoidal behavior, Figure S8A, inset). The tendency in both refolding and unfolding curves, followed by fluorescence and CD, was the same (Figure S8); however, the refolding CD results indicate that the final ellipticity, acquired by the native state, was not the same as that in the unfolding experiments. Moreover, the fluorescence refolding curves indicate that the value of the $<1/\lambda>$ was different to that in the unfolding ones, even though we used the same protein concentration. Then, we conclude that there was a hysteresis behavior, and chemical denaturations were also irreversible, as it could be expected for a protein composed of several domains (see below, Section 2.4.).

In addition, comparison of the unfolding CD and fluorescence results suggest that the unfolding of LrtA was not a simple two-state process, as the denaturation curves by both techniques were different. Whereas fluorescence curves showed two transitions, CD reported a sole one, whose apparent midpoint did not overlap with that of the fluorescence (Figure S8). Fluorescence denaturation curves at several protein concentrations (in the range from 1.9 to 19 μM (in protomer units)) indicate that the first transition monitored corresponded to a protein-concentration-dependent process (Figure S8B inset), as at high protein concentrations (19 μM) two transitions were observed, with apparent midpoints around 1.0 and 2.0 M GdmCl. This result further confirms that LrtA was an oligomeric protein.

2.4. Sequence Properties and Molecular Modeling of LrtA

The primary structure of LrtA possesses a relatively large fraction of charged residues, both acidic and basic. Due to their high hydrophilicity, these residues tend to hamper the hydrophobic collapse and increase disorder in the protein backbone. Predictors of local disorder [26–29] based on a variety of physical properties (Figure S9) were used to estimate the propensity of the protein sequence to fold. There is a consensus indicating the region around residues 100–130 is highly disordered, together with a few other residues at both protein termini. Since the experimentally determined radius of monomeric LrtA has a value close to that of a compact protein (see Section 2.1., DOSY-NMR results), this may indicate that the region 100–130 was either a long, disordered loop within a single domain protein or a coil region separating two distinct but spatially close domains.

The overall propensity of LrtA to fold into a well-structured protein was also explored by mapping its properties in terms of charge and hydropathy (Figure 3). In particular, Figure 3A compares the location of LrtA sequence within an Uversky diagram [30], which provides indication on the possibility that the protein belongs to the IDP (intrinsically disordered protein) class through the identification of a boundary hydropathy that separates folded and unfolded polypeptides. When the overall primary structure was considered, LrtA fell into the region of the diagram that is mostly populated by well-folded proteins [31], although also accessible to a few IDPs. However, when the sequence of LrtA was divided into two separate portions, they had distinct features. The first-half of LrtA sequence (residues 1–100) more distinctly belonged to the region occupied by ordered polypeptides. In contrast, the second-half (residues 100–191) fell in the region of the diagram that is populated by IDPs, although also accessible to some well-folded proteins. On the other hand, a Das–Pappu diagram [32] showed that LrtA should be considered a so-called 'Janus sequence' in between weak and strong polyampholytes (Figure 3B), independently whether the whole protein or just the two halves of its sequence are considered. This observation strongly suggests that the structure of LrtA is context-dependent, and may easily become more expanded/collapsed or structured/unstructured according to the environment (such as solution conditions or the presence of biomolecular partners). From our experimental results, LrtA had folded regions at pH 8.0 in aqueous solution (Figure S2 and Figure 2B, inset), although with a small stability, as suggested by thermal denaturations (Figure 2C and Figure S7A).

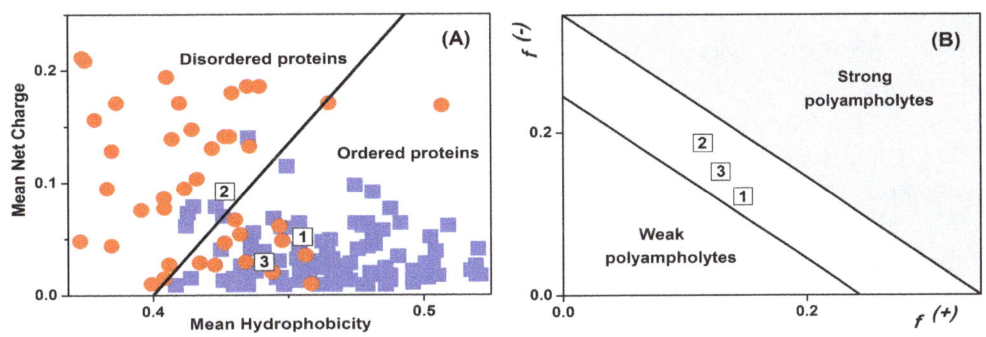

Figure 3. Location of LrtA in the diagram of state for charged polypeptides: Symbols "1", "2", and "3" indicate, respectively, the first two halves of the LrtA sequence (residues 1–100 and 101–191) and the whole protein. (**A**) Uversky plot based on the absolute mean net charge as a function of the mean scaled hydropathy, as obtained with PONDR [31]; well-folded (blue squares) and disordered proteins (red circles) are shown. (**B**) Das–Pappu plot based on the fraction f(+) and f(−) of positively and negatively charged residues, respectively [32].

With the aim of building a model for the secondary and tertiary structures of LrtA, the protein sequence was submitted to full-chain protein structure prediction servers [33–36]. A particularly interesting result was obtained by using I-TASSER [33], which is one of the most popular and accurate software for generating high-quality model predictions of tridimensional protein structures. The best models predicted by I-TASSER (Figure S10) all included a well-structured domain spanning the first 100 residues, followed by a collapsed and poorly-structured region. The well-structured domain consisted of two parallel α-helices, and a β-sheet formed by four anti-parallel β-strands. These models were remarkable because they predicted a degree of order in the structure of LrtA that is in reasonable agreement with our expectations based on the CD experimental results (see above, Section 2.2.2), whereas in most cases the algorithms tend to overestimate the amount of secondary structure when applied to intrinsically unfolded polypeptides. Furthermore, the absence of a defined folding topology for the second half of LrtA sequence is consistent with the theoretical predictions discussed above. It is worth mentioning that the C-terminal region of HPF of *S. aureus*, another member of the long sub-family of HPF, in EM (electron microscopy) preparations was folded [5], in contrast with our model; then, it seems that in LrtA from *Synechocystis* sp. PCC 6803, the C-terminal region has specific features, which might be related to protein function. Finally, the conformations predicted for the first-half of LrtA sequence were in common with those obtained with the other algorithms of structural modeling that we used (i.e., FALCON [34], SWISS-MODEL [35], and Robetta [36]), although details of the geometry and orientation of the α-helices and β-strands were in some models different. This was particularly intriguing, especially because a four-strand motif is typical of many RNA-binding proteins ([5] and references therein).

Our theoretical predictions are not difficult to reconcile with the findings show by NMR, at physiological pH, where the spectrum of LrtA was that of a folded molecule (Figure S2). In fact, the signals of the proton nuclei in the unfolded and folded halves of the protein had a different behavior. The amide protons of the unfolded half of the protein would appear between 8.0 and 8.5 ppm [37], where they would be probably obscured, although they should be sharper than the rest of the signals, by many of the amide resonances of the folded half (those of the residues connecting the α-helices and the β-strands); it is interesting to note, however, the presence of a higher intensity at 8.2–8.3 ppm (Figure S2B), which could be due to the sharper resonance of the unfolded region of LrtA. Furthermore, the majority, if not all, of the amide protons of the unfolded half will be broadened and exchanged with the solvent at pH 8.0 [37], as it has been observed to occur in other intrinsically disordered regions, when the pH is raised and even when the temperature is decreased at the highest explored pH [38]. However, we tried to acquire a 1D ^{1}H NMR spectrum at pH 6.9 (in the presence of 0.5 M NaCl) and 15 °C; under these conditions (Figure S11) some amide signals appearing between 7.8 and 8.3 became sharper, as expected for a disordered region that has a fast molecular tumbling. The methyl region, under these solvent conditions, was similar to that acquired at higher pH and temperature (Figure S2). In addition, all the methyl peaks corresponding to the side-chains of Val, Ile, and Leu residues of the disordered half of the protein under any of the conditions explored (pH 8.0, 20 °C or pH 6.9, 15 °C) would appear at basically the same chemical shifts as those of the corresponding folded region, i.e., around 0.8 ppm [37].

The models predicted by I-TASSER provided a static picture of LrtA that does not take into account its dynamics, which could be expected to be significant to determine the properties of such a chameleonic protein. Furthermore, the main difference between the predicted models and our experimental findings is the presence of a larger amount of β-structures in the former. Thus, we suspected that the structure predicted corresponded to the most stable structure that LrtA can assume, e.g., under ideal conditions in solution or when bound to a partner molecule, although, the experimental evidence (Figure 2C, Figures S7 and S8) suggests that this structure was not very stable. For those reasons, we used MD simulations to study the behavior of the protein structure both at room and high temperatures. The latter case corresponds to the simplest and most direct way to investigate the dynamics of a protein under non-native conditions [39], speeding up the sampling by overcoming the energetic barriers that restrain the structure in a given conformation.

The MD results showed that the region including residues 1–100 is stable and maintains its folding topology and structure when simulated at room temperature (Figure S12). In particular, as shown in Figure 4, Tyr19 and Tyr77 interact to fix the two α-helices, whereas the other two Tyr residues are on the opposite face of the protein (Figure 4B). In contrast, when the temperature was raised, Tyr19 and Tyr77 lost their coordination and the first N-terminal β-strand immediately started losing its anchoring with the rest of the protein and the β-sheet scaffold (Figure 4B), increasing the amount of coil and helical structure. The anchoring was not recovered in annealing runs performed by reducing back the temperature, unless they were started at the earliest step of the local unfolding process. This finding suggests that the folding of the N-terminal region of LrtA was possibly assisted by interactions with other biomolecules, which may include other monomers of LrtA or binding partners, such as RNA. In contrast, the rest of the protein was very stable, and did not lose its folding topology even under the most extreme simulation conditions (Figure 4C).

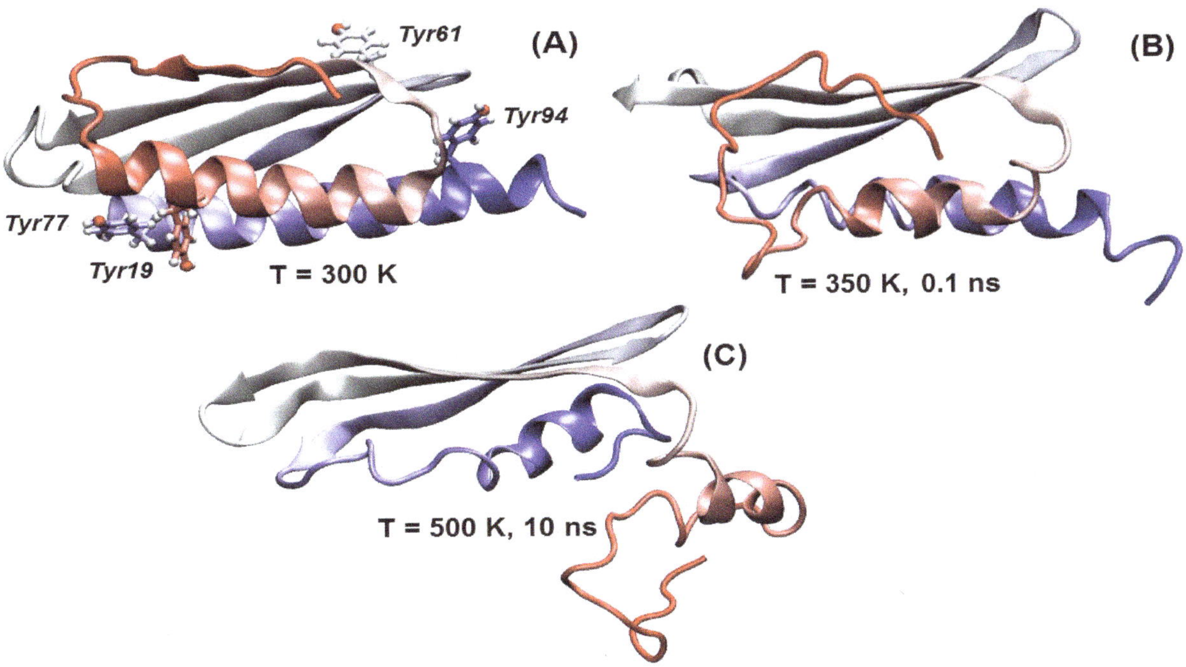

Figure 4. Dynamic behavior of the predicted folded domain of LrtA: The region comprising residues 1–100 of LrtA is shown in cartoon representation (colored from red (N terminus) to silver-white (mid-sequence regions) up to blue (C terminus of the domain)). (**A**) Structure at room temperature, with Tyr residues indicated. (**B**) Simulation under unfolding conditions: in the N-terminal region (indicated in red), the first β-strand loses it structure and coordination with the rest of the β-sheet. (**C**) Structure under extreme conditions: the folding topology is maintained. VMD [40] is used for the protein displays.

3. Discussion

LrtA seemed to acquire a native-like conformation from pH 6.0 to 9.0. Changes in secondary (far-UV CD) and tertiary (intrinsic fluorescence) structures, and in the burial of hydrophobic residues (ANS fluorescence) occurred concomitantly at low pH. Under acidic conditions (pH 3.0), species with a higher amount of secondary structure (as indicated by a larger (in absolute value) ellipticity, Figure 2B) appeared to be populated, although they had hindered solvent-accessibility towards I^- and acrylamide quenchers (Table 1). Therefore, at low pH, LrtA had non-native conformations with non-stable secondary and tertiary structures (as judged by the absence of a sigmoidal shape in thermal denaturation curves). These results are further supported by the NMR spectra acquired at low pH, where there was no dispersion of amide or methyl signals (Figure S3), suggesting the presence of conformations with an unfolded structure. Besides, the broadening observed in the methyl and amide regions suggested the presence of aggregation, which was further confirmed by the far-UV CD spectra at pH 4.5 at different protein concentrations. Then, the protein at acidic pH had a larger tendency to associate than at physiological pH. The increase of ANS fluorescence at low pH, indicating a large solvent-accessible hydrophobic surface area, may appear difficult to reconcile with the results of I^- or acrylamide quenching, suggesting a smaller solvent accessibility towards Tyr (Table 1). However, the larger amount of hydrophobic surface area (monitored by ANS), that became solvent-exposed at acidic pH values, could involve several of the Val, Leu, and Ile residues, which are highly abundant in LrtA (46 out of 197 amino acids).

LrtA had well-folded regions in the pH range from 6.0 to 9.0, as indicated by: (i) the sigmoidal curves in the thermal and chemical denaturations (fluorescence and far-UV CD) (Figure 2C, Figure S7 and S8); and (ii) the 1D ^{1}H-NMR spectrum at pH 7.0 and 8.0 (Figure S2). Moreover, the protein was capable of binding homogenous yeast RNA (from Sigma) with an affinity of ~1 μM (Figure S13), and therefore the purification protocol did not affect the conformational features of the protein. However, this structure was not highly rigid, as judged from the apparent thermal midpoint obtained from the irreversible denaturation curves (~40 °C, Figure 2C); these findings agree with the MD results in this work. As the secondary and tertiary structures of LrtA are only stable in a narrow thermal range, an increase in the temperature environment reduces the availability of well-folded and active protein, and therefore, a larger amount of protein is needed to carry out the cyanobacterial functions. The LrtA structure under native conditions, as suggested by the deconvolution of CD data, had a smaller percentage of α-helix structure than other HPF members (30% vs. 45%), as well as a lower percentage of β-sheet (17% vs. 27%). These experimental percentages were confirmed by the results of our MD simulations.

LrtA was an oligomeric protein at physiological pH. We showed that some of the Tyr residues seemed to be involved in the self-associating interface, as judged by the changes in the fluorescence lifetimes (Table 1) or the protein-concentration dependence of the curve denaturations midpoints (Figure S8B inset). In addition, our MD simulations at room temperature suggest that Tyr19 and Tyr77 in the folded domain of the protein were key in anchoring amino acids of the β-sheet, and the loss of such anchoring during the high-temperature simulations caused a partial disruption of the protein β-sheet. Then, Tyr residues were important for quaternary and secondary scaffolding in LrtA. Recently, it has been observed that, in EM preparations, the HPF of *S. aureus* (another member of the long HPF subfamily) is forming domain-swapped dimeric species [5]. Furthermore, crystals of the short HPF from *Vibrio cholerae* show the presence of dimers mediated by Co (II) anchoring residues of the β-sheets of two monomers [11], further pinpointing to the crucial importance of residues in the β-sheet for a possible quaternary arrangement of any HPF member. However, the importance of oligomerization for the function of all those proteins (included LrtA) remains to be elucidated, as it could be an adaptive mechanism of regulation to interact with other proteins or even with RNA.

4. Materials and Methods

4.1. Materials

Deuterium oxide and IPTG was obtained from Apollo Scientific (Stockport, UK). Sodium trimethylsilyl [2,2,3,3-$^2H_4$] propionate (TSP), imidazole, DNase, Trizma base and acid, yeast RNA, glutaraldehyde (25% *w*/*v* solution), ANS, deuterated acetic acid, its sodium salt and His-Select HF nickel resin were from Sigma-Aldrich (Madrid, Spain). The β-mercaptoethanol (β-ME) was from BioRad (Madrid, Spain). Triton X-100 and protein marker, PAGEmark-tricolor (G Biosciences) were from VWR (Barcelona, Spain). Dialysis tubing, with a molecular weight cut-off of 3500 Da, was from Spectrapor (Spectrum Laboratories, Breda, The Netherlands). Amicon centrifugal devices with a cut-off molecular weight of 3000 Da were from Millipore (Barcelona, Spain). Standard suppliers were used for all other chemicals. Water was deionized and purified on a Millipore system.

4.2. Protein Expression and Purification

Expression of LrtA was carried out in BL21(DE3) or C41 [41] strains with a final ampicillin concentration of 100 mg/mL at 37 °C. The cells were cultured in 1 L flasks. Protein expression was induced with a final concentration of 1.0 mM IPTG when the absorbance of the cell culture at 600 nm was 0.4–0.9, and the cells were grown for 15–16 h at 37 °C. Cells were harvested at 8000 rpm in a JA-10 rotor (Beckman Coulter, Miami, FL, USA) for 15 min. The pellet from 5 L of culture was re-suspended in 50 mL of buffer A (500 mM NaCl, 5 mM imidazole, 20 mM Tris buffer (pH 8), 0.1% Triton X-100 and 1 mM β-ME), supplemented with a tablet of Sigma Protease Cocktail EDTA-free and 2 mg of DNase (per 5 L of culture). After being incubated with gentle agitation at 4 °C for 10 min, cells were disrupted by sonication (Branson sonicator, 750 W, Richmond, VA, USA), with 10 cycles of 45 s at 55% of maximal power output and an interval of 15 s between the cycles. All the sonication steps and the interval waits were carried out in ice. The lysate was clarified by centrifugation at 18,000 rpm for 40 min at 4 °C in a Beckman JSI30 centrifuge with a JA-20 rotor (Beckman Coulter, Miami, FL, USA).

The clarified lysate from such first centrifugation did not contain a large amount of LrtA, and thus, we suspected that most of the protein was present in the cell debris precipitate. Therefore, the precipitate was treated with buffer A supplemented with 8 M urea and a tablet of Sigma Protease Cocktail EDTA-free and 2 mg of DNase (per 5 L of culture). The re-suspended sample was treated with another 10 cycles of sonication in ice, and the sample was clarified by centrifugation at 20,000 rpm for 30 min at 4 °C. LrtA was in the supernatant and was purified by immobilized affinity chromatography (IMAC). The supernatant was added to 5 mL of Ni-resin previously equilibrated in buffer A supplemented with 8 M urea. The mixture was incubated for 20 min at 4 °C, and afterwards, the lysate was separated from the resin by gravity. On-column refolding was carried out during the washing step with 20 mL of buffer B (20 mM Tris buffer (pH 8.0), 500 mM NaCl, 1 mM β-ME, and 20 mM imidazole); the protein was eluted by gravity from the column with buffer C (20 mM Tris buffer (pH 8.0), 500 mM NaCl, 1 mM β-ME, and 500 mM imidazole). The eluted LrtA was extensively dialyzed against buffer D (100 mM sodium phosphate buffer (pH 8.0) with 500 mM NaCl). Precipitate in the dialysis tubing after five dialysis steps in buffer D was removed by centrifugation at 20,000 rpm for 30 min at 4 °C. The final yield of protein was 4.5–6.5 mg/L of culture (with both cellular strains assayed), and the protein was 90–95% pure as judged by SDS gels (Figure S14). This purity percentage takes into account the possible contamination due to the presence of deoxyribonucleotides, as judged by their absorbance at 260 nm (see below).

We attempted to re-purify the protein recovered from IMAC by using gel filtration chromatography in a Superdex 16/600, 75 pg column (GE Healthcare, Barcelona, Spain) connected to an AKTA FPLC system (GE Healthcare) by monitoring the absorbance at 280 nm; nevertheless, the protein was bound to the column and did not come out within its bed volume. Binding to the column has been also observed during purification of the recombinant HPF from *Staphylococcus aureus* [5], another member of the long subfamily of HPF.

The eluted protein from IMAC showed absorbance at 260 nm, suggesting that it was probably contaminated with di- or tri-deoxyribonucleotides (from the cleavage of the DNase used during purification and even though sample was dialyzed against 500 mM of NaCl). Presence of deoxyribonucleotides has been also observed in the recombinant HPF from *S. aureus* after its purification [5]. We tried to remove the deoxyribonucleotides by using different concentrations of polyethylenimine (PEI), ranging from 0.2 to 1% (v/v) [42], but most of LrtA co-precipitated with the oligonucleotides. The total protein concentration, P_c (in mg/mL) was determined by using the expression [43]: $P_c = 1.55\ A_{280} - 0.75\ A_{260}$, where A_{280} and A_{260} are the absorbance of the dialyzed protein solution at 280 and 260 nm, respectively. However, it is important to note that the presence of deoxyribonucleotides did not affect the spectroscopic signals either of both fluorescence and far-UV

CD, because DNA is spectroscopically silent in fluorescence, and in the far-UV region of CD spectra between 210–240 nm, deoxyribonucleotides do not absorb [44,45]. It is also interesting to note that deoxyoligonucleotides in aqueous solution or intact DNA show a small or slightly positive ellipticity around 222 nm [45–49] where we have carried out the study of the CD biophysical properties of the protein (see Results section).

4.3. Fluorescence

Fluorescence spectra were collected on a Cary Varian spectrofluorimeter (Agilent, Foster City, CA, USA), interfaced with a Peltier, at 25 °C. LrtA concentration in the pH- or chemical-denaturation experiments was 9.8 μM (in protomer units). For experiments with ANS, a final probe concentration of 100 μM was added. A 1-cm-pathlength quartz cell (Hellma, Mullheim, Germany) was used.

In the pH-induced unfolding curves, the pH was measured after completion of the experiments with an ultra-thin Aldrich electrode in a Radiometer pH-meter (Madrid, Spain). The acids and salts used were: pH 2.0–3.0, phosphoric acid; pH 3.0–4.0, formic acid; pH 4.0–5.5, acetic acid; pH 6.0–7.0, NaH_2PO_4; pH 7.5–9.0, Tris acid; pH 9.5–11.0, Na_2CO_3; pH 11.5–13.0, Na_3PO_4. Chemical and pH denaturations were repeated three times with new samples at any of the concentrations assayed. Appropriate blank corrections were made in all spectra both in pH- and chemical-denaturation experiments.

For GdmCl-denaturation experiments the samples were prepared the day before from a 7 M GdmCl concentrated stock and left overnight to equilibrate; before experiments, samples were left to 25 °C for 1 h. For the refolding experiments, the sample was exchanged in 7 M GdmCl by using Amicon centrifugal devices; protein concentration was the same as in the unfolding experiments.

The emission intensity weighted average of the inverse wavelengths (also called the spectrum mass center, or the spectral average energy of emission), <1/λ>, was calculated as described [50]. Briefly, we define <1/λ> as: $\langle 1/\lambda \rangle = \sum_{1}^{n} \frac{1}{\lambda_i} I_i / \sum_{1}^{n} I_i$, where I_i is the intensity at wavelength λ_i. We shall report <1/λ> in units of μm^{-1}.

Steady-State Spectra—The experimental set-up for the intrinsic and ANS fluorescence pH-denaturation experiments has been described previously [50]. Briefly, protein samples were excited at 278 nm, for the intrinsic fluorescence, and 380 nm for the ANS experiments. In all cases, excitation and emission slits were 5 nm. The experiments were recorded between 300 and 400 nm (for the intrinsic fluorescence) and between 400 to 600 for the ANS experiments. The signal in all cases was acquired for 1 s and the increment of wavelength was set to 1 nm. For the chemical denaturations, following intrinsic fluorescence, several protein concentrations were used in the range from 1.9 to 19.6 μM (in protomer units) at 100 mM phosphate buffer (pH 7.0) and 50 mM NaCl.

Thermal Denaturations—Thermal denaturations of isolated LrtA at different pH values were carried out with the same experimental set-up described [50] and protein concentrations of 9.8 and 5 μM (in protomer units). Briefly, these experiments were performed at constant heating rates of 60 °C/h and an average time of 1 s. Thermal scans were collected at 308 nm after excitation at 278 nm from 25 to 95 °C and acquired every 0.2 ° C.

Fluorescence Quenching—Quenching by iodide and acrylamide was examined at different solution conditions, with an LrtA concentration of 9.8 μM (in protomer units): pH 3.5 (formic buffer, 50 mM), pH 7.0 (phosphate buffer, 50 mM), and pH 11.0 (boric buffer, 50 mM). Experiments were also carried out in the presence of 6 M GdmCl at pH 7.0 (50 mM, phosphate buffer). The experimental set-up for both quenchers was the same described above for the intrinsic fluorescence experiments. The data for KI were fitted to [12]

$$\frac{F_0}{F} = 1 + K_{sv}[KI] \quad (1)$$

where K_{sv} is the Stern-Volmer constant for collisional quenching; F_0 is the fluorescence intensity in the absence of KI; and F is that at any KI concentration. The range of KI concentrations explored was 0–0.7 M. For experiments with acrylamide, the data were fitted to [12]

$$\frac{F_0}{F} = (1 + K_{sv}[acrylamide])e^{\upsilon[acrylamide]} \tag{2}$$

where υ is the dynamic quenching constant. Fittings to Equations (1) and (2) were carried out by using Kaleidagraph (Synergy software, Dubai, United Arab Emirates).

4.4. Fluorescence Lifetimes

Lifetimes were measured on an EasyLife V™ lifetime fluorometer (Madrid, Spain) with the stroboscopic technique, by using as excitation source a pulsed light of a diode LED operating at 278 nm. The number of channels used for each scan was 500, and the integration time was 1 s. Three scans were averaged in each experiment, and they were repeated twice at 25 °C in 50 mM phosphate buffer (pH 8.0) and 500 mM NaCl.

The experimental fluorescence decays ($D(t)$) were fitted to a sum of exponential functions: $D(t) = \sum_{i=1}^{n} a_i \exp(-t/\tau_i)$, where τ_i is the the lifetime of the electronic excited states of the fluorescent species present in solution, and a_i the pre-exponential factor of those electronic states. The pre-exponential factors can be interpreted not only in terms of the populations of the corresponding species, but also in terms of the radiative probability constants of Tyr residues, in the case of LrtA. We also determined the mean lifetime, $<\tau>$, as: $< \tau >= \sum_{i=1}^{n} f_i\tau_i$, where the f_is are defined as: $f_i = a_i\tau_i/\sum_j a_j\tau_j$. The fitting procedure of the experimental fluorescence lifetime curves used an iterative method based on the Levenberg–Marquardt algorithm [51]. The temporal width of the excitation pulse, which distorted the observed decay, was taken into account through the instrument response function (IRF), which was determined by using a scattered solution of Ludox. Goodness of the fittings was tested by using a reduced χ^2, that was calculated by measuring the spectral noise at time t, and determining the measurement uncertainties [52].

4.5. CD

The far-UV CD spectra were collected on a Jasco J815 spectropolarimeter (Jasco, Tokyo, Japan) fitted with a thermostated cell holder, and interfaced with a Peltier unit, at 25 °C. The instrument was periodically calibrated with (+)-10-camphorsulphonic acid. Several protein concentrations of LrtA (4.5, 9.8 and 19.6 μM, in protomer units) were used to test for concentration-dependent changes in the shape and intensity of the steady-state spectrum at two different pH values; differences were not observed at these concentrations (9.8 and 19.6 μM) at pH 8.0. However, protein-concentration-dependent changes (in the concentration range of 4.5 and 9.8 μM, in protomer units) were observed in the shape and ellipticity at pH 4.5, suggesting that at this pH the protein showed a larger tendency to self-associate (see above, Section 2.1, Figure S4) than at pH 8.0. Molar ellipticity was calculated as described [44,45,50].

Steady-State Spectra—Experiments were acquired with the same experimental set-up described previously [50]. Typically, spectra were acquired at a scan speed of 50 nm/min with a response time of 2 s and averaged over six scans with a bandwidth of 1 nm. Spectra were corrected by subtracting the baseline in all cases. Protein concentrations were 9.8 μM (in protomer units) for pH- and chemical-denaturation experiments (for both unfolding and refolding). The refolding samples of LrtA were prepared as described above for fluorescence.

Thermal Denaturations—Thermal denaturations were carried out with the same experimental set-up described previously [50] and a protein concentration of 9.8 μM. Briefly, thermal denaturations were performed at a constant heating rate of 60 °C/h, a response time of 8 s, a band width of 1 nm, and acquired every 0.2 °C. Thermal scans were collected in the far-UV region at 222 nm from 25 to 85 °C.

4.6. NMR Spectroscopy

The NMR experiments were acquired at 20 °C or 15 °C, when stated, on a Bruker Avance DRX-500 spectrometer equipped with a triple resonance probe and z-pulse field gradients.

1D ^{1}H-NMR experiments—Homonuclear 1D-^{1}H-NMR experiments were performed with LrtA at a concentration of 30 μM (in protomer units) in 0.5 mL, in 100 mM phosphate buffer (pH 8.0 or pH 6.9) and 500 mM NaCl in H_2O/D_2O (90%/10%, v/v), uncorrected for deuterium isotope effects. The 1D-^{1}H spectra under these conditions were acquired with 16 K data points, with 2 K scans and a 6000 Hz spectral width (12 ppm), with the WATERGATE sequence [53]. Baseline correction and zero-filling were applied before processing. We also acquired a spectrum at pH 4.5 (100 mM acetate) with 64 K scans, while the other experimental parameters were as above. During buffer exchange in Amicon centrifugal devices the sample precipitated and we had to increase the number of scans to have a good signal-to-noise ratio. All spectra were processed and analyzed by using TopSpin 2.1 (Bruker GmbH, Karlsruhe, Germany). TSP was used as the external chemical shift reference, taking into account the pH-dependence of its resonance [15].

Translational Diffusion Measurements—The DOSY experiments at pH 8.0 and 4.5 were performed with the pulsed-gradient spin-echo sequence, as described previously [50], with 16 gradient strengths ranging linearly from 2 to 95% of the total power of the gradient unit. At both pH values, the duration of the pulse gradient was 2.5 ms, and the time between the two gradients was 150 ms. Samples were exchanged in D_2O buffer (100 mM phosphate buffer (pH 8.0, not corrected by isotope effects) or 100 mM deuterated acetate buffer (pH 4.5, not corrected by isotope effects) with 500 mM NaCl at both pH values) by using Amicon centrifugal devices during 4 to 6 h. Precipitation was observed after the buffer exchange (from pH 8.0 to pH 4.5), and the sample was centrifuged at 13,000 rpm for five minutes, before putting it into the NMR tube. Further and more severe precipitation was observed during the D_2O buffer exchange at pH 4.5 and attempts to acquire a DOSY by increasing the number of scans (until 8 K) at each pulse field gradient strength failed. The methyl region was used for intensity measurements in the experiment acquired at pH 8.0.

4.7. DLS

DLS measurements were performed at 25 °C in 100 mM phosphate buffer (pH 8.0) with 500 mM NaCl at different protein concentrations. Experiments were performed at fixed angle (θ = 173°) in a Zetasizer nano instrument (Malvern Instrument Ltd., Malvern, UK) equipped with a 10-mW helium-neon laser (λ = 632.8 nm) and a thermoelectric temperature controller. Experiments were analyzed with Zetasizer software (Malvern Instrument Ltd., Malvern, UK), and they were carried out as described [50]. Each sample was measured 10 times with 10 runs of 30 s each. The Z-average size was obtained by fitting the autocorrelation function with the cumulants method. The R_S was calculated by applying the Stokes–Einstein equation: $D = kT/6\pi\eta R_S$, where k is the Boltzmann constant and T is the temperature (in K). Experiments were also carried out at pH 4.5 (100 mM acetic acid) and 500 mM NaCl, but the poor signal obtained precluded any reliable conclusion.

4.8. Molecular Modelling

Disorder propensity was estimated by using scoring functions according to independent algorithms [26–29], as calculated through submission to their respective web servers. Three-dimensional models of the protein were obtained using the on-line structure predictors I-TASSER [33], FALCON [34], SWISS-MODEL [35], and Robetta [36]. Folding stability in the LrtA structure predicted by I-TASSER was assessed in MD simulations at room temperature performed with the GROMACS package [54]. The AMBER ff99SB-ILDN force field [55] and TIP3P model [56] were used for the protein and water, respectively, and other simulation conditions were as previously described [57,58]. High temperature simulations were performed at 400 and 500 K, according to a protocol previously adopted [59].

4.9. SEC

SEC experiments were carried out in a Superose 12 10/300 GL column at pH 8.0 (50 mM Tris) and 0.7 M NaCl, connected to an AKTA FPLC (GE Healthcare), and monitoring the absorbance at 280 nm. Flow rate was 0.7 mL/min, and the protein was loaded in a volume of 100 μL at a concentration of 97 μM (in protomer units). Baselines of the chromatograms were strongly curved and therefore the UNICORN software 5.0.2 (GE Healthcare, Barcelona, Spain) was used for automatic baseline correction. The markers used to calibrate the column were albumin, bovine RNase A and blue dextran (GE Healthcare); the isolated markers were loaded in the column in the above described buffer in three independent measurements.

4.10. Glutaraldehyde Cross-Linking

Glutaraldehyde cross-linking was carried out at pH 8.0 (50 mM Tris buffer) in the presence of 0.5 M NaCl at room temperature. Sample volume was 200 μL. Protein concentration was 97 μM (in protomer units). The protocol used was that described previously [60], taking aliquots of a fresh 25% (*w*/*v*) glutaraldehyde solution (Sigma-Aldrich) to yield a final concentration of the cross-linker in the protein solution of 1%. Aliquots of 30 μL were extracted at defined times and the cross-linking reaction was stopped by adding an identical volume of SDS-loading buffer to the aliquot. The extracted samples were run in a 12% SDS-PAGE gel immediately. Experiments were repeated twice.

5. Conclusions

Our results show that LrtA from cyanobacteria, a member of the long HPF subfamily, shows a clear propensity to self-associate and has an apparent low conformational stability. The structure was predicted to be formed by two domains, one of which was well-folded, whereas the other was disordered. These conformational features, together with the presence of a relatively high number of charged residues, provide an overall structural plasticity.

Supplementary Materials:
There are 14 figures, showing: the intensity curve of the methyl groups from the DOSY experiment at pH 8.0 and 500 mM NaCl (Figure S1); the spectroscopic characterization of LrtA by NMR at pH 8.0 and 500 mM NaCl and 20 °C (Figure S2); the 1D 1H-NMR spectrum of LrtA at pH 4.5 and 500 mM NaCl and 20 °C (Figure S3); far-UV CD spectra of LrtA at pH 4.5 at two different concentrations (Figure S4); the SDS-PAGE gel with the glutaraldehyde cross-linking results (Figure S5); the fluorescence lifetime at pH 8.0 and 500 mM NaCl of LrtA at 0.98 μM (in protomer units) (Figure S6); the thermal denaturations followed by fluorescence at different pH values (Figure S7); the chemical (urea and GdmCl) denaturations of LrtA followed by fluorescence and CD (Figure S8); the predictions of disorder probability along the sequence (Figure S9); several models of the LrtA structure as obtained from homology modeling (Figure S10); the spectroscopic characterization of LrtA by NMR at pH 6.9 and 500 mM NaCl at 15 °C (Figure S11); the secondary structure of the protein obtained in MD simulations (Figure S12); the titration of LrtA to yeast RNA (Figure S13); and the SDS-gel of purified LrtA (Figure S14).

Author Contributions: Conceptualization, L.M.C., P.S., A.C.-A., J.G.H.-C., B.R., and J.L.N.; Methodology, L.M.C., P.S., A.C.-A., J.G.H.-C., B.R., and J.L.N.; Software, B.R.; Investigation, L.M.C., P.S., A.C.-A., J.G.H.-C., B.R., and J.L.N.; Resources, F.J.F. and M.I.M.-P.; Writing—Original Draft Preparation, L.M.C., P.S., A.C.-A., J.G.H.-C., B.R., F.J.F., M.I.M.-P., J.G.d.l.T., and J.L.N.; Writing—Review and Editing, B.R., A.C.-A., and J.L.N.; Supervision, J.G.d.l.T. and J.L.N.

Abbreviations

ANS	8-anilino-1-naphthalene sulfonic acid
β-ME	β-mercaptoethanol
CD	circular dichroism
DLS	dynamic light scattering

DOSY	diffusion ordered spectroscopy
EM	electron microscopy
GdmCl	guanidine hydrochloride
HPF	hibernating promoting factor
IDP	intrinsically disordered protein
IMAC	immobilized affinity chromatography
IRF	instrument response function
ITC	isothermal titration calorimetry
MD	molecular dynamics
NMR	nuclear magnetic resonance
PEI	polyethylenimine
RMF	ribosome modulation factor
RNase	ribonuclease
SDS-PAGE	sodium dodecyl sulfate polyacrylamide gel electrophoresis
SEC	size exclusion chromatography
UV	ultraviolet

References

1. Tan, X.; Varughese, M.; Widger, W.R. A light-repressed transcript found in *Synechococcus* sp. PCC 7002 is similar to a chloroplast-specific small subunit ribosomal protein and to a transcription modulator protein associated with sigma 54. *J. Biol. Chem.* **1994**, *269*, 20905–20912. [PubMed]
2. Samartzidou, H.; Widger, W.R. Transcriptional and post-transcriptional control of mRNA from lrtA, a light-repressed transcript in *Synechococcus* sp. PC 7002. *Plant Physiol.* **1998**, *117*, 225–234. [CrossRef] [PubMed]
3. Galmozzi, C.V.; Florencio, F.J.; Muro-Pastor, M.I. The cyanobacterial ribosomal-associated protein LrtA is involved in post-stress survival in *Synechocystis* sp. PCC 6803. *PLoS ONE* **2016**. [CrossRef] [PubMed]
4. Yoshida, H.; Wada, A. The 100S ribosome: Ribosomal hibernation induced by stress. *Wiley Interdiscip. Rev. RNA* **2014**, *5*, 723–732. [CrossRef] [PubMed]
5. Khusainov, I.; Vicens, Q.; Ayupov, R.; Usachev, K.; Myasnikov, A.; Simonetti, A.; Validov, S.; Kieffer, B.; Yuspova, G.; Yusupov, M.; et al. Structures and dynamics of hibernating ribosomes from *Staphylococcus aureus* mediated by intermolecular interactions of HPF. *EMBO J.* **2017**, *36*, 2073–2087. [CrossRef] [PubMed]
6. Starosta, A.L.; Lasak, J.; Jung, K.; Wilson, D.N. The bacterial translation stress response. *FEMS Microbiol. Rev.* **2014**, *38*, 1172–1201. [CrossRef] [PubMed]
7. Agafonov, D.E.; Spirin, A.S. The ribosome-associated inhibitor A reduces translation errors. *Biochem. Biophys. Res. Commun.* **2004**, *320*, 354–358. [CrossRef] [PubMed]
8. Polikanov, Y.S.; Blaha, G.M.; Steitz, T.A. How hibernation factors RMF, HPF and YfiA turn off protein synthesis. *Science* **2012**, *336*, 915–918. [CrossRef] [PubMed]
9. Ueta, M.; Yoshida, H.; Wada, C.; Baba, T.; Mori, H.; Wada, A. Ribosome binding proteins YHbH and YfiA have opposite functions during 100S formation in the stationary phase of *Escherichia coli*. *Genes Cells* **2005**, *10*, 1103–1112. [CrossRef] [PubMed]
10. Ueta, M.; Ohniwa, R.L.; Yoshida, H.; Maki, Y.; Wada, C.; Wada, A. Role of HPF (hibernation promoting factor) in translational activity in *Escherichia coli*. *J. Biochem.* **2008**, *143*, 425–433. [CrossRef] [PubMed]
11. De Bari, H.; Berry, E.A. Structure of *Vibrio cholerae* ribosome hibernation factor. *Acta Crystallogr. Sect. F* **2013**, *69*, 228–236. [CrossRef] [PubMed]
12. Lakowicz, J.R. *Principles of Fluorescence Spectroscopy*, 2nd ed.; Plenum Press: New York, NY, USA, 1999.
13. Albani, J.R. *Principle and Applications of Fluorescence Spectroscopy*; Blackwell Publishing: Oxford, UK, 2007.
14. Cantor, C.R.; Schimmel, P.R. *Biophysical Chemistry*; W. H. Freeman: New York, NY, USA, 1980.
15. Cavanagh, J.F.; Wayne, J.; Palmer, A.G., III; Skelton, N.J. *Protein NMR Spectroscopy: Principles and Practice*, 1st ed.; Academic Press: San Diego, CA, USA, 1996.
16. Wilkins, D.K.; Grimshaw, S.B.; Receveur, V.; Dobson, C.M.; Jones, J.A.; Smith, L.J. Hydrodynamic radii of native and denatured proteins measured by pulse field gradient NMR techniques. *Biochemistry* **1999**, *38*, 16424–16431. [CrossRef] [PubMed]

17. Nobbmann, U.; Connah, M.; Fish, B.; Varley, P.; Gee, C.; Mulot, S.; Chen, J.; Zhou, L.; Lu, Y.; Sheng, F.; et al. Dynamic light scattering as a relative tool for assessing the molecular integrity and stability of monoclonal antibodies. *Biotechnol. Genet. Eng. Rev.* **2007**, *24*, 117–128. [CrossRef] [PubMed]
18. Pace, C.N.; Grimsley, G.R.; Scholtz, J.M. Protein ionizable groups: p*K* values and their contribution to protein stability and solubility. *J. Biol. Chem.* **2009**, *284*, 13285–13289. [CrossRef] [PubMed]
19. Grimsley, G.R.; Scholtz, J.M.; Pace, C.N. A summary of the measured p*K* values of the ionizable groups in folded proteins. *Protein Sci.* **2009**, *18*, 247–251. [PubMed]
20. Lux, B.; Gerard, D.; Laustriat, G. Tyrosine fluorescence of S8 and S15 *Escherichia coli* ribosomal proteins. *FEBS Lett.* **1977**, *80*, 66–70. [CrossRef]
21. Soengas, M.S.; Mateo, C.R.; Salas, M.; Acuña, A.U.; Gutiérrez, C. Structural features of φ29 single-stranded DNA-binding protein. Environment of tyrosines in terms of complex formation with DNA. *J. Biol. Chem.* **1997**, *272*, 295–302. [CrossRef] [PubMed]
22. Whitmore, L.; Wallace, B.A. Protein secondary structure analysis from circular dichroism spectroscopy: Methods and reference databases. *Biopolymers* **2008**, *89*, 392–400. [CrossRef] [PubMed]
23. Whitmore, L.; Wallace, B.A. DICHROWEB, an online server for protein secondary structure analyses from circular dichroism spectroscopic data. *Nucleic Acids Res.* **2004**, *32*, W668–W673. [CrossRef] [PubMed]
24. Woody, R.W. Circular dichroism. *Methods Enzymol.* **1995**, *246*, 34–71. [PubMed]
25. Kelly, S.M.; Jess, T.J.; Price, N.C. How to study proteins by circular dichroism. *Biochim. Biophys. Acta* **2005**, *1751*, 119–139. [CrossRef] [PubMed]
26. Dosztányi, Z.; Csizmók, V.; Tompa, P.; Simon, I. IUPred: Web server for the prediction of intrinsically unstructured regions of proteins based on estimated energy content. *Bioinformatics* **2005**, *21*, 3433–3434. [CrossRef] [PubMed]
27. Yang, Z.R.; Thomson, R.; McNeil, P.; Esnouf, R.M. RONN: The bio-basis function neural network technique applied to the detection of natively disordered regions in proteins. *Bioinformatics* **2005**, *21*, 3369–3376. [CrossRef] [PubMed]
28. Ishida, T.; Kinoshita, K. PrDOS: Prediction of disordered protein regions from amino acid sequence. *Nucleic Acids Res.* **2007**, *35*, W460–W464. [CrossRef] [PubMed]
29. Lobanov, M.Y.; Sokolovskiy, I.V.; Galzitskaya, O.V. IsUnstruct: Prediction of the residue status to be ordered or disordered in the protein chain by a method based on the Ising model. *J. Biomol. Struct. Dyn.* **2013**, *31*, 1034–1043. [CrossRef] [PubMed]
30. Uversky, V.; Gillespie, J.; Fink, A. Why are "natively unfolded" proteins unstructured under physiological conditions? *Proteins* **2000**, *41*, 415–427. [CrossRef]
31. Dunker, A.K.; Lawson, J.D.; Brown, C.J.; Williams, R.M.; Romero, P.; Oh, J.S.; Oldfield, C.J.; Campen, A.M.; Ratliff, C.M.; Hipps, K.W.; et al. Intrinsically disordered proteins. *J. Mol. Graph. Model.* **2001**, *19*, 26–59. [CrossRef]
32. Das, R.K.; Pappu, R.V. Conformations of intrinsically disordered proteins are influenced by linear sequence distributions of oppositely charged residues. *Proc. Natl. Acad. Sci. USA* **2013**, *110*, 13392–13397. [CrossRef] [PubMed]
33. Yang, J.; Yan, R.; Roy, A.; Xu, D.; Poisson, J.; Zhang, Y. The I-TASSER Suite: Protein structure and function prediction. *Nat. Methods* **2015**, *12*, 7–8. [CrossRef] [PubMed]
34. Wang, C.; Zhang, H.; Zheng, W.M.; Xu, D.; Zhu, J.; Wang, B.; Ning, K.; Sun, S.; Li, S.C.; Bu, D. FALCON@home: A high-throughput protein structure prediction server based on remote homologue recognition. *Bioinformatics* **2016**, *32*, 462–464. [CrossRef] [PubMed]
35. Biasini, M.; Bienert, S.; Waterhouse, A.; Arnold, K.; Studer, G.; Schmidt, T.; Kiefer, F.; Cassarino, T.G.; Bertoni, M.; Bordoli, L.; et al. SWISS-MODEL: Modelling protein tertiary and quaternary structure using evolutionary information. *Nucleic Acids Res.* **2014**, *42*, W252–W258. [CrossRef] [PubMed]
36. Kim, D.E.; Chivian, D.; Baker, D. Protein structure prediction and analysis using the Robetta server. *Nucleic Acids Res.* **2004**, *32*, W526–W531. [CrossRef] [PubMed]
37. Wüthrich, K. *NMR of Proteins and Nucleic Acids*; Wiley and Sons: New York, NY, USA, 1986.
38. Pantoja-Uceda, D.; Neira, J.L.; Saelices, L.; Robles-Rengel, L.; Florencio, F.J.; Muro-Pastor, M.I.; Santoro, J. Dissecting the binding between glutamine synthetase and its tow natively unfolded protein inhibitors. *Biochemistry* **2016**, *55*, 3370–3382. [CrossRef] [PubMed]

39. Rizzuti, B.; Daggett, V. Using simulations to provide the framework for experimental protein folding studies. *Arch. Biochem. Biophys.* **2013**, *531*, 128–135. [CrossRef] [PubMed]
40. Humphrey, W.; Dalke, A.; Schulten, K. VMD: Visual molecular dynamics. *J. Mol. Graph. Model.* **1996**, *14*, 33–38. [CrossRef]
41. Miroux, B.; Walker, J.E. Over-production of proteins in *Escherichia coli*: Mutant hosts that allow synthesis of some membrane proteins and globular proteins at high levels. *J. Mol. Biol.* **1996**, *260*, 289–298. [CrossRef] [PubMed]
42. Burgess, R.R. Protein precipitation techniques. *Methods Enzymol.* **2009**, *463*, 331–342. [PubMed]
43. Dunn, M.J. Initial planning: Determination of total protein concentration. In *Protein Purification Methods*; Harris, E.L.V., Angal, S., Eds.; Oxford University Press: Oxford, UK, 1995; pp. 10–20.
44. Neira, J.L. Fluorescence, circular dichroism and mass spectrometry as tools to study virus structure. *Subcell. Biochem.* **2013**, *68*, 177–202. [PubMed]
45. Tinoco, I., Jr.; Sauer, K.; Wang, J.C. *Physical Chemistry: Principles and Applications in Biological Sciences*, 3rd ed.; Prentice-Hall: New York, NY, USA, 1995; pp. 585–588.
46. Bokma, J.T.; Johnson, W.C., Jr.; Blok, J. CD of the Li-salt of DNA in ethanol/water mixtures: Evidence for the B- to C-form transition in solution. *Biopolymers* **1987**, *26*, 893–909. [CrossRef] [PubMed]
47. Jin, X.; Johnson, W.C., Jr. Comparison of base inclination in Ribo-AU and Deoxyribo-AT polymers. *Biopolymers* **1995**, *36*, 303–312. [CrossRef] [PubMed]
48. Charnavets, T.; Nunvar, J.; Necasova, I.; Völker, J.; Breslauer, K.J.; Schneider, B. Conformational diversity of the single-stranded DNA from bacterial repetitive extragenic palindromes: Implications for the DNA recognition elements of transposases. *Biopolymers* **2015**, *103*, 585–596. [CrossRef] [PubMed]
49. Atkins, P.; De Paula, J. *Physical Chemistry for the Life Sciences*, 1st ed.; W. H. Freeman: Oxford, UK, 2006; pp. 565–566.
50. Neira, J.L.; Hornos, F.; Bacarizo, J.; Cámara-Artigas, A.; Gómez, J. The monomeric species of the regulatory domain of Tyrosine Hydroxylase has a low conformational stability. *Biochemistry* **2016**, *55*, 6209–6220. [CrossRef] [PubMed]
51. Bevington, P.R.; Robinson, K.D. *Data Reduction and Error Analysis for the Physical Sciences*, 3rd ed.; McGraw-Hill: New York, NY, USA, 2003.
52. James, D.R.; Siemiarczuk, A.; Ware, W.R. Stroboscopic optical boxcar technique for the determination of fluorescence lifetimes. *Rev. Sci. Instrum.* **1992**, *63*, 1710–1716. [CrossRef]
53. Piotto, M.; Saudek, V.; Sklenar, V. Gradient-tailored excitation for single-quantum NMR spectroscopy of aqueous solutions. *J. Biomol. NMR* **1993**, *2*, 661–665. [CrossRef]
54. Abraham, M.J.; Murtola, T.; Schulz, R.; Páll, S.; Smith, J.C.; Hess, B.; Lindahl, J. GROMACS: High performance molecular simulations through multi-level parallelism from laptops to supercomputers. *SoftwareX* **2015**, *1–2*, 19–25. [CrossRef]
55. Lindorff-Larsen, K.; Piana, S.; Palmo, K.; Maragakis, P.; Klepeis, J.L.; Dror, R.O.; Shaw, D.E. Improved side-chain torsion potentials for the Amber ff99SB protein force field. *Proteins* **2010**, *78*, 1950–1958. [CrossRef] [PubMed]
56. Jorgensen, W.L.; Chandrasekhar, J.; Madura, J.D.; Impey, R.W.; Klein, M.L. Comparison of simple potential functions for simulating liquid water. *J. Chem. Phys.* **1983**, *79*, 926–935. [CrossRef]
57. Evoli, S.; Guzzi, R.; Rizzuti, B. Molecular simulations of β-lactoglobulin complexed with fatty acids reveal the structural basis of ligand affinity to internal and possible external binding sites. *Proteins* **2014**, *82*, 2609–2619. [CrossRef] [PubMed]
58. Rizzuti, B.; Bartucci, R.; Sportelli, L.; Guzzi, R. Fatty acid binding into the highest affinity site of human serum albumin observed in molecular dynamics simulation. *Arch. Biochem. Biophys.* **2015**, *579*, 18–25. [CrossRef] [PubMed]
59. Guglielmelli, A.; Rizzuti, B.; Guzzi, R. Stereoselective and domain-specific effects of ibuprofen on the thermal stability of human serum albumin. *Eur. J. Pharm. Sci.* **2018**, *112*, 122–131. [CrossRef] [PubMed]
60. Rudolph, R.; Böhm, G.; Lilie, H.; Jaenicke, R. Folding proteins. In *Protein Function: A Practical Approach*, 2nd ed.; Creighton, T.E., Ed.; Oxford University Press: Oxford, UK, 1997; pp. 57–99.

8

The Melting Diagram of Protein Solutions and its Thermodynamic Interpretation

Kálmán Tompa [1], Mónika Bokor [1] and Péter Tompa [2,3,*]

1 Institute for Solid State Physics and Optics, Wigner RCP of the HAS, Konkoly-Thege út 29-33, H-1121 Budapest, Hungary; tompa.kalman@wigner.mta.hu (K.T.); bokor.monika@wigner.mta.hu (M.B.)
2 Institute of Enzymology, Research Centre for Natural Sciences of the HAS, Magyar Tudósok körút. 27, H-1117 Budapest, Hungary
3 VIB Center for Structural Biology, Vrije Universiteit Brussel, Building E, Pleinlaan 2, 1050 Brussel, Belgium
* Correspondence: peter.tompa@vub.be

Abstract: Here we present a novel method for the characterization of the hydration of protein solutions based on measuring and evaluating two-component wide-line ^{1}H NMR signals. We also provide a description of key elements of the procedure conceived for the thermodynamic interpretation of such results. These interdependent experimental and theoretical treatments provide direct experimental insight into the potential energy surface of proteins. The utility of our approach is demonstrated through the examples of two proteins of distinct structural classes: the globular, structured ubiquitin; and the intrinsically disordered ERD10 (early response to dehydration 10). We provide a detailed analysis and interpretation of data recorded earlier by cooling and slowly warming the protein solutions through thermal equilibrium states. We introduce and use order parameters that can be thus derived to characterize the distribution of potential energy barriers inhibiting the movement of water molecules bound to the surface of the protein. Our results enable a quantitative description of the ratio of ordered and disordered parts of proteins, and of the energy relations of protein–water bonds in aqueous solutions of the proteins.

Keywords: protein; hydration; wide-line ^{1}H NMR

---

1. Introduction

Wide-line ^{1}H NMR is an accepted method to delineate the structures of hydrogen-containing molecules determined primarily by X-ray and, to a lesser extent, by neutron-scattering. This way, it can provide information on the location and structural environment of hydrogen atoms in proteins. It has a unique capability, on the other hand, in the direct observation of translational and rotational movements of molecules in the condensed phase.

NMR characteristics of aqueous solutions rapidly frozen and then slowly thawed through equilibrium thermal states provide direct information on the immobile and partially or fully mobile parts of the molecules. We have previously reviewed relevant features of this approach in our works "Hydrogen skeleton, mobility and protein architecture" [1] and "Studying molecular motions in solid states by NMR" [2].

Based on these studies, we state that molecular motions in the sample result in narrowing of the wide-line NMR spectrum. This phenomenon is known as motional narrowing in the literature [3]. Our goal is to advance from this observation to arrive at the thermodynamic characterization of protein systems.

In Figure 1a, we show the typical ^{1}H NMR free-induction decay (FID) signal of a set of spins containing proton–proton pairs of different mobilities. In Figure 1b, we present the deduced NMR spectrum. Similar FID signals and NMR spectra are observed at certain temperatures when studying the aqueous solution of a protein that contains hydrogen pairs.

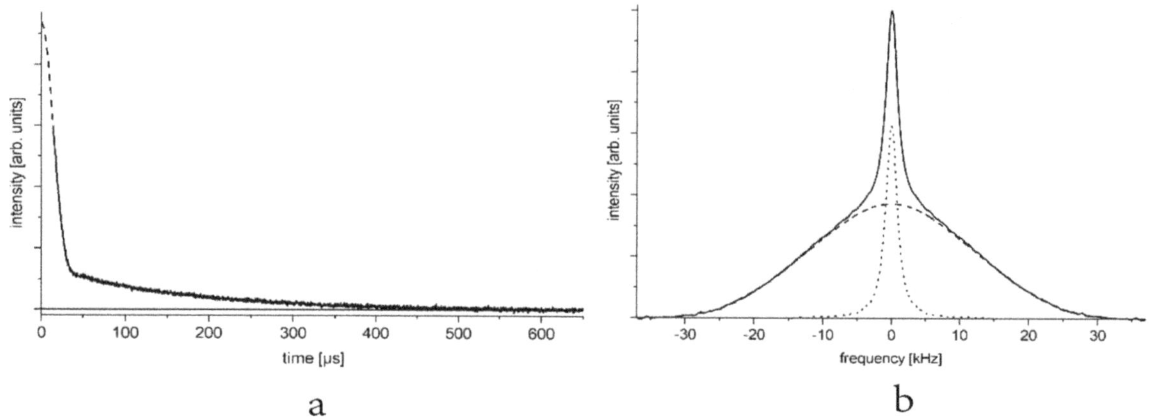

Figure 1. Free induction decay (FID, panel (**a**)) and spectrum (panel (**b**)) of a motionally two-state spin system. (We focus on the slow component of the FID, the initial part of which is lost in the dead time of the spectrometer, marked by dashed line, and can be disregarded).

Time domain (Figure 1a) and frequency or energy domain (Figure 1b) representation of the spectra are linked through Fourier transformation, yet it may be useful to consider both, as they provide information on different practical utilities. The amplitude of the FID signal (even considering its slow component) extrapolated to time zero gives the number of relevant nuclei (spins) through nuclear magnetization. The amplitude of response to the 90° radiofrequency pulse is proportional to the relevant x-y component of nuclear magnetization that is further proportional to $M_0 \approx (nB_0)/T$, in which B_0 is the constant magnetic induction, T is the absolute temperature, and n is the number of resonant nuclei (in our case, it equals the number of protons in water). On the other hand, the width of the spectrum gives direct information on the motional characteristics of proton spin pairs. In a system of two components (e.g., one that contains both mobile and immobile spin-pairs), it is important to have direct information on both parameters.

It is questionable whether such a simple approach can give significant novel information on the dynamics of a complex system, such as a protein and its environment in an aqueous solution. The independent measurement over a broad temperature range of the two parameters of the slow FID component (FID amplitude extrapolated to $t = 0$ and the spectral width) is debatable.

Therefore, here we address the behavior of the slow-FID, and the narrow-spectral component. Our working hypothesis (that we already partially proved) is that the narrow spectral component comes from water molecules bound to the protein, termed bound water-molecules [4]. One may ask a range of relevant questions about their number, their strength of binding to the protein vs. the neighboring water molecules, and about their potential field following molecular changes of the protein, etc. Similar questions can also be asked for the broad-spectrum component, which we have already addressed before [1,2].

In earlier studies [5–8], we have addressed in detail the behavior of globular and intrinsically disordered proteins (IDPs) in aqueous solutions and provided an initial and partial interpretation of experimental observations. As relevant examples, we refer to results with proteins, such as ubiquitin, bovine serum albumin, α-synuclein (and its point mutants), calpastatin, ERD10 (early response to dehydration 10), and lysozyme. Here, we demonstrate our point by focusing on two proteins, ubiquitin (Ubq) and ERD10, as they have been thoroughly studied earlier; one (Ubq) is a globular/structured protein and the other (ERD10) is intrinsically disordered, i.e., they are representatives of these distinct structural classes. We show the temperature dependence of the slowly-decaying component of the FID extrapolated to $t = 0$ (which gives directly the ratio of relevant mobile water protons). We show the observed behavior in the form of a melting diagram (*MD*). In Figure 2, we show the *MD* of three studied systems (bulk water and the aqueous solution of two proteins, ubiquitin and ERD10) in the usual °C scale.

The melting process of inhomogeneous systems (such as the protein solutions we study), basically differs from the first-order phase transition of homogeneous, single component material, such as the melting of ice at a given transition temperature.

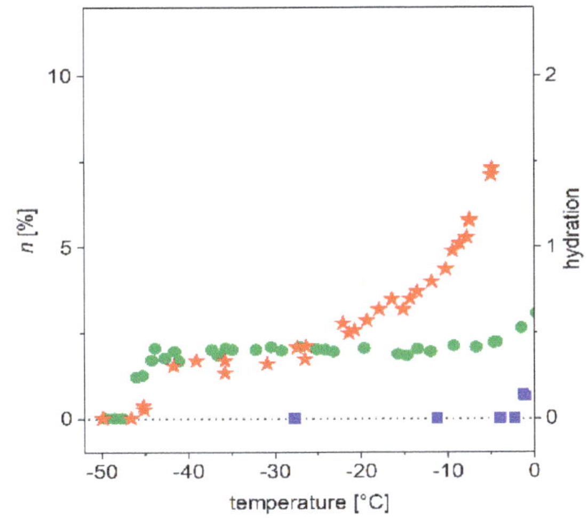

Figure 2. "Old fashioned" melting diagrams, i.e., the total number of mobile water molecules (through protons) normalized to the total number of water molecules, as a function of temperature (blue squares: bulk water, green circles: ubiquitin, red stars: ERD10 proteins in aqueous solutions). The data are given for 50 mg/mL protein concentration.

We consider melting as the process of the beginning of movement of a component of the mixture (such as a bound water-molecule, or a fragment of the protein of high symmetry, e.g., a methyl group or other terminal moiety), in which either translation or rotation begins. In our case, these (individual) events of initial movements show a temperature distribution characteristic of the given molecule, and the derived *MD*s link the well-defined, directly measurable NMR characteristics with atomic/molecular motions.

These characteristics can thus also give direct information on molecular interactions. The water molecules associated with the protein molecule constitute an integral part of the system. Thus, their nuclei, rather than large energy particles applied in scattering techniques (such as X-ray crystallography), monitor the potential energy surface of the protein as built-in probes. In our previous works [5–8], however, we only drew qualitative conclusions from the *MD*s.

These were as follows. In aqueous solutions, melting (that is, beginning of molecular motions) of protein-bound water molecules begins at a much lower temperature than the melting of bulk ice. Each protein has a unique *MD* (individual profile or fingerprint) that results from its individual thermodynamic characteristics. The *MD* of globular and ID proteins vastly differ. They can be characterized by temperature-independent FID amplitudes—a plateau (globular protein)—or they can lack a plateau (IDP) or can have a plateau of small temperature extension (partly IDP).

1.1. Energetic Interpretation of Melting Diagrams

We have made significant advances in several respects of interpreting our results [9,10] since we last addressed these questions [5–8]. Key steps are detailed in chapters 4–6 of ref [9]; here, we add a new element and summarize these steps in more detail, following the logical order of the application.

As a reminder, we are following the beginning of the movement—probably the rotation—of water molecules bound to the surface of the protein, by observing motional narrowing in wide-line ^{1}H NMR spectroscopy. For the first time in the field—following the seminal work of Kittel and Kroemer [11]—we introduced the concept of fundamental temperature, T_f, and also introduced here the idea suggested by Waugh and Fedin [12] for connecting the thermal excitation energy, V_0, in which molecular motions begin with the temperature, T, as V_0 = constant $\times$ T.

In some detail, the key steps taken are as follows.

1.1.1. Fundamental Temperature

As a first step, we introduced the use of the scale of fundamental temperature, i.e., thermal excitation energy scale, T_f, and its version normalized to the melting point of ice, T_{fn}. By definition, $T_f = k_B T$, in which $k_B = 1.381 \cdot 10^{-16}$ erg/K ($k_B = 1.381 \cdot 10^{-23}$ J/K) is the Boltzmann constant, and T is the absolute temperature in K. We can also use the equation of $T_f = RT$, in which $R = 8.317$ J/mol·K, the universal gas constant. If we need a dimensionless scale, it is expedient to use the normalized fundamental temperature scale, T_{fn}, nor malized to the melting temperature of bulk water formally as $T_{fn} = k_B \cdot T/(k_B \times 273.15) = T/273.15$. This way, it becomes possible to characterize the events of the beginning of molecular motion on an energy scale.

1.1.2. Energy Scale and the Heterogeneity of the Protein Surface

As a next step, we invoked the formula of Waugh and Fedin, after the improvement of placing it on a fundamental temperature scale of the right dimension. The formula can then be used for aqueous solutions. The equation at atomic/molecular level is

$$E_{0a}\ [\text{erg}] = ck_B T\ [\text{erg}], \tag{1a}$$

or applied to molar quantities it is

$$E_{0m}\ [\text{kJ/mol}] = cRT\ [\text{kJ/mol}]. \tag{1b}$$

In these equations, c is a dimensionless quantity, i.e., a number, the value of which was determined by applying Equation (1b) to the melting of bulk ice, considering the melting heat of ice (6.01 kJ/mol [13]). The fundamental temperature equivalent with 273.15 K is $T_f = RT = 2.272$ kJ/mol. In Equation (1b), the c proportionality constant is 2.65. When comparing Equation (1a) with the energy pertaining to one degree of freedom by the equation of equipartition ($1/2\ k_B T$), we may deduce the degree of freedom of a water molecule as 5.3, which seems to be in the right range for a rotating (and not translating) electric water dipole.

In addition, we introduced dynamic parameters for the quantitative characterization of the ordered/disordered state of protein molecules, which goes beyond their static structural description. Before formalizing the definitions, let us take a look at Figure 2 (and for details, Figures 3 and 4). There is a marked difference between the globular and intrinsically disordered proteins. On the melting diagram of the globular protein Ubq one can see a broad, temperature- (or excitation energy-) independent region (plateau). On the other hand, the plateau of the IDP ERD10 is significantly smaller. (A similar behavior was also seen for other proteins [5–10]). Significantly more information is provided by the initial (T_{fno}) and the ending temperature (T_{fne}) values of the plateau. The region between these two temperatures shows homogeneous bond (potential energy barrier) distribution, whereas the region above T_{fne} shows a heterogeneity in terms of protein–water-bond energy distribution. After this introduction, the following quantities can be defined.

Heterogeneity ratio, *HeR*. According to our observations [5–10] and the literature quoted therein, protein molecules can be characterized and categorized by the homogeneity/heterogeneity of the energy distribution of water binding. The basis of the classification is the measurement of the ratio, for which we suggest the relation

$$HeR = (1 - T_{fne})/(1 - T_{fno}), \tag{2}$$

in which $(1 - T_{fne})$ and $(1 - T_{fno})$ give the measured distances from the melting point of ice. These values can be easily read from the novel *MDs*. *HeR* is 1 (one) for systems showing heterogeneous water binding (lacking a plateau) and 0 (zero) for homogeneous binding systems (e.g., bulk water), and is between 0 and 1 for partially heterogeneous systems. *HeR* therefore gives the order parameter type specification for what extent of the surface of the protein molecule can be regarded as showing heterogeneous potential energy distribution (disordered) in terms of water binding. It must be emphasized that this correlation measures the heterogeneity ratio based on the comparison of the extent of the two possible regions and does not measure the number of actual protein–water bonds in them.

1.1.3. An Analytical Description of *n*

The introduction of fundamental temperature or energy scale makes it possible to describe *MD* by power series in the form

$$n = A + B(T_{fn} - T_{fn1}) + C(T_{fn} - T_{fn2})^2 + \ldots \quad (3)$$

That is, we can define the total number of water molecules, n, that are moving at a given thermal energy (temperature), as well as the change of *MD* on a normalized fundamental energy scale, i.e., the differential form of melting diagram, *DMD*

$$\Delta n / \Delta T_{fn} = B + 2C(T_{fn} - T_{fn2}) + \ldots, \quad (4)$$

which defines the number of water molecules that begin to move at the given excitation energy. T_{fnx} (with $x = 1, 2, \ldots, n$) is fitting parameter in Equations (3) and (4), in which x is equal to the exponent in each term (in the other terms too, with $n \geq 3$ not given here in detail). The present form of equations calls attention to the validity of any term in a given temperature range. It should be emphasized that all quantities and coefficients are dimensionless in these formulae.

Number of protein–water bonds, HeR_n. We can make a statement about the homogeneity/heterogeneity of bonds (potential barriers) if we ask about the exact number of protein–water bonds in the given excitation energy range. Parameters that fit the power series provide the answer. In the simplest cases (including, in our experience, aqueous solutions with distilled water), in which there is only a wider heterogeneous range in *MD*, the number of water bonds in the heterogeneous region depends on the number of fitting members, $B/(1\ 2212\ T_{fne})$, and $2C/(1 - T_{fne})$; if both, then it depends on the sum of the two members. As simplification of the determination of the number of protein–water bonds (the degree of hydration), it can be directly read from the *DMDs*, i.e., the value or the sum of the areas colored in the figures enter (in principle, the definite integrals within the region T_{fne} to $T_{fn} \approx 1$). Let n_{ho} be the number of water molecules in the first hydrate shell and n_{he} the total number of water molecules in the entire heterogeneous region. In this case, the second relation suggested for the ratio of heterogeneity is

$$HeR_n = n_{he} / (n_{he} + n_{ho}). \quad (5)$$

The value of n_{ho} (approximately) is given by the area of the rectangle at the lowest excitation energy region, whereas n_{he} in our case is given by the areas of triangles (in general, those described by members of higher exponents; see Figures 3 and 4).

The numbers in the equation can be measured directly based on *MD*. n_{ho} can be determined with high accuracy as the average of all n points measured on the plateau, and $(n_{he} + n_{ho})$ as an approximate value by the n value reliably measured at a temperature close to the highest temperature, $T_{fn} \approx 1$. The process has a self-checking potential and thus improves the reliability of the data.

The measure of heterogeneity is *HeM*. We suggested [9] to introduce this as the parameter

$$HeM = (B + 2C)/(1 - T_{fne}). \tag{6}$$

This relationship is also correct in terms of dimensions, and the *HeM* value is generally a positive number. Its value is zero for proteins of almost equipotential molecular surface, so it can be considered as a quasi-order parameter. The denominator, $(1 - T_{fne})$ designates the energy range in which there are varying protein–water bonds (of heterogeneous distribution), and $B + 2C$ (going till the second term of non-zero exponent) is the number of bonds within this range. The fraction is thus a kind of slope of the *MD* function; its values cannot be limited to the range of 0 to 1, just like for the tangent function. Non-heterogeneously binding proteins, such as globular proteins by our experience, have a *HeM* value, by definition, which applies to the region above the plateau. It is not unfounded to suggest that there is a similar dynamic difference in the hydrogen mobility (*HM* [1]) and *HeM* parameters, i.e., in the mobility of all proton–proton pairs, and in the degree of heterogeneity of protein–water bonds.

In the power series of n Equation (3), we only went till the first two members of non-zero exponent (which is enough to interpret the results presented in most of our examples). Heterogeneity and homogeneity can be observed in both the nature and the magnitude of the respective potentials and their distance dependence. Variants of theoretical possibilities are found in the literature [14–17].

The determination of the *MD* function and its differential form that can also be described analytically allows for the unique and individual mapping of the energy distribution of the potential barriers that inhibit the motion of water molecules bound to the protein. Using the elements required for the interpretation of measured *MD*s we have introduced, the purpose of our present work can be easily formulated. Specifically, it is a deeper, thermodynamic interpretation of our results. The examples that illustrate this statement are presented through the analysis of the *MD* of the globular standard protein ubiquitin, and the intrinsically disordered ERD10.

2. Results and Discussion

In Figures 3 and 4, we show the *MD*s determined for the two proteins, dissolved in double-distilled water, with a panel (a) showing measurements on reference water too, and panel (b) the derived curves, *DMD*s (that is $\Delta n/\Delta T_{fn}$ the potential distribution of protein–water bonds). The information on the origin of the samples, the measuring equipment, and the details of the measurements is described in our above-mentioned articles and in book chapters [4,5].

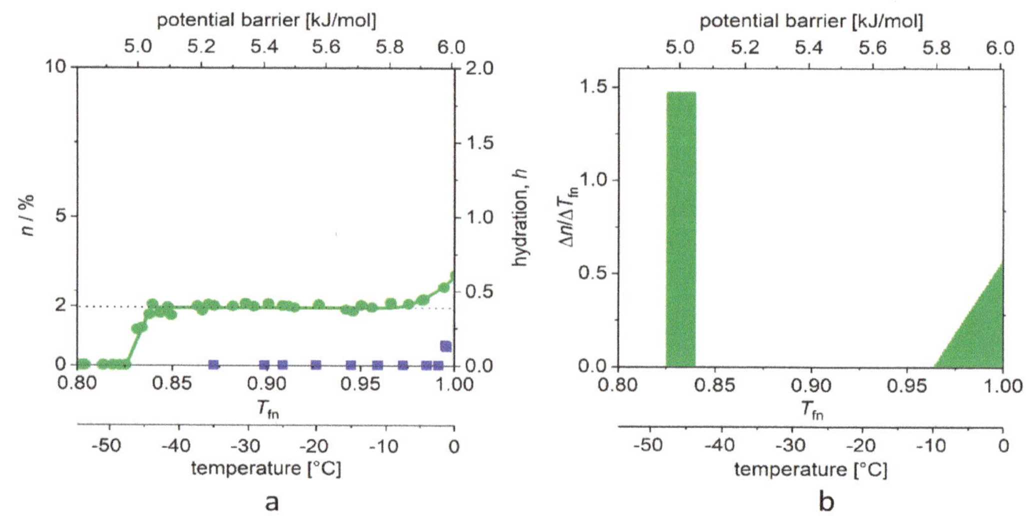

Figure 3. (**a**) Melting diagram (*MD*, green circles) of ubiquitin dissolved in double distilled water and that of frozen water under identical conditions (blue squares). (**b**) *DMD* curves (that is, the potential barrier distribution of protein–water bonds). There is no reliable measured data in the range −1–0 °C (0.995–1.00 T_{fn}). The data are given for 50 mg/mL protein concentration.

Perhaps it is not unnecessary to repeat that the amplitude value of the slow component of the measured FID signal extrapolated to $t = 0$ gives directly the number n of resonant protons (i.e., the protein-bound water molecules), whereas the temperature dependence of *MD* gives the dependence of n on thermal excitation energy.

The information can be read from Figures 3 and 4 as follows. Bulk water (blue squares) show the microscopic image of the ice-water phase transition. What would we expect of an absolute pure water sample of infinite size (in theory, one having a periodic boundary condition)? A single step of infinite slope at $T_{fn} = 1.00$ and $E_a = 6.01$ kJ/mol excitation energy (at 0.00 °C), in which all four bonds of the water molecule in the tetrahedral bond symmetry environment "melt" simultaneously. Instead moving water molecules are detected already below 0 °C. There are several reasons for this. The sample is not of infinite size, and the environment of the water molecules on the surface of the small sample is not the same as of those in the bulk environment. Secondly, the sample is not of absolute purity, so the environment of pollutions is not the same as in the clean environment. Third, the temperature of the sample in the measurement can be controlled and determined with limited accuracy only, especially at 0 °C.

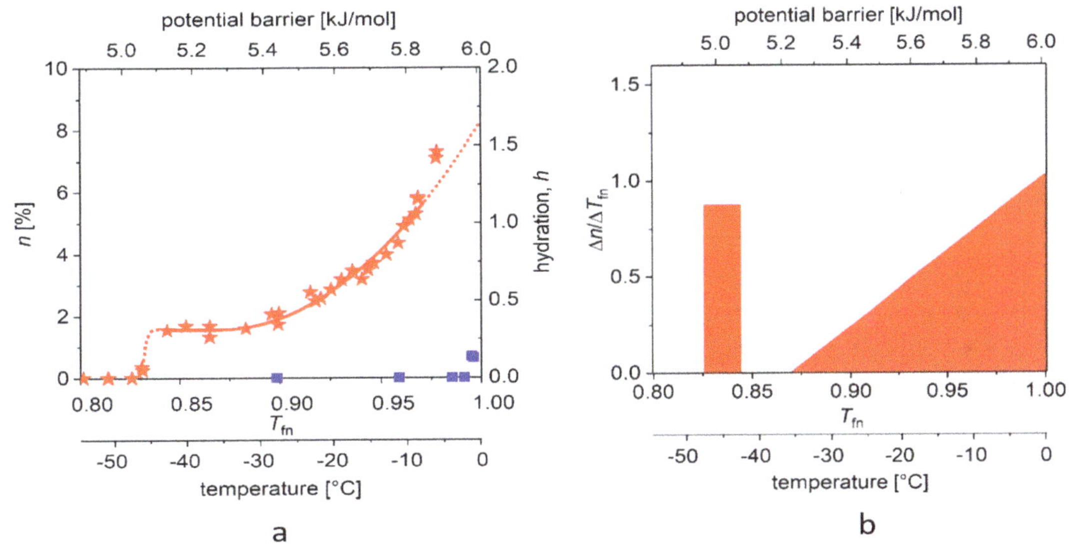

Figure 4. (**a**) The melting diagram (*MD*, red stars) of ERD10 dissolved in double-distilled water and the melting curve (blue squares) of the solvent (water). (**b**) *DMD* curves are shown (that is, the potential barrier distribution of protein–water bonds). There is no reliable measured data in the range −1–0 °C (0.995–1.00 T_{fn}). The data are given for 50 mg/mL protein concentration.

In Figure 3a, the "melting point" (−46 (1) °C (for definition of error, see Table 1) of the aqueous solution of ubiquitin shows the thermal energy investment (ΔQ) that is required to start to move the water molecules that are bound to the protein. The steep step (with a narrow, ≈0.01 kJ/mol energy range) shows that there are water molecules in the first hydrate shell that are bound almost identically. It is a reasonable approximation to consider these energies nearly the same, and the relevant molecular surface equipotential. This potential field of nearly identical elements resembles the feature of the H-bridges [16–19] and is largely different from strongly distant dependent potentials (the variants can be find in the text-books [16–19]). The number of protein–water bonds in the actual region is given by the area of the rectangle. As a self-check, the same quantity can be more accurately determined from the average of all n points on the *MD* plateau.

The next wide region is the plateau. (This region begins at $T_{fno} = 0.832$ (4), in Figure 3b, in which the value of $\Delta n/\Delta T_{fn}$ is zero). The plateau carries very important information. No new water molecules begin to move in this excitation energy region, because there are no water molecules that are bound by corresponding energy to the protein. We can suggest that the H-bridges here, which link the bulk of the protein molecule to a globule. Thus, this can be an ancestral form of a higher order structure, which is represented not only by geometry but also by a bonding network of a certain energy. The heat invested within the plateau region does not start to move new water molecules; rather, it increases the specific heat, and the rotational speed of already rotating water molecules, as we have seen in a previous work on differential scanning calorimetry (DSC) measurements and data interpretation [7]. (Based on this interpretation, these statements can be made to be more accurate, which we intend to do in a short notice.)

T_{fne} is the end of the plateau, and here begins the energy region where there are binding energies close to the binding energy of water–water bonds, presumably on parts of the protein molecule that are better exposed to water. In principle, the temperature dependence of n can be described by the higher exponents of the power function; the quadratic member was sufficient in this case. All data are available; we summarize the values and the order parameters introduced by us in Table 1.

In Figure 4a, we show the "melting point" (approximately −42 (2) °C) for ERD10 (red stars). We also repeat the above procedure for ERD10 with different parameters. The steep step (with narrow, ≈0.01 kJ/mole energy range) shows the presence of water molecules of nearly identical binding energy in the first hydrate shell, but this region is followed by a plateau, which is significantly narrower than that observed in the case of globular proteins. We then observe a phase of continuous rise in *MD*, which can be well approximated by the quadratic (or even higher) component of the summation. A much larger part of the molecular surface is exposed to water than in the case of ubiquitin, i.e., about 69–77% of the protein molecule can be described as disordered. The range $(1 - T_{fne})$ of energy barriers inhibiting water movements (which can be defined as disorder) is more than three times broader than for the selected globular protein.

Figure 3b and Figure 4b depict the changes (differential quotient) of the mobile water fractions by normalized functional temperature, i.e., they are the graphical representations of Equation (4). As outlined, the bars at low temperature (around −45 °C) correspond to the relatively high differential quotient values describing the first few data points greater than zero. The fraction of mobile hydration water increases here within a few degrees to the level of $n(E_{a,o})$ or A while the first mobile hydration layer forms, which gives the high differential quotient values.

Table 1. Characteristic thermal quantities for two sample proteins. T_{fno} end T_{fne} give the start and the end points of the plateau in *MD*s, respectively, as normalized fundamental temperature. n_{ho} and n_{he} values are given as the mobile hydration water fraction and as the number of mobile hydration water per protein molecule. *HeR*, *HeR*$_n$, and *HeM* are dynamic parameters describing heterogeneity from various aspects (see text).

Protein	T_{fno}	T_{fne}	*HeR* (4) *	n_{ho}	n_{he} **	*HeR*$_n$ (6) *	*HeM*
UBQ	0.832 (4)	0.961 (5)	0.23 (2)	0.019 (1) 226 (3)	>0.009 (3) >102 (33)	0.3 (1)	241 (147)
ERD10	0.835 (3)	0.889(2)	0.73 (4)	0.0157 (4) 514 (13)	>0.098 (8) >3200 (275)	0.9 (1)	415 (60)

* The number in parentheses is the measurement error in the order of magnitude of the last number; the heterogeneity ratio is defined by the relation (4) or (6); ** Lower limit estimate due to the uncertainty of measured data is close to $T_{fn} = 1$; at T_{fno} value given in Table 1 (−43 °C), the excitation energy is 5.06 (4) kJ/mol for both proteins; at T_{fne} for ubiquitin, the excitation energy is 5.798 (2) kJ/mol at −9.9 °C; and for ERD10, it is 5.31 (3) kJ/mol at −36 °C.

A comparison of *HeM* values (analogous with the tangent function) shows that in globular proteins the realization of the two extreme values, conditions in the first hydrate shell and water-water bonding, are very close. For ERD10, a much wider distribution of potential energy barriers is characteristic of structural disorder. The typical data are summarized in Table 1. The ordered/disordered state of the two protein molecules only approximates the ideal limiting values, *HeR* = 0 and *HeR* = 1.

The reality of the n_{ho} number of protein–water bonds in the homogeneous binding energy region (in other words, in the first hydrate shell) is better appreciated by reference to our knowledge of the hydration of protein-forming amino acids [18]. The sum of the numbers of the possible H-bridges of ubiquitin molecule gives 211. According to our measurements, the number of water molecules bound in the first hydration shell by similar binding energies is $n_{ho} = 226$ (3).

The summation of possible H-bridges within ERD10 yields 986. According to our measurements, the number of water molecules bound in the first hydration shell by similar binding energies is $n_{ho} = 514$ (13). The difference between measured and estimated values is unsurprising, especially in light of the good agreement found for ubiquitin. It is reasonable to ask the question whether approximately half of the H-bridges does not link with other water molecules, but realize some other type of bond.

Among the quantities given in Table 1, it is necessary to emphasize the determination of the relative number of bonds that fall into the heterogeneous region ($n_{he}/(n_{he} + n_{ho})$). The result is surprising if one is thinking in terms of a globular protein molecule, because for ubiquitin, the protein is in contact with an additional $n_{he} > 102$ (33) water molecules, which is approximately 36% of all bound water-molecules. The bonds of these water molecules are dominated by water–protein bonds, which are close in energy to the of water–water bonds. In the case of ERD10, the protein surface is in contact with an additional $n_{he} > 2200$ (220) water molecules, which is approximately 73% of all bound water-molecules. In the bonds of the latter water-molecules, water–protein bonds similar to water–water bonds dominate with a substantially wider energy distribution for this initially disordered protein.

It is maybe unnecessary to emphasize that the values we suggest are derived from direct measurements, i.e., they do not rely on assuming any hypothesis or model! They allow to determine the number of first-neighbor water molecules per amino acid (n_{ho}/amino acid), which is $226/76 \cong 3.0$ for UBQ and $514/260 \cong 2.0$ for ERD10. The round value within an error of 1% is surprising, as well as the close match of 2.0 with other values observed for other globular proteins (casein, lysozyme, and BSA, to be published).

Therefore, the measured number of bound water-molecules for ubiquitin is 328 (30). Molecular dynamics simulation estimation from the literature [19] gives a value of 379. For ERD10, the numbers per protein molecule is 2714 (263) (measured) and 881 (estimated by molecular dynamics simulation [19]). The difference between the two proteins and the reverse ratio raise many questions about the nature of protein–water bonds that are still difficult to answer.

3. Materials and Experimental Methods

3.1. Selection of Proteins

We have selected these proteins, as we and others have collected ample evidence for their function depending on their particular structural class, which is folded and intrinsically disordered. Ubiquitin (UBQ) is a small, 76-amino acid globular protein that is found ubiquitously in the cells of all eukaryotic organisms, carrying out basic and indispensable functions in regulating protein function [20]. That is, proteins targeted for degradation are covalently modified by a mono- or poly-ubiquitin chain and are directed for degradation by the 26S proteasome, whereas other proteins labeled with a mono-ubiquitin chain enter regulatory interactions in transcription regulation, for example.

ERD10 [21], on the other hand, is a plant dehydrin that has its cellular protection function strictly linked with structural disorder [22]. Its length is 260 amino acids; it is structurally disordered by a broad range of biophysical techniques, and it functions by protecting the structural integrity of client proteins under the conditions of dehydration and other stresses.

3.2. Expression and Purification of Proteins

Lyophilized ubiquitin (UBQ) was obtained from Sigma Chemical (St. Louis, MO, USA), whereas plant late embryogenesis abundant protein early response to dehydration 10 (ERD10, UniProt P42759) was produced via recombinant expression in *Escherichia coli* BL21(DE3) Star expression strain and purified as described previously [22]. In short, purification was carried out through three chromatographic steps: an ion exchange on HiTrap Q FF at pH 9.5 with gradient elution, followed by two gel-filtration steps on Superdex 200 and Superdex 75 columns, on an AKTA Avant (GE Healthcare, Little Chalfont, UK) FPLC system. The purity of the proteins was checked by SDS-PAGE and was found to be at least 98%.

3.3. Wide-Line NMR

We have reviewed the varieties of radiofrequency excitations applied in wide-line NMR in two book chapters [23,24], and here we only address the simplest excitation protocol that uses a 90° ($\pi/2$) radiofrequency pulse. NMR measurements and data acquisition were performed with a Bruker AVANCE III NMR pulse spectrometer at a frequency of 82.4 MHz with a stability of better than 10^{-6}. The $\pi/2$ pulse was 3–4 μs, and the dead time of the spectrometer was 6–8 μs. The inhomogeneity of the magnetic field was 2 ppm. The accumulated repeat number of the measurements was between 50 and 80.

The temperature was controlled by an open-cycle Janis cryostat with an uncertainty better than 0.5 K. The system was complemented by an adequate NMR head and by a closed sample holder.

4. Conclusions

By reinterpreting our previous results, we have determined the energy distribution of the potential barriers inhibiting the movement of water molecules bound to two protein molecules in aqueous solution. Based on our results, we could deduce quantitative conclusions about the ratios of the globular/ordered and more solvent exposed/disordered regions of the protein molecules and the extent of the latter, as well as the energy relations of the protein–water bonds. We suggest that short range forces (H-bonds) play a dominant role in the formation of the first hydrate shell.

The mapping of the water-binding characteristics of protein molecules is certainly not the only area of the application of wide-line NMR measurements and this novel interpretational procedure. The rapid, non-disruptive measurement and the data interpretation had already opened a novel avenue to study molecular interactions and to determine the moisture content of solid phase samples. In the outline of our previous work [9], we listed some additional possibilities. Three of these are also mentioned here: (i) the possibility to directly demonstrate the interaction between different molecules (e.g., protein and drug), (ii) the possibility of direct, non-destructive measurement of the different bonds between identical molecules, and (iii) the possibility to determine the effect of a standard (often NaCl containing) solvent on the structure and properties of protein molecules.

Author Contributions: Conceptualization, K.T., M.B., and P.T.; Methodology, M.B.; Validation, K.T., M.B. and P.T.; Formal Analysis, M.B.; Investigation K.T. and M.B.; Data Curation, M.B.; Writing-Original Draft, K.T. Writing—Review & Editing K.T., M.B. and P.T.; Visualization M.B.; Supervision, P.T. Funding Acquisition P.T.

Acknowledgments: We would like to thank all colleagues, who contributed to previous studies that resulted in the results on which this work is based.

Abbreviations

FID	free induction decay
IDP	intrinsically disordered protein
MD	melting diagram
DMD	differential form of the melting diagram
DSC	differential scanning calorimetry

References

1. Tompa, K.; Bokor, M.; Han, K.H.; Tompa, P. Hydrogen skeleton mobility and protein architecture. *Intrinsically Disord. Proteins* **2013**, *1*, 77–86. [CrossRef] [PubMed]
2. Grüner, G.; Tompa, K. Molekuláris mozgások vizsgálata szilárdtestekben NMR módszerrel. *Kémiai Közlemények* **1968**, *30*, 315–356.
3. Slichter, C.P. *Principles of Magnetic Resonance*, 3rd ed.; Springer: Berlin/Heidelberg, Germany, 1990; ISBN 978-642080692.
4. Cooke, R.; Kuntz, I.D. The properties of water in biological systems. *Annu. Rev. Biophys. Bioeng.* **1974**, *3*, 95–124. [CrossRef] [PubMed]
5. Tompa, K.; Bánki, P.; Bokor, M.; Kamasa, P.; Lasanda, G.; Tompa, P. Interfacial water at protein sur-faces: Wide-line NMR and DSC characterization of hydration in ubiquitin solutions. *Biophys. J.* **2009**, *96*, 2789–2798. [CrossRef] [PubMed]
6. Csizmók, V.; Bokor, M.; Bánki, P.; Klement, E.; Medzihradszky, K.F.; Friedrich, P.; Tompa, K.; Tompa, P. Primary contact sites in intrinsically unstructured proteins: The case of calpastatin and microtubule-associated protein 2. *Biochemistry* **2005**, *44*, 3955–3964. [CrossRef] [PubMed]
7. Hazy, E.; Bokor, M.; Kalmar, L.; Gelencser, A.; Kamasa, P.; Han, K.H.; Tompa, K.; Tompa, P. Distinct hydration properties of wild-type and familial point mutant A53T of α-synuclein associated with parkinson's disease. *Biophys. J.* **2011**, *101*, 2260–2266. [CrossRef] [PubMed]
8. Tompa, P.; Bánki, P.; Bokor, M.; Kamasa, P.; Kovács, D.; Lasanda, G.; Tompa, K. Protein–water and protein-buffer interactions in the aqueous solution of an intrinsically unstructured plant dehydrin: NMR intensity and DSC aspects. *Biophys. J.* **2006**, *91*, 2243–2249. [CrossRef] [PubMed]
9. Tompa, K.; Bokor, M.; Verebélyi, T.; Tompa, P. Water rotation barriers on protein molecular sur-faces. *Chem. Phys.* **2015**, *448*, 15–25. [CrossRef]
10. Tompa, K.; Bokor, M.; Ágner, D.; Iván, D.; Kovács, D.; Verebélyi, T.; Tompa, P. Hydrogen mobility and protein–water interactions in proteins in the solid state. *ChemPhysChem* **2017**, *18*, 677–682. [CrossRef] [PubMed]
11. Kittel, C.; Kroemer, H. *Thermal Physics*, 2nd ed.; W. H. Freeman and Company: San Francisco, CA, USA, 1980; p. 445, ISBN 9780716710882.
12. Waugh, J.S.; Fedin, E.I. On the determination of hindered rotation barriers in solids. *Leningrad* **1962**, *4*, 2233–2237.
13. *CRC Handbook of Chemistry and Physics*, 97th ed.; Haynes, W.M. (Ed.) CRC Press, Taylor & Francis: Boca Raton, FL, USA, 2016; ISBN 9781498784542.
14. Atkins, P.; de Paula, J. *Physical Chemistry*, 9th ed.; Oxford University Press: Oxford, UK, 2010; ISBN 9781429218122.
15. Chaplin, M. Water's Hydrogen Bond Strength. Available online: https://arxiv.org/abs/0706.1355 (accessed on 19 October 2018).
16. Steiner, T. The hydrogen bond in the solid state. *Angew. Chem. Int. Ed. Engl.* **2002**, *41*, 49–76. [CrossRef]
17. Stone, A. *The Theory of Intermolecular Forces*, 2nd ed.; Oxford University Press: New York, NY, USA, 2013; ISBN 0198558848.
18. Kuntz, I.D. Hydration of macromolecules. III. Hydration of polypeptides. *J. Am. Chem. Soc.* **1971**, *93*, 514–516. [CrossRef]
19. Bizzarri, A.R.; Wang, C.X.; Chen, W.Z.; Cannistraro, S. Hydrogen bond analysis by MD simulation of copper plastocyanin at different hydration levels. *Chem. Phys.* **1995**, *201*, 463–472. [CrossRef]
20. Braten, O.; Livneh, I.; Ziv, T.; Admon, A.; Kehat, I.; Caspi, L.H.; Gonen, H.; Bercovich, B.; Godzik, A.; Jahandideh, S.; et al. Numerous proteins with unique characteristics are degraded by the 26S proteasome following monoubiquitination. *Proc. Natl. Acad. Sci. USA* **2016**, *113*, E4639–E4647. [CrossRef] [PubMed]
21. UniProtKB-P42759. Available online: https://www.uniprot.org/uniprot/P42759 (accessed on 19 October 2018).
22. Kovacs, D.; Kalmar, E.; Torok, Z.; Tompa, P. Chaperone activity of ERD10 and ERD14, two disordered stress-related plant proteins. *Plant Physiol.* **2008**, *147*, 381–390. [CrossRef] [PubMed]

23. Tompa, K.; Bokor, M.; Tompa, P. Hydration of Intrinsically Disordered Proteins from Wide-Line NMR. In *Instrumental Analysis of Intrinsically Disordered Proteins: Assessing Structures and Conformation*; Uversky, V., Longhi, S., Eds.; Wiley & Sons Inc.: Hoboken, NJ, USA, 2010; pp. 345–368, ISBN 9780470343418.
24. Tompa, K.; Bokor, M.; Tompa, P. Wide-Line NMR and Protein Hydration. In *Intrinsically Disordered Protein Analysis*; Uversky, V.N., Dunker, A.K., Eds.; Humana Press: New York, NY, USA, 2012; Volume 895, ISBN 10643745.

9

Hydrodynamic Behavior of the Intrinsically Disordered Potyvirus Protein VPg, of the Translation Initiation Factor eIF4E and of their Binary Complex

Jocelyne Walter *, Amandine Barra, Bénédicte Doublet, Nicolas Céré, Justine Charon † and Thierry Michon *

UMR 1332 Biologie du Fruit et Pathologie, INRA, Université de Bordeaux, CS 20032, 33140 Villenave d'Ornon, France; amandine.barra@inra.fr (A.B.); benedicte.doublet@inra.fr (B.D.); nicolas.cere@hotmail.fr (N.C.); justine.charon@sydney.edu.au (J.C.)

* Correspondence: Jocelyne.walter@inra.fr (J.W.); thierry.michon@inra.fr (T.M.)

† Present address: School of Life & Environmental Sciences, University of Sydney, Sydney, NSW 2006, Australia.

Abstract: Protein intrinsic disorder is involved in many biological processes and good experimental models are valuable to investigate its functions. The potyvirus genome-linked protein, VPg, displays many features of an intrinsically disordered protein. The virus cycle requires the formation of a complex between VPg and eIF4E, one of the host translation initiation factors. An in-depth characterization of the hydrodynamic properties of VPg, eIF4E, and of their binary complex VPg-eIF4E was carried out. Two complementary experimental approaches, size-exclusion chromatography and fluorescence anisotropy, which is more resolving and revealed especially suitable when protein concentration is the limiting factor, allowed to estimate monomers compaction upon complex formation. VPg possesses a high degree of hydration which is in agreement with its classification as a partially folded protein in between a molten and pre-molten globule. The natively disordered first 46 amino acids of eIF4E contribute to modulate the protein hydrodynamic properties. The addition of an N-ter His tag decreased the conformational entropy of this intrinsically disordered region. A comparative study between the two tagged and untagged proteins revealed the His tag contribution to proteins hydrodynamic behavior.

Keywords: intrinsically disordered protein; plant virus; eIF4E; VPg; potyvirus; molten globule; protein-protein interaction; fluorescence anisotropy; protein hydrodynamics

---

1. Introduction

Many biologically functional protein regions do not fold spontaneously. This class of proteins, termed intrinsically disordered proteins (IDP), contains intrinsically disordered regions (IDR) which are devoid of stable secondary and tertiary structures under physiological conditions and rather, exist as dynamic ensembles of inter-converting conformers [1]. Many of these proteins gain a stable 3D structure only when they interact with their target molecules [2]. The ability to exert specific biological functions and to interact with various partners in spite of the lack of a precise 3D scaffold, challenges the classic paradigm according to which specificity can only be achieved through surface complementation between structured and conserved domains. It is now well accepted that intrinsic disorder is involved in a large spectrum of functional properties modulated through multi-partnership interactions with proteins and nucleic acids. With between 7.3% and 77% of residues being disordered, the proteome of viruses on the whole presents the highest variability of intrinsic disorder in the living world [3–5]. The genus *Potyvirus* represents one of the largest and most economically damaging genus of plant-infecting viruses [6]. These viruses possess a single-stranded, polyadenylated, positive-sense genomic RNA which is covalently linked at its 5′ end to a viral protein, the viral protein genome-linked (VPg) [7,8]. The VPgs from several potyviruses, namely lettuce mosaic virus (LMV) [9], potato virus Y (PVY) [10], and Potato virus A (PVA) [11] have been experimentally characterized as intrinsically disordered. The potyviral VPg has been shown to interact with several viral and host factors. It is assumed to be a multifunctional protein involved in essential steps of the virus infectious cycle, translation, replication, and movement [12,13]. The VPg recruits the host eukaryotic translation initiation factor 4E (eIF4E), or its isoform eIF(iso)4E, in an interaction that is crucial for virus infection [14–16]. Mutations in the central region of VPg (residues 80–125) are associated with host resistance breakdown, [17–19]. In the LMV VPg, this central region interacts with eIF4E [20]. This region has been predicted to be an IDR for VPg from twelve potyviral species [9,21]. The study reported here analyzes the contribution of VPg and eIF4E flexible regions to the hydrodynamic properties of the two proteins, either monomeric, or associated in a binary complex.

2. Results and Discussion

2.1. Hydrodynamic Behavior of the Histidine Tagged Forms Assessed by Size Exclusion Chromatography

The secondary and tertiary structures of a protein involve non-covalent interactions. The more the amino-acid residues will interact, the more compact (globular) the protein will be. IDRs will involve fewer interactions between residues and lower compaction. Consequently, the hydration sphere of an intrinsically disordered protein is often larger than what would be expected for globular proteins of a similar molecular weight. A simple way to assess the degree of compactness of a protein in solution is to measure its hydrodynamic radius (R_h). A commonly used method for measuring R_h is size exclusion chromatography, SEC (Figure 1).

When submitted to SEC (Figure 1A), the hydrodynamic behavior of His_6 eIF4E, *M* 28,550 Da (MALDI-TOF spectrometry of the purified recombinant protein), was in agreement with that of a 30 kDa globular protein. R_h values were deduced from experimentally determined apparent molecular weights (M_{app}) (see Section 4). The elution profile of purified recombinant His_6 VPg, *M* 26,250 Da (mass spectrometry), featured two populations, with the major one (69%) suggesting that of a 40 kDa globular protein, and the minor species (27%) averaging 90 kDa. Clearly, His_6 VPg did not behave like a globular protein. The trypsin hydrolysis kinetics of His_6 VPg shows a moderate proteolytic resistance profile comparable to that of α-casein, a disordered protein (Figure S1). A method was developed to classify IDPs according to the relationship between their apparent molecular density (ρ) and their true molecular weight (*M*) [22]. From Figure 1C,D it can be deduced that His_6 VPg shares hydrodynamic features of molten and pre-molten globules. The binary complex His_6 VPg-His_6 eIF4E showed a more complex elution profile (Figure 1A). Its major component (peak 2) displayed a M_{app} of 56kDa, which is close to the weight (*M* 54,772 Da) calculated by adding the two partners molecular weights. SDS-PAGE analysis showed that this elution fraction contained equal amounts of His_6 VPg and His_6 eIF4E indicating a possible compaction of VPg upon its association with eIF4E. The major component eluted under peak 1, M_{app} = 110 kDa, corresponds to oligomeric forms of His_6 VPg (Figure 1B). The VPg propensity to aggregate was previously described [10].

Using PONDR-VLXT, intrinsic disorder was predicted for the tagged and untagged proteins. The His_6 tag potentially brings disorder to the N-terminus of all proteins. After the His_6 tag removal, the first 46 amino acids segment of the native eIF4E was still predicted as unstructured. This prediction was validated by previously reported structural data showing that the first 40 amino acids of eIF4E were intrinsically disordered and fold upon binding with eIF4G [23]. In addition, upon His_6 tag removal, the first 25 amino acids of VPg were predicted as disordered (Figure S2). This region was recently reported as being a conformational switch [24]. Therefore, the contribution of these regions, predicted as disordered, to the hydrodynamic properties of the proteins was assessed.

The hydrodynamic behavior of VPg, eIF4E and their binary complex was deduced from fluorescence anisotropy measurements.

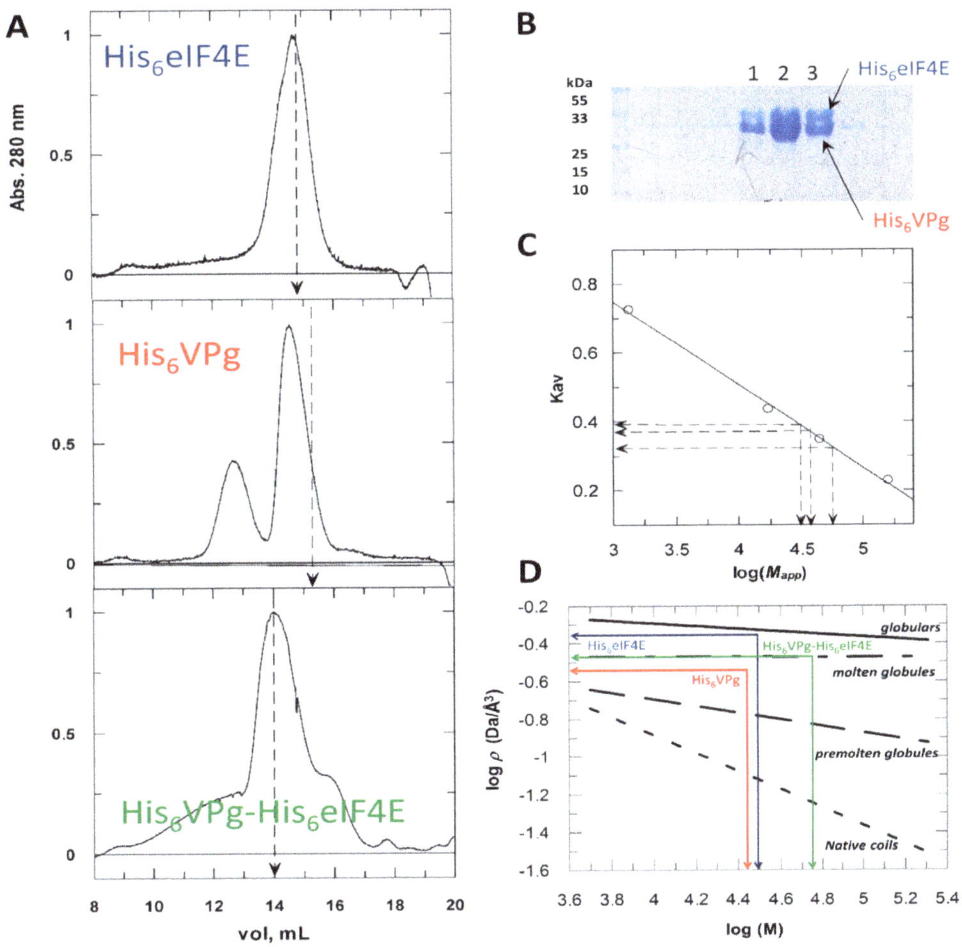

Figure 1. Size exclusion chromatography of His_6 tagged VPg, His_6 tagged 4E (eIF4E) and His_6 VPg-His_6 eIF4E, their binary complex. (**A**) Separated runs of the purified monomers (1 mL of 0.7–1 mg/mL) were performed. Vertical dashed lines refer to the elution volumes expected for globular proteins of 28,549 Da (upper panel) and 26,137 Da (middle panel), the molecular weights of recombinant His_6 eIF4E and His_6 VPg respectively (MALDI-TOF spectrometry determinations). The His_6 VPg-His_6 eIF4E binary complex was pre-formed by mixing His_6 eIF4E (0.7 mg/mL with an excess of His_6 VPg (1.2 mg/mL) and loaded up to the column (lower panel). A mass of 54,772 Da was estimated for the binary complex (summing His_6 VPg and His_6 eIF4E molecular weights) vertical dashed line. For comparison, absorbance values were standardized to the maximum value of each peak. (**B**) Distribution of the various molecular species through the size exclusion chromatography (SEC) separation of a His_6 VPg-His_6 eIF4E mix. Upon elution, fractions 1. 2 and 3 were recovered and submitted to SDS-PAGE analysis. (**C**) Determination of the three molecular species apparent molecular weights (M_{app}) deduced from standard calibration with a set of known globular proteins. K_{av} is a mean value determined from at least three independent SEC runs. (**D**) Apparent molecular densities (ρ) of the three molecular species were deduced from their experimentally determined hydrodynamic radius (R_h) values (see material and methods). For each protein, the intersection between logρ and log(M), M being the true molecular mass, allows to deduce the conformational families to which they belong to.

Because SEC leads to proteins diluting and to complexes partly dissociating during the chromatography process, it required substantial amount of proteins at concentration above 0.5 mg/mL. We experienced difficulties to obtain isolated untagged VPg and eIF4E at the concentrations suited for SEC experiments. Indeed, in vitro enzymatic tag cleavage resulted in a mixture of molecular species, which after separation (0.1–0.2 mg/mL), were not concentrated enough for SEC. Consequently, hydrodynamic parameters were deduced from fluorescence anisotropy measurements, which return a more detailed analysis of hydrodynamic behavior and can be operated at much lower concentrations. Anisotropy measurements give access to the rotational correlation time (θ) of the proteins. This parameter is strongly related to their hydrodynamic properties, as it depends on the protein shape and it is linked to the effective solvent shell accompanying the protein rotational diffusion. For that purpose, a fluorescent probe, *N*-acetyl-*N′*-(5-sulfo-1-naphtyl)ethylenediamine (AEDANS) was linked to the VPg single cysteine with an efficiency of 0.9 AEDANS moiety per VPg. In another set of experiments, AEDANS was also coupled to His_6 eIF4E, eIF4E and $eIF4E^{\Delta 1-46}$ to evaluate the first 1–46 disordered residues contribution to eIF4E compaction and also, more generally, the effect of the His_6 tag on the compaction of monomeric forms. There are four cysteine residues within lettuce eIF4E, among which two are strictly conserved in plant orthologues. The modification resulted in a mean of 1.7 AEDANS moieties per molecule. The addition of the fluorophore did not alter the binding properties of His_6 eIF4E, eIF4E, and $eIF4E^{\Delta 1-46}$ to VPg (Figure S3). This result was not surprising as wheat eIF4E, either reduced, oxidized, or with a cysteine-to-serine mutation do not undergo structural changes and are functional, all binding m7GTP in a similar and labile manner [25]. In addition, in the structure of pea eIF4E, the two sulfur atoms are in close proximity but are clearly not bridged [26].

2.2. *His Tagging Modulates Proteins Hydrodynamic Parameters*

As expected, upon addition of His_6 eIF4E, the fluorescence anisotropy of His_6 VPg*, increased proportionally to the amount of His_6 VPg*-His_6 eIF4E complex formed. It reached a plateau value indicating a saturation, Figure 2A inset.

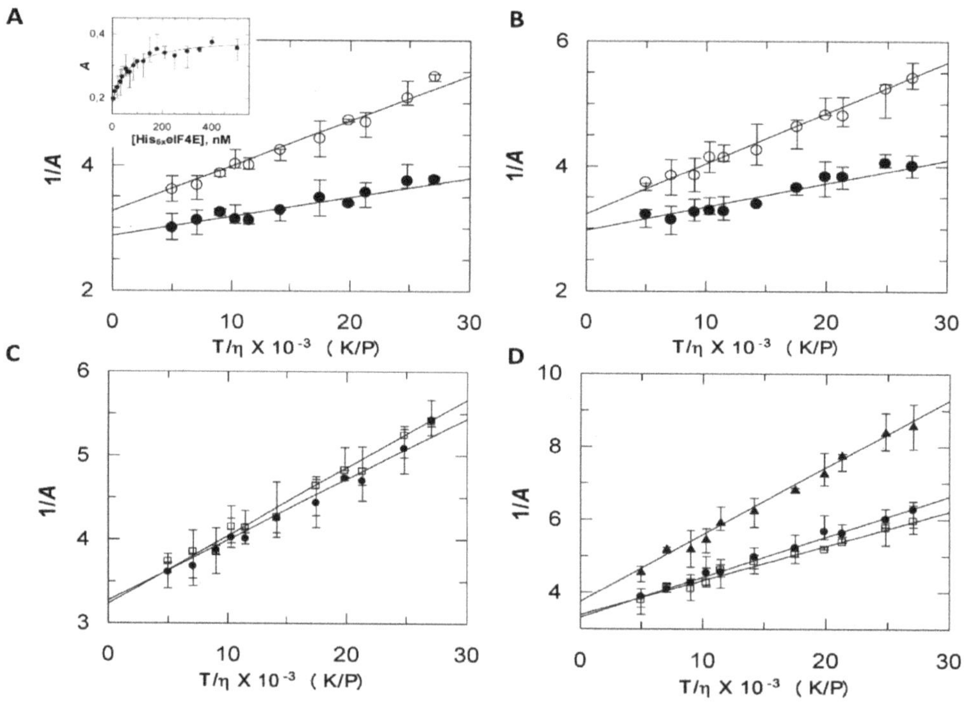

Figure 2. Fluorescence anisotropy of the various molecular forms of eIF4E and VPg. (**A**) The fluorescence anisotropy of His_6 VPg* (300 nM, open circles) and His_6 VPg*-His_6 eIF4E complex (mix of 300 nM His_6 VPg* and 2 μM His_6 eIF4E filled circles) was recorded as a function of the viscosity increase at 25 °C. The reciprocal of the emitted light anisotropy (Perrin's plot), $1/A$, is plotted as a function of T/η. V, the apparent molecular volume of the proteins and their complexes and A_0, the fundamental anisotropy were obtained respectively from the slope and intercept at infinite viscosity. Inset, the fluorescence anisotropy increase upon association of His_6 eIF4E with His_6 VPg* (300 nM). (**B**) Perrin's plots of untagged VPg* (open circles) and VPg*-eIF4E complex (filled circles) in the same conditions as (**A**). (**C**) Perrin's plots of His_6 VPg* (open squares), VPg* (filled circles) and D. His_6 eIF4E* (filled circles), eIF4E* (open squared) and $eIF4E^{\Delta 1-46}$* (filled triangles). Experimental conditions were the same as in B. All measurements were obtained using VPg from lettuce mosaic virus (LMV) AF199 strain.

The modification by the probe did not significantly change the binding strength as comparable dissociation constants (K_D) values were found for probed and unprobed proteins. A K_D value of 63 nM could be extracted from the data, in agreement with intrinsic fluorescence measurements (Figure S3). The hydrodynamic molar volume (V) and the rotational correlation time (θ_{exp}) of the various molecular species were determined from steady state fluorescence anisotropy measurements as described in the experimental procedure section.

2.3. Discrepancies between SEC and Fluorescence Anisotropy Suggest a Contribution of Tags in Proteins Hydrodynamic Behavior

The parameters V and θ_{exp} were derived for the various monomeric species (Figure 2C,D, and Table 1). The more the ratio $\theta_{exp}/\theta_{calc}$ differs from unit, the more asymmetric the protein is. The $\theta_{exp}/\theta_{calc}$ ratio of His_6 VPg (2.41) and His_6 eIF4E (1.62) indicated that their shape differed from a sphere. In addition, although the His_6 VPg molecular weight was 4295 Da less than His_6 eIF4E, its V value was 1.37 times larger, indicating that it was significantly less compact, a pre-molten globule feature. The untagged form of VPg displayed an expected decrease of its hydrodynamic molar volume with respect to its tagged form. Interestingly, the eIF4E* untagged form showed a higher V value (46.5 L/mol) and hence, a significant decrease in compaction compared to His_6 eIF4E* (42.9 L/mol). This could be due to interactions between the His_6 tag and the eIF4E N-ter IDR [23], an effect previously discussed [27]. Because of the intrinsic conformational entropy and the positive charges cluster of the disordered His_6 tag, its interactions with other parts of the protein are more likely to occur. This could account for the more compact hydrodynamic behavior of His_6 eIF4E. For most 3D structures solved from tagged proteins, the peptide tag is too disordered to be resolved. However, it is worth mentioning that, in most of the cases, comparisons of tagged with corresponding untagged structures determined from X ray diffraction data revealed only minor structural differences of the type that might be observed when comparing two identical sequences solved in different space groups [28]. This tends to show that these purification tags generally had no significant effect on the structure of the native protein. However, the importance of the proteins hydrodynamics properties cannot be understated as it accounts for the polypeptide chain dynamics, which drives most of biological functions. This is exemplified through the richness of the functional interpretations provided by NMR data related to disordered segments [29].

A comparison of V values between tagged and untagged species allows for the discussing of tag contribution to compaction. A decrease of the Perrin's plot slope was observed for His_6 VPg*-His_6 eIF4E binary complex when compared to the slope of His_6 VPg* (Figure 2A). Upon analysis of this data, it was shown that the hydrodynamic molar volume V linked to the labeled VPg enlarged from 58 L/mol to 121 L/mol, (Table 1). The later, attributed to the complex, was 20% larger than could be expected by adding the V value of His_6 VPg* and His_6 eIF4E monomers.

Hence, the anisotropy approach revealed a possible contribution of the two flexible His tags conformational entropy to the proteins hydrodynamic behavior. By contrast, the V value experimentally determined for the native untagged binary complex was close to that of the sum of the monomer values (Table 1). Compaction can be estimated by ρ, the molecular density value. Interestingly, the compaction within the binary complex suggested by SEC was not observed by anisotropy. In the SEC experiments, elution is concomitant to dilution, and hence to a modification of the species distribution. This could contribute to average the observed apparent molecular weight of each species. As stated in the experimental procedures section, anisotropy measurements on binary complexes as a function of various viscosity values were performed after the molecular species in presence have reached an equilibrium. The ρ value is directly obtained from V the experimental molecular volume, and thus, the anisotropy approach provides a likely more accurate way to estimate compaction than SEC does through the use of M_{app} in the empirical linear Equation (4). More generally, ρ values provided by anisotropy were 1.5 time higher than those deduced from the SEC (Table 1).

Table 1. Physical parameters of the various molecular forms of VPg, eIF4E and of their binary complexes.

Parameter	Units	HV	V	HE	E	EΔ(1–46)	HVHE	VE
MW	g/mol	26,222 †	21,781 †	28,550 †	26,076 †	21,246 †	54,772 ‡	47,857 ‡
SEC								
V_R	mL	14.4	nd	14.8	nd	nd	13.7	nd
M_{app}	kg/mol	38.0 ± 1.0	nd	30.2 ± 0.8	nd	nd	56.2 ± 1.5	nd
R_h	Å	28.1 ± 0.7	nd	26.4 ± 1.1	nd	nd	33.0 ± 0.9	nd
ρ	g/mol·Å ^{3}	0.28	nd	0.40	nd	nd	0.38	nd
Conformation		MG/PMG		F/MG			MG	
1 **Fluorescence Anisotropy**								
V_{app}	L/mol	58.7 ± 1.8	51.6 ± 0.2	42.9 ± 0.8	46.5 ± 1.0	24.3 ± 0.5	121.3 ± 3.2	102.4 ± 0.9
θ_{exp}	s ($\times 10^9$)	23.7	20.9	17.3	18.8	9.8	49 [3]	41.4
θ_{calc}	s ($\times 10^9$)	9.8	8.9	10.7	9.8	7.9	20.6	18.0
$(\theta_{exp})/(\theta_{calc})$		2.41	2.55	1.62	1.92	1.23	2.38	2.30
h	g water/g	1.44	1.57	0.72	1	0.38 (0.53) [2]	1.41	1.35
R_h	Å	24	23	22	22	18	31	29
ρ	g/mol·Å ^{3}	0.45	0.42	0.67	0.56	0.88	0.45	0.47

HV, His_6 VPg; HE, His_6 eIF4E; V, His_6 VPg; E, eIF4E; EΔ(1–46), 1–46 residues deleted eIF4E; HVHE, His_6 VPg-His_6 eIF4E binary complex; VE, VPg-eIF4E binary complex. MW (molecular weight), V_R Retention volume), M_{app} (apparent molecular mass), R_h (hydrodynamic radius), ρ (apparent molecular density), V_{app} (apparent molecular volume), θ_{exp} (experimental rotational correlation time), θ_{calc} (calculated rotational correlation time), h (estimated hydration degree). [1] Determined for η = 0.01 P (water) at 298 K; τ, AEDANS fluorescence lifetime: 15.6 ns; [2] Calculated from pea eIF4E structure (2WMC) using the Hydropro software (http://leonardo.inf.um.es/macromol/programs/hydropro/hydropro.htm); [3] value obtained from His_6 VPg*-His_6 eIF4E binary complex; † Mass spectrometry (MALDI-TOF); ‡ Calculated from amino acid sequence.

2.4. His_6 VPg and His_6 VPg*-His_6 eIF4E Binary Complex Shapes Differ from the Globular State*

The deduced experimental rotational correlation time, θ_{exp}, increased from 23.7 ns for His_6 VPg* to 49 ns for the His_6 VPg*-His_6 eIF4E binary complex. The θ_{calc} values for hydrated rigid spheres of 26 kDa and 54.7 kDa were 9.8 ns and 20.6 ns (His_6 VPg* and His_6 VPg*-His_6 eIF4E respectively), indicating that His_6 VPg* and His_6 VPg*-His_6 eIF4E shapes differ from the compact globular state. On an indicative basis, the compact EΔ(1–46) [26] displayed an θ_{exp} value (7.9 ns) which was close to the θ_{calc} value (9.8 ns) calculated for a globular protein of comparable molecular weight. The $\theta_{exp}/\theta_{calc}$ ratio for the His_6 VPg*-His_6 eIF4E complex and His_6 VPg* were close, suggesting that the complex formation has no effect on the probe segmental motion within the VPg [30].

2.5. Compaction and Hydration are Experimentally Correlated

Because of their non-compact structures and more solvent-accessible surface area, disordered regions tend to display a higher hydration water density as compared to more ordered regions [31]. The degree of hydration (h) of each species was estimated from θ_{exp} (see experimental section). For compact globular proteins, this value is usually in between 0.2 and 0.4 g water/g protein. Our data suggests that His_6 VPg and VPg forms are more solvated than standard folded proteins (1.44 and 1.57 g/g respectively). This is in agreement with previous hydrodynamic experiments on IDPs. For instance, in the absence of calcium, the adenylate cyclase toxin calcium binding domain is intrinsically disordered and displays a high hydration propensity [32]. Moreover, molecular dynamics simulations show that partially disordered proteins like VPg have a higher capacity to bind hydration water as compared to globular proteins [33]. Interestingly, His_6 eIF4E and eIF4E, which both include a long disordered N-terminal region, have a rather high hydration degree (0.72 and 1 g/g respectively) although the presence of the His_6 tag seems to slightly decrease the hydration degree in accordance to its compacting effect discussed above. Conversely, eIF4E$^{\Delta 1-46}$, which corresponds to the globular part of eIF4E, possesses a hydration value expected for compact proteins (0.38 g/g). Using the PDB coordinates of eIF4E$^{\Delta 1-46}$ from pea (PDB file 2WMC), we calculated the protein rotational diffusion coefficient (D) [34] from Equation (12). From this value (1.49×10^7 s^{-1}) the rotational coefficient time was deduced and, in turn a h value of 0.53 g/g, which is in fairly good agreement with our estimation

(Table 1). Finally, the binary complex formation is not associated with a significant modification of hydration.

3. Conclusions

As opposed to fully disordered proteins, which display a random coil state, VPg possesses a significant content in secondary structure. These more compact intermediates have led to the concept of molten globule (MG) and the somewhat less compact pre-molten globule (PMG) states. A MG is characterized by a large internal flexibility of its side chains and backbone, with a R_h 1.5–2.0 times larger than that of globular proteins. As they are usually distributed in solutions between a limited number of conformers, these proteins prove to be more complex to analyze than fully disordered polypeptides. Modern NMR approaches provide an excellent way to study such proteins [35,36]. However the concentrations required [37], usually from 2 to 10 mg/mL, are far beyond what can be stabilized in solution in the present case. Because it enables the development of low-resolution structural models, taking into account the contribution of intrinsically disordered regions, a hydrodynamic analysis can provide useful data. Hydrodynamic parameters are usually assessed using SEC, AUC (analytical ultracentrifugation), and DLS (dynamic light scattering). However, these techniques also require protein concentrations above 1 mg/mL. We propose an elegant way to deal with especially difficult proteins. Although less used, fluorescence anisotropy can prove quite resolutive and especially suitable when protein concentration is the limiting factor. One can argue that this method gives access to the rotational mobility of the reporter fluorophore and not of the macromolecule itself. However, if, as it is the case here, the probe motion displays a low degree of freedom within the macromolecule, the extracted parameters reflect the hydrodynamic properties of the macromolecule well. The data obtained by SEC and fluorescence spectroscopy were in good agreement. Because analytical ultracentrifugation (AUC) is based on equilibrium and non-equilibrium thermodynamics, it is referred as a gold standard for characterizing the hydrodynamic properties. The expected sedimentation coefficients ($s_{20,w}$) for the LMV VPg and lettuce eIF4E were derived from their experimentally measured diffusion coefficients (see Equation (13) in the experimental section). They were in agreement with values reported for their homologous counterparts PVY VPg (46.7% identity, 76.4% similar with LMV VPg) and human eIF4E (43.2% identity, 67.6% similarity with lettuce eIF4E), (Table 2). A fine analysis of potyviral VPg conformers distribution on the basis of their sedimentation properties was recently reported [24]. A $s_{20,w}$ value of 3.2 Svedberg was determined for the major VPg molecular species (70%). [10]. This suggests that most of the VPg was present as a dimer. Indeed, this value is in accordance with the value obtained for non-reduced VPg from PVY [10], Table 2.

Table 2. VPgs and eIF4E sedimentation coefficients in water at 20 °C.

Molecular Species	MM, kDa *	$s_{20,w}$, s ($\times 10^{13}$)	Ref.	Method
VPg (LMV)-R	21.8	1.8	This study	Fluorescence
VPg (PVY)-R	22	1.7	[10]	UAC
VPg (PVY)	22	3.0	[10]	UAC
VPg (PVBV)	21.8	3.2	[24]	UAC
eIF4E (lettuce)	26.1	2.2	This study	Fluorescence
eIF4E (human)	24.3	2.03	[38]	UAC

LMV (lettuce mosaic virus), PVY (potato virus Y), PVBV (Pepper vein banding virus). * theoretical molecular weight. VPg (LMV)-R and VPg (PVY)-R, VPg under reduced conditions.

4. Materials and Methods

4.1. Protein Preparation

The gene coding for eIF4E initiation factor from the lettuce (Lactuca sativa cultivar Salinas GenBank AF 530162) and its derived molecular species were cloned into the vector pENTR/D-TOPO®

(Invitrogen, Carlsbad, CA, USA). They were transferred into pDEST™17 using the Gateway® recombinant Technology to allow production of N-terminal fusions with an hexahistidine tag (Invitrogen). In addition, full length eIF4E and eIF4E $^{(\Delta 1–46)}$ were cloned into the vector pENTR/SD/D-TOPO® (Invitrogen) and transferred into the Gateway pDEST™14 expression vector according to manufacturer's instructions to obtain untagged full length eIF4E and eIF4E$^{(\Delta 1–46)}$. The constructs were introduced into *E. coli* (BL21-AI strain), expression and purification of the His-tagged proteins were performed on ion metal affinity chromatography followed by m7GTP-Sepharose 4B (GE Healthcare, Amersham, UK) as previously described [39]. Untagged proteins were obtained by one step affinity purification on m7GTP sepharose 4B. The Lettuce mosaic virus VPg (isolate AF199GenBank AJ2 78854) coding sequence was cloned into the pTrcHis C expression vector downstream from an hexahistidine tag (Invitrogen). The vector contains a specific enterokinase cleavage site in frame with the protein for proteolytic tag removal. The protein was produced and purified as previously reported [39] except that the final monoQ chromatographic step was replaced by a size exclusion chromatography on a Superdex 75 HR 10/30 column (GE Healthcare) in 20 mM Hepes pH 8, 300 mM NaCl, and 2 mM DTT. For His-tag removal, this last step was omitted and the protein was diluted twice in the same buffer with reduced ionic strength (150 mM NaCl) containing 1 mM $CaCl_2$, 0.1% Tween-20. His tagged enterokinase (0.02 mg for 0.2 mg VPg) was added and the protein mix was dialyzed overnight at 4 °C against the same buffer. The protease and uncleaved tagged VPg were subsequently trapped on a NiNTA resin and the pure free VPg form was recovered in the flow through (30–40% yield).

Protein labelling with *N*-(iodoacetyl)-*N*′-(5-sulfo-1-naphtyl)ethylenediamine, (IAEDANS).

Cysteine residues were reduced before labelling. The VPg or eIF4E solutions (0.2 mg/mL) were dialyzed overnight in 25 mM HEPES, pH 7.5, 6 mM βMe, 2 mM AcNa, 5 mM EDTA, 25 mM DTT, NaCl 0.3 M at 4 °C. The protein solutions (1 mL) were loaded on a 5 mL G25 column equilibrated in the coupling buffer (25 mM HEPES, pH 7.5, NaCl 0.3 M). A 10 times molar excess of IAEDANS was added in the same buffer and the mix was incubated in the dark at 25 °C for 2 h. The reaction was stopped by addition of DTT (25 mM final concentration). A buffer exchange was performed over a G25 equilibrated in 25 mM HEPES, pH 7.5, 150 mM NaCl. About 1 and 2 AEDANS molecules were bound per VPg and eIF4E respectively.

4.2. *Size Exclusion Chromatography*

A superose 12 HR 10/30 column (GE Healthcare) was calibrated with separate 500 μL injections of the following native globular proteins: γ-globulin (bovine, 158 kDa), Ovalbumin (Chiken) 44 kDa, myoglobin (horse, 17 kDa and vitamin B_{12} (1.35 kDa). Excluded volume (V_0, 8.74 mL) was determined with blue dextran. The volume of the column, V_t was 24 mL. Chromatography conditions were 20 mM HEPES/KOH, 0.4M KCl and 1.4 mM 2-Mercaptoethanol, flowrate 0.5 mL/min.

4.3. *Hydrodynamic Radius Measure*

The Stokes radius, also termed R_h, is the radius of a hard sphere that diffuses at the same rate as the protein. A commonly used method for measuring R_h is SEC. The size exclusion column is calibrated using the elution volume (V_e) of standard folded proteins (i.e., Globular) of known molecular weight. The apparent molecular weight of the protein of interest is then deduced from V_e. R_h is determined as the R_h expected for a globular protein of that apparent molecular weight, for which simple relations exist [22,40].

The retention factor K_{av} of the proteins was determined as follows:

$$K_{av} = \frac{V_e - V_0}{V_t - V_0} \tag{1}$$

There was a linear relationship between K_{av} and logM_{app} (Figure 1C):

$$K_{av} = -0.24 \log M_{app} + 1.47 \tag{2}$$

$$M_{app} = 10^{\frac{1.47-K_{av}}{0.24}} \quad (3)$$

It was assumed that the protein considered has the hydrodynamic behavior of its equivalent rigid sphere of R_h and an M_{app}. A linear relationship exists between $\log M_{app}$ and $\log R_h$ which, knowing the SEC derived apparent molecular weight of a protein, allows the determination of its R_h [22]. For a globular protein, the expression is:

$$\log(R_h) = -0.2 + 0.36 \log(M_{app}) \quad (4)$$

4.4. Structural Feature Estimation

The apparent molecular density or compaction (ρ) of a globular protein is:

$$\rho = \frac{M}{4/3\pi R_h^3} \quad (5)$$

with M, the molecular weight calculated from the protein amino acid composition. A plot of log(ρ) vs. log(M) allowed us to estimate the structural family; ordered globular, molten globule, pre-molten globule, or native coil-like protein (Figure 1D). Straight lines that define the different groups of conformational states were calculated from [41].

Fluorescence measurements. All steady state fluorescence acquisitions were obtained at 25 °C in 20 mM HEPES pH 7.5, 0.25 mM NaCl and 1 mM DTT using a SAFAS Xenius spectrofluorimeter (Monaco) equipped with a Peltier temperature controller. For optical characteristics of the instrument, see [39]. Affinity constants were deduced from steady state eIF4E tryptophan intrinsic fluorescence decrease upon titration by VPg as previously described [39]. In order to ensure that the system reached an equilibrium before measurements, a thorough mixing (gentle back and forth syringe flushing) of the various molecular species, was followed by a 10 min incubation. Then, an average value was collected during another 10 min, both for the acquisition of eIF4E steady state intrinsic fluorescence and anisotropy measurements.

Fluorescence anisotropy measurement. The fluorescent probe bound to the VPg was chosen so that τ its fluorescent lifetime be of the order of magnitude of θ, the rotational correlation time of the VPg in solution. A, the measured anisotropy is defined as:

$$\frac{A_0}{A} = 1 + \frac{\tau}{\theta} \quad (6)$$

A_0 is the fundamental anisotropy observed in the absence of other depolarizing processes such as rotational diffusion or energy transfer. If $\theta \gg \tau$ then the measured anisotropy is equal to A_0 (infinite viscosity, no motion of the macromolecule). If $\theta \ll \tau$ then the anisotropy is zero. For the AEDANS group, a fluorescence lifetime of 15.6 ns was determined by phase-modulation fluorimetry on a SLM 4800 fluorimeter. This value is in the range of θ values for proteins (15–70 kDa).

The anisotropy of the AEDANS labelled VPg either free or associated with eIF4E was measured at 25 °C in solutions of various viscosity (η). The dependency of the anisotropy on viscosity is given by the Perrin equation:

$$\frac{1}{A} = \frac{1}{A_0} + \alpha\frac{T}{\eta} \quad (7)$$

with

$$\alpha = \frac{\tau R}{A_0 V} \quad (8)$$

Plotting $1/A$ versus T/η gives usually a straight line. The viscosity was experimentally increased by addition of sucrose in the buffer. V the hydrodynamic molar volume was determined from α, the slope value:

$$V = \frac{\tau R}{A_0 \alpha} \quad (9)$$

The experimental rotational correlation time θ_{exp} was deduced from V:

$$\theta_{exp} = \frac{\eta V}{RT} \tag{10}$$

The rotational correlation time of an equivalent rigid sphere of the same molecular weight M was calculated as follows:

$$\theta_{calc} = \frac{\eta M}{RT}(\overline{v} + h) \tag{11}$$

where $\overline{v}$ is the protein partial specific volume (usually 0.73 cm^3/g) and h is the degree of hydration (g H_2O per g of protein; usually $0.2 < h < 0.4$); $R = 8.31 \times 10^7$ erg $mol^{-1} \cdot K^{-1}$. From Equation (13), replacing the calculated rotational time by rotational correlation times experimentally determined for each molecular species, leads to an estimation of h their degree of hydration (Table 1). Alternatively, knowing D, the rotational diffusion coefficient of the protein, its rotational correlation time can be obtained:

$$\theta = \frac{1}{6D} \tag{12}$$

The expected sedimentation coefficient can be deduced as follows:

$$s = \frac{MD(1 - \rho_{20}\overline{v})}{RT} \tag{13}$$

with M, molecular weight, ρ_{20} solvent density (water).

Author Contributions: T.M. and J.W. conceived and designed the experiments. A.B., N.C., B.D., J.C., J.W., and T.M. performed the experiments. T.M. and J.W. analyzed the data. T.M. wrote the manuscript. T.M. and J.W. discussed the results and commented on the manuscript.

Acknowledgments: We thank Sonia Longhi (CNRS-Marseille, France) for fruitful discussions. We are indebted to Stephane Claverol (centre de genomique fonctionnelle, Bordeaux) for mass spectrometry analysis.

Abbreviations

LMV	Lettuce mosaic virus
PVY	Potato virus Y
TEV	Tobacco etch virus
TuMV	Turnip mosaic virus
IAEDANS	*N*-(iodoacetyl)-*N′*-(5-sulfo-1-naphtyl)ethylenediamine
AEDANS	*N*-(acetyl)-*N′*-(5-sulfo-1-naphtyl)ethylenediamine
His_6 eIF4E	N-ter hexahistidine tagged lettuce eIF4E
His_6 VPg	N-ter hexahistidine tagged LMV VPg
$eIF4E^{\Delta 1-46}$	untagged lettuce eIF4E deleted from its first 46 N-ter amino acid
His_6 VPg*	AEDANS labelled N-ter hexahistidine tagged LMV VPg
His_6 eIF4E	AEDANS labelled N-ter hexahistidine tagged lettuce eIF4E
eIF4E*	AEDANS labelled untagged lettuce eIF4E
$eIF4E^{\Delta 1-46*}$	AEDANS labelled untagged lettuce eIF4E deleted from its first 46 N-ter amino acids
[His_6 VPg*-His_6 eIF4E]	AEDANS labelled binary complex

References

1. Boehr, D.D.; Wright, P.E. Biochemistry. How do proteins interact? *Science* **2018**, *320*, 1429–1430. [CrossRef]
2. Habchi, J.; Tompa, P.; Longhi, S.; Uversky, V.N. Introducing Protein Intrinsic Disorder. *Chem. Rev.* **2014**, *114*, 6561–6588. [CrossRef]
3. Xue, B.; Blocquel, D.; Habchi, J.; Uversky, A.V.; Kurgan, L.; Uversky, V.N.; Longhi, S. Structural Disorder in Viral Proteins. *Chem. Rev.* **2014**, *114*, 6880–6911. [CrossRef]
4. Pushker, R.; Mooney, C.; Davey, N.E.; Jacqué, J.-M.; Shields, D.C. Marked Variability in the Extent of Protein Disorder within and between Viral Families. *PLoS ONE* **2013**, *8*, e60724. [CrossRef]
5. Xue, B.; Dunker, A.K.; Uversky, V.N. Orderly order in protein intrinsic disorder distribution: Disorder in 3500 proteomes from viruses and the three domains of life. *J. Biomol. Struct. Dyn.* **2012**, *30*, 137–149. [CrossRef]
6. Adams, M.; Zerbini, F.; French, R.; Rabenstein, F.; Stenger, D.; Valkonen, J. *Potyviridae. Virus Taxonomy, 9th Report of the International Committee for Taxonomy of Viruses*; Elsevier Academic Press: San Diego, CA, USA, 2011.
7. Murphy, J.F.; Klein, P.G.; Hunt, A.G.; Shaw, J.G. Replacement of the tyrosine residue that links a potyviral VPg to the viral RNA is lethal. *Virology* **1996**, *220*, 535–538. [CrossRef]
8. Murphy, J.F.; Rychlik, W.; Rhoads, R.E.; Hunt, A.G.; Shaw, J.G. A tyrosine residue in the small nuclear inclusion protein of tobacco vein mottling virus links the VPg to the viral RNA. *J. Virol.* **1991**, *65*, 511–513.
9. Hebrard, E.; Bessin, Y.; Michon, T.; Longhi, S.; Uversky, V.N.; Delalande, F.; Van Dorsselaer, A.; Romero, P.; Walter, J.; Declerck, N.; et al. Intrinsic disorder in Viral Proteins Genome-Linked: Experimental and predictive analyses. *Virol. J.* **2009**, *6*, 23–36.
10. Grzela, R.; Szolajska, E.; Ebel, C.; Madern, D.; Favier, A.; Wojtal, I.; Zagorski, W.; Chroboczek, J. Virulence Factor of Potato Virus Y, Genome-attached Terminal Protein VPg, Is a Highly Disordered Protein. *J. Biol. Chem.* **2008**, *283*, 213–221. [CrossRef] [PubMed]
11. Rantalainen, K.I.; Uversky, V.N.; Permi, P.; Kalkkinen, N.; Dunker, A.K.; Makinen, K. Potato virus A genome-linked protein VPg is an intrinsically disordered molten globule-like protein with a hydrophobic core. *Virology* **2008**, *377*, 280–288. [CrossRef]
12. Jiang, J.; Laliberte, J.F. The genome-linked protein VPg of plant viruses-a protein with many partners. *Curr. Opin. Virol.* **2011**, *1*, 347–354. [CrossRef] [PubMed]
13. Martínez, F.; Rodrigo, G.; Aragonés, V.; Ruiz, M.; Lodewijk, I.; Fernández, U.; Elena, S.F.; Daròs, J.-A. Interaction network of tobacco etch potyvirus NIa protein with the host proteome during infection. *BMC Genom.* **2016**, *17*, 87. [CrossRef]
14. Charron, C.; Nicolaï, M.; Gallois, J.L.; Robaglia, C.; Moury, B.; Palloix, A.; Caranta, C. Natural variation and functional analyses provide evidence for co-evolution between plant eIF4E and potyviral VPg. *Plant J.* **2008**, *54*, 56–68. [CrossRef] [PubMed]
15. Leonard, S.; Plante, D.; Wittmann, S.; Daigneault, N.; Fortin, M.G.; Laliberte, J.F. Complex formation between potyvirus VPg and translation eukaryotic initiation factor 4E correlates with virus infectivity. *J. Virol.* **2000**, *74*, 7730–7737. [CrossRef]
16. Robaglia, C.; Caranta, C. Translation initiation factors: A weak link in plant RNA virus infection. *Trends Plant Sci.* **2006**, *11*, 40–45. [CrossRef]
17. Moury, B.; Charron, C.; Janzac, B.; Simon, V.; Gallois, J.L.; Palloix, A.; Caranta, C. Evolution of plant eukaryotic initiation factor 4E (eIF4E) and potyvirus genome-linked protein (VPg): A game of mirrors impacting resistance spectrum and durability. *Infect. Genet. Evol.* **2014**, *27*, 472–480. [CrossRef] [PubMed]
18. Ayme, V.; Souche, S.; Caranta, C.; Jacquemond, M.; Chadoeuf, J.; Palloix, A.; Moury, B. Different mutations in the genome-linked protein VPg of potato virus Y confer virulence on the pvr2(3) resistance in pepper. *Mol. Plant Microbe Interact.* **2006**, *19*, 557–563. [CrossRef]
19. Ayme, V.; Petit-Pierre, J.; Souche, S.; Palloix, A.; Moury, B. Molecular dissection of the potato virus Y VPg virulence factor reveals complex adaptations to the pvr2 resistance allelic series in pepper. *J. Gen. Virol.* **2007**, *88*, 1594–1601. [CrossRef]
20. Roudet-Tavert, G.; Michon, T.; Walter, J.; Delaunay, T.; Redondo, E.; Le Gall, O. Central domain of a potyvirus VPg is involved in the interaction with the host translation initiation factor eIF4E and the viral protein HcPro. *J. Gen. Virol.* **2007**, *88*, 1029–1033. [CrossRef]

21. Charon, J.; Theil, S.; Nicaise, V.; Michon, T. Protein intrinsic disorder within the Potyvirus genus: From proteome-wide analysis to functional annotation. *Mol. BioSyst.* **2016**, *12*, 634–652. [CrossRef]
22. Uversky, V.N. What does it mean to be natively unfolded? *Eur. J. Biochem.* **2002**, *269*, 2–12. [CrossRef]
23. von der Haar, T.; Oku, Y.; Ptushkina, M.; Moerke, N.; Wagner, G.; Gross, J.D.; McCarthy, J.E. Folding transitions during assembly of the eukaryotic mRNA cap-binding complex. *J. Mol. Biol.* **2006**, *356*, 982–992. [CrossRef]
24. Sabharwal, P.; Srinivas, S.; Savithri, H.S. Mapping the domain of interaction of PVBV VPg with NIa-Pro: Role of N-terminal disordered region of VPg in the modulation of structure and function. *Virology* **2018**, *524*, 18–31. [CrossRef]
25. Monzingo, A.F.; Dhaliwal, S.; Dutt-Chaudhuri, A.; Lyon, A.; Sadow, J.H.; Hoffman, D.W.; Robertus, J.D.; Browning, K.S. The structure of eukaryotic translation initiation factor-4E from wheat reveals a novel disulfide bond. *Plant Physiol.* **2007**, *143*, 1504–1518. [CrossRef]
26. Ashby, J.A.; Stevenson, C.E.M.; Jarvis, G.E.; Lawson, D.M.; Maule, A.J. Structure-based mutational analysis of eIF4E in relation to sbm1 resistance to Pea seed-borne mosaic virus in Pea. *PLoS ONE* **2011**, *6*, e15873. [CrossRef]
27. Marsh, J.A.; Forman-Kay, J.D. Sequence determinants of compaction in intrinsically disordered proteins. *Biophys. J.* **2010**, *98*, 2374–2382. [CrossRef]
28. Carson, M.; Johnson, D.H.; McDonald, H.; Brouillette, C.; Delucas, L.J. His-tag impact on structure. *Acta Crystallogr. D Biol. Crystallogr.* **2007**, *63*, 295–301. [CrossRef]
29. Mollica, L.; Bessa, L.M.; Hanoulle, X.; Jensen, M.R.; Blackledge, M.; Schneider, R. Binding Mechanisms of Intrinsically Disordered Proteins: Theory, Simulation, and Experiment. *Front. Mol. Biosci.* **2016**. [CrossRef]
30. Granon, S.; Kerfelec, B.; Chapus, C. Spectrofluorimetric investigation of the interactions between the subunits of bovine pancreatic procarboxypeptidase A-S6. *J. Biol. Chem.* **1990**, *265*, 10383–10388.
31. Aggarwal, L.; Biswas, P. Hydration Water Distribution around Intrinsically Disordered Proteins. *J. Phys. Chem. B.* **2018**, *122*, 4206–4218. [CrossRef]
32. Chenal, A.; Guijarro, J.I.; Raynal, B.; Delepierre, M.; Ladant, D. RTX calcium binding motifs are intrinsically disordered in the absence of calcium: Implication for protein secretion. *J. Biol. Chem.* **2009**, *284*, 1781–1789. [CrossRef]
33. Rani, P.; Biswas, P. Local Structure and Dynamics of Hydration Water in Intrinsically Disordered Proteins. *J. Phys. Chem. B.* **2015**, *119*, 10858–10867. [CrossRef]
34. Ortega, A.; Amoros, D.; Garcia de la Torre, J. Prediction of hydrodynamic and other solution properties of rigid proteins from atomic- and residue-level models. *Biophys. J.* **2011**, *101*, 892–898. [CrossRef]
35. Jensen, M.R.; Ruigrok, R.W.H.; Blackledge, M. Describing intrinsically disordered proteins at atomic resolution by NMR. *Curr. Opin. Struct. Biol.* **2013**. [CrossRef]
36. Wells, M.; Tidow, H.; Rutherford, T.J.; Markwick, P.; Jensen, M.R.; Mylonas, E.; Svergun, D.I.; Blackledge, M.; Fersht, A.R. Structure of tumor suppressor p53 and its intrinsically disordered N-terminal transactivation domain. *Proc. Natl. Acad. Sci. USA* **2008**. [CrossRef]
37. Kosol, S.; Contreras-Martos, S.; Cedeño, C.; Tompa, P. Structural characterization of intrinsically disordered proteins by NMR spectroscopy. *Molecules* **2013**, *18*, 10802–10828. [CrossRef]
38. Modrak-Wojcik, A.; Gorka, M.; Niedzwiecka, K.; Zdanowski, K.; Zuberek, J.; Niedzwiecka, A.; Stolarski, R. Eukaryotic translation initiation is controlled by cooperativity effects within ternary complexes of 4E-BP1, eIF4E, and the mRNA 5′ cap. *FEBS Lett.* **2013**. [CrossRef]
39. Michon, T.; Estevez, Y.; Walter, J.; German-Retana, S.; Le Gall, O. The potyviral virus genome-linked protein VPg forms a ternary complex with the eukaryotic initiation factors eIF4E and eIF4G and reduces eIF4E affinity for a mRNA cap analogue. *FEBS J.* **2006**, *273*, 1312–1322. [CrossRef]
40. Uversky, V.N. Natively unfolded proteins: A point where biology waits for physics. *Protein Sci.* **2002**, *11*, 739–756. [CrossRef]
41. Uversky, V.N.; Santambrogio, C.; Brocca, S.; Grandori, R. Length-dependent compaction of intrinsically disordered proteins. *FEBS Lett.* **2012**, *586*, 70–73. [CrossRef]

10

The Effect of FG-Nup Phosphorylation on NPC Selectivity: A One-Bead-Per-Amino-Acid Molecular Dynamics Study

Ankur Mishra [1], Wouter Sipma [1], Liesbeth M. Veenhoff [2], Erik Van der Giessen [1] and Patrick R. Onck [1,*]

1 Zernike Institute for Advanced Materials, University of Groningen, Groningen, 9747 AG, The Netherlands; a.mishra@rug.nl (A.M.); woutersipma@gmail.com (W.S.); e.van.der.giessen@rug.nl (E.V.G.)
2 European Research Institute for the Biology of Ageing, University of Groningen, University Medical Centre, Groningen, 9713 AV, The Netherlands; l.m.veenhoff@rug.nl
* Correspondence: p.r.onck@rug.nl

Abstract: Nuclear pore complexes (NPCs) are large protein complexes embedded in the nuclear envelope separating the cytoplasm from the nucleoplasm in eukaryotic cells. They function as selective gates for the transport of molecules in and out of the nucleus. The inner wall of the NPC is coated with intrinsically disordered proteins rich in phenylalanine-glycine repeats (FG-repeats), which are responsible for the intriguing selectivity of NPCs. The phosphorylation state of the FG-Nups is controlled by kinases and phosphatases. In the current study, we extended our one-bead-per-amino-acid (1BPA) model for intrinsically disordered proteins to account for phosphorylation. With this, we performed molecular dynamics simulations to probe the effect of phosphorylation on the Stokes radius of isolated FG-Nups, and on the structure and transport properties of the NPC. Our results indicate that phosphorylation causes a reduced attraction between the residues, leading to an extension of the FG-Nups and the formation of a significantly less dense FG-network inside the NPC. Furthermore, our simulations show that upon phosphorylation, the transport rate of inert molecules increases, while that of nuclear transport receptors decreases, which can be rationalized in terms of modified hydrophobic, electrostatic, and steric interactions. Altogether, our models provide a molecular framework to explain how extensive phosphorylation of FG-Nups decreases the selectivity of the NPC.

Keywords: Nuclear pore complex; FG-Nups; phosphorylation

---

1. Introduction

Eukaryotic cells are characterized by the presence of the nuclear envelope (NE), a lipid bilayer membrane that separates the cells into two compartments, i.e., the nucleus and the cytoplasm. The NE contains many nuclear pore complexes (NPCs), which are the sole gateway for the exchange of essential biomolecules between the two compartments. NPCs are large protein complexes, with a molecular mass of ~55–66 MDa [1,2] in yeast and ~ 125 MDa in vertebrates [3]. The NPC is composed of 30 different types of proteins called nucleoporins (Nups) [4,5]. One third of these Nups are intrinsically disordered proteins (IDPs), which are anchored to the inner wall of the NPC and are rich in phenylalanine-glycine (FG) repeats. Inside the NPC, these FG-Nups form a central meshwork that provides a permeability barrier for translocating molecules. Various studies have revealed that the NPC allows rapid transport of small molecules (30 kDa or ~5 nm in diameter), but drastically slows down the translocation of larger molecules from one compartment to the other [6–8]. It also has been found that FG Nups bind to nuclear transport receptors (NTRs) [9,10] by means of hydrophobic interactions, which facilitates the translocation of NTRs by lowering the permeability barrier [11]. Cargoes of diameter up to 40 nm are known to translocate by this facilitated transport mechanism [12,13]. Therefore, the FG-Nups are considered to be crucial in establishing the selective permeability barrier of the NPC.

Nucleocytoplasmic trafficking can be altered by a change in the surface properties of translocating molecules [14], by the deletion of FG-Nups [6,8], and by the change in cohesiveness of the FG-Nups [15]. For example, mutation of the hydrophobic F residues of Nsp1, a representative yeast FG-Nup, into the hydrophilic Serine S reduces the propensity of Nsp1 to form a hydrogel [16]. These experiments revealed that the hydrogels exclude inert molecules, but allow hydrophobic NTRs to enter, which is explained in terms of a local disruption of the cohesive gel network [16,17]. In a separate study [15], the Nsp1 molecules tethered onto the inner surface of solid state NPC mimics formed a dense phase (over 100 mg/mL) and enabled transport selectivity. Kap95 (a yeast NTR) traversed the pore whereas the translocation of tCherry (an inert molecule of similar size) was inhibited. The F, I, L, V to S mutation of Nsp1 resulted in a remarkably less dense FG-Nup network inside the pore, which led to a loss of selectivity, as both tCherry and Kap95 were able to translocate. Taken together, these studies show that the transition from a dense, hydrophobic phase to a dispersed, hydrophilic phase results in the nanopores losing their selective barrier function.

The hydrophobicity of FG-Nups can also be altered through phosphorylation, one of the most abundant protein modifications inside the cell [18,19]. Phosphorylation is catalyzed by kinases and can be reversed by phosphatases. It has been shown that extracellular signal-regulated kinase (ERK), a phosphorylating agent, can directly interact with FG-Nups [20,21], causing FG-Nups to phosphorylate [22–24]. Several in vitro studies have revealed that specifically Nup62, Nup98, Nup153, Nup214, and Nup358 can undergo phosphorylation [25,26]. Furthermore, there is evidence which confirms that FG-Nups undergo phosphorylation in vivo as well [21,23,27,28]. Transport studies demonstrated that the phosphorylation of nucleoporins results in decreased kinetics of active transport of Kap95 [25,27,29] and Kap-cargo complexes [30,31], and increased kinetics of passive transport [32]. These studies indicate that phosphorylation can modulate the selective permeability of the NPCs. However, the molecular mechanism behind the alteration in nucleocytoplasmic transport due to phosphorylation is not well understood.

Molecular dynamics (MD) simulations have proved to be a powerful tool to study the disordered protein structure inside the NPC and the transport through native and biomimetic nanopores [8,15,33–35]. Therefore, in order to understand the molecular mechanism behind the phosphorylation-induced alteration in transport kinetics, we carried out MD studies using our earlier developed one-bead-per-amino-acid (1BPA) coarse grained (CG) model for FG-Nups [35], extended here for phosphorylated FG-Nups. This 1BPA model has been successfully applied to probe the (doughnut-like) density distribution of the disordered domain of yeast NPCs [35], the facilitated transport of NTRs through yeast [33] and biomimetic [15] NPCs, and the size selectivity for passive transport [6], in good agreement with experiments. Although the transport experiments on phosphorylated NPCs cited above were carried out on mammalian NPCs, we here used the yeast NPC model, which has structural and functional similarities to the vertebrate NPC [36].

In the current study, we extended our 1BPA model to phosphorylated FG-Nups and carried out MD simulations of FG-Nups in isolation, as well as within the NPC. We studied the impact of phosphorylation on the structure of the disordered phase and the transport across the NPC in two scenarios. In the first (referred to as the Phos_N scenario), we used the NetPhosYeast 1.0 server [37] to obtain the phosphorylated residues of the yeast FG-Nups (yielding phosphorylated serine (S) and threonine (T) residues only), and in the second (referred to as the Phos_Max scenario), we assumed that all phosphorylatable residues (serine (S), histidine (H), threonine (T), and tyrosine (Y)) were phosphorylated. We investigated the changes in conformation of phosphorylated FG-Nups compared to FG-Nups in their native state by using the Stokes radius (R_S) as a measure for their size (see Section 2.1). We found that phosphorylation causes FG-Nups to extend by an amount that depends on the fraction of phosphorylatable residues and positively charged residues. In Section 2.2, we present a study on the collective interaction of phosphorylated FG-Nups inside the confined environment of the NPC. We found that phosphorylation drastically reduces the FG-Nup density inside the NPC. Finally, in Section 2.3 we report on simulation results of the phosphorylation-affected transport of inert particles and Kap95, and discuss these results in light of the various contributions to the interaction energy inside the NPC. Our transport simulations are in qualitative agreement with the experimentally-observed increase and decrease in transport rate of the passive and active transport pathways, respectively. Note that the Phos_N scenario predicts more phosphorylation sites than other phosphorylation databases, such as the fungi phosphorylation database (FPD), which provides a comprehensive list of experimentally validated phosphorylation sites [38]. The prediction from the FPD database is incorporated in the Supplementary Materials (see the section "Sensitivity analysis") to provide a scenario for experimentally validated phosphorylation sites. It is important to note that it is unclear which phosphosites predicted in either scenario are phosphorylated simultaneously in vivo, and hence the predictions provided in this study are not meant to mimic specific biological conditions, but rather to shed light on the fundamental mechanisms underlying the changes in transport kinetics of phosphorylated NPCs.

2. Results

In order to study the effect of phosphorylation on FG-Nups, we started with our previously developed MD model for intrinsically disordered proteins (IDPs) in their native state, coarse-grained at a resolution of one bead per amino acid (1BPA) [35]. This 1BPA model accounts for non-bonded hydrophobic and electrostatic interactions between the amino acids, including the effect of solvent polarity and ionic screening to mimic the solvent conditions inside the NPC. The model is accurate (within 20% error) in predicting the Stokes radius R_S [35] for a range of FG-Nups and FG-Nup segments [39]. In the current study, we extended the model for phosphorylation by accounting for the change in hydrophobicity and charge of four amino acids: S, H, T, and Y. We used a weighted average scheme of five predictor programs KOWWIN, ClogP, ChemAxon, ALOGPS, and miLogP [40–43] to predict the change in hydrophobicity due to the change in the chemical structure. For details on the model development for phosphorylated FG-Nups we refer to the Materials and Methods section (Section 4.2). The new parameters for phosphorylated amino acids are summarized in Table 1.

Table 1. Parameters in the 1BPA forcefield for phosphorylated amino acids. Note that: ε_{1BPA} and $\varepsilon_{weighted}$ are the normalized hydrophobicity values (between 0 and 1) from the 1BPA model [35] and the weighted average scheme (see Section 4) for the amino acids in their native state, respectively; ε_P is the hydrophobicity of the phosphorylated amino acid; and q and q_P denote the charge of the amino acids in their native and phosphorylated conditions, respectively.

AA	ε_{1BPA}	$\varepsilon_{weighted}$	ε_P	q	q_P
Ser (S)	0.45	0.41	0.07	0	−2e
His (H)	0.53	0.44	0.06	0	−2e
Thr (T)	0.51	0.52	0.23	0	−2e
Tyr (Y)	0.82	0.83	0.67	0	−2e

2.1. *Effect of Phosphorylation on Isolated FG Nups*

We used our newly developed parametrization for phosphorylation and performed MD simulations to study the effect of phosphorylation on the conformation of isolated FG-Nup segments [35]. The simulated trajectories were analyzed to determine the time averaged R_S using the Hydro program [44,45]. The predicted R_S values for the phosphorylated FG-Nups are compared with that of FG-Nups in their native state (from experiments [39] and simulations [35]) in Figure 1. The error bars for the simulation data represent the standard deviation in time for R_S (See Table S1 in the supplementary data for the source data).

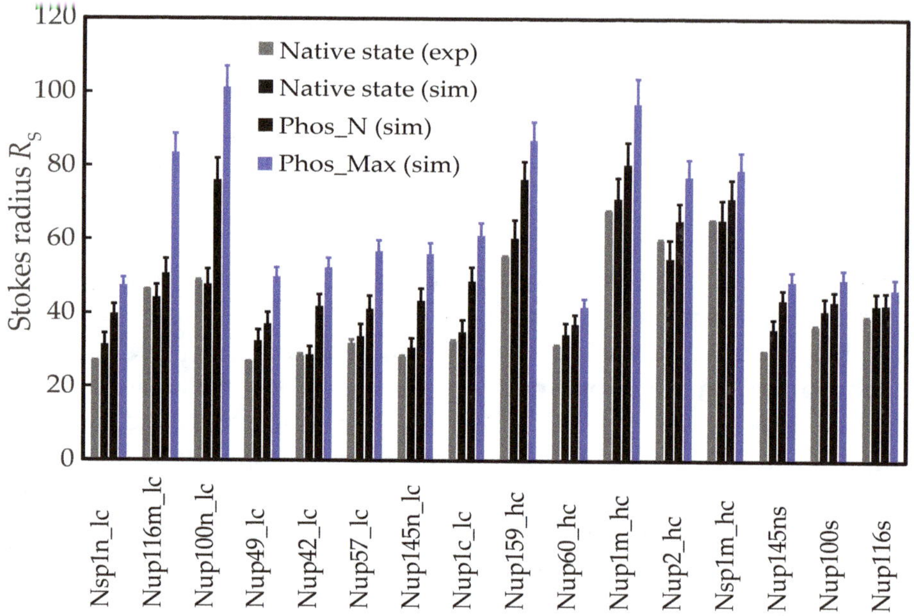

Figure 1. Phosphorylation-induced extension of FG-Nup segments. The Stokes radius R_S (in Angstrom) is depicted for a range of FG-Nup segments in their native and phosphorylated states. The suffix *lc* denotes low charge, *hc* high charge and *s* refers to the stalk region of the Nup. The grey and black bars represent the data in the native state from experiments [39] and simulations (results reproduced from [35]), respectively, and the prediction for the phosphorylated states are plotted in red for Phos_N and blue for Phos_Max. For the simulation data, the error bars represent the standard deviation in time (see Table S1 for the source data).

As a result of phosphorylation, the amino acids become more hydrophilic and negatively charged (see Table 1). Thus, compared to the native state, the phosphorylated FG-Nups exhibit enhanced electrostatic repulsion and reduced hydrophobic attraction leading to an overall decrease in intra-molecular cohesion and thus a more extended configuration (see Figure 1). In Table S2, we have summarized the number of amino acids that can be phosphorylated in each FG-Nup segment. The FG-segments are grouped as low charged (*lc*), high charged (*hc*), and stalk (*s*) domains, following the definition of Yamada et al. [39]. We found that the relative abundance of phosphorylatable residues in all FG-Nup segments ranges from ~15% (for Nup116s) to ~33% (for Nsp1n_lc) for the maximally phosphorylated (Phos_Max) condition, whereas for the Phos_N scenario the range is from ~4% (Nup116s) to ~17% (Nup159_hc). In order to quantify the change in Stokes radius in terms of the number of residues undergoing phosphorylation, we plot the normalized change in R_S as a function of the percentage of phosphorylatable residues (n) for the low charged, high charged, and stalk groups in blue, red, and green data points, respectively, for the Phos_Max (Figure 2a) and Phos_N (Figure 2b) scenarios. The change ΔR_S is normalized as $\Delta R_S/(N-1)b$, where N is the total number of residues of the FG-Nup segment and b is the coarse-grained bond length (3.8 Angstrom). We fitted the data points for individual groups to a straight line passing through the origin, represented as colored lines in Figure 2a,b. We note that for both Phos_Max and Phos_N, the FG-segments from the *lc* group show the highest normalized change in R_S (blue line in Figure 2a,b), whereas phosphorylation has a smaller effect on size for the *hc* and *s* groups (red and green lines in Figure 2a,b).

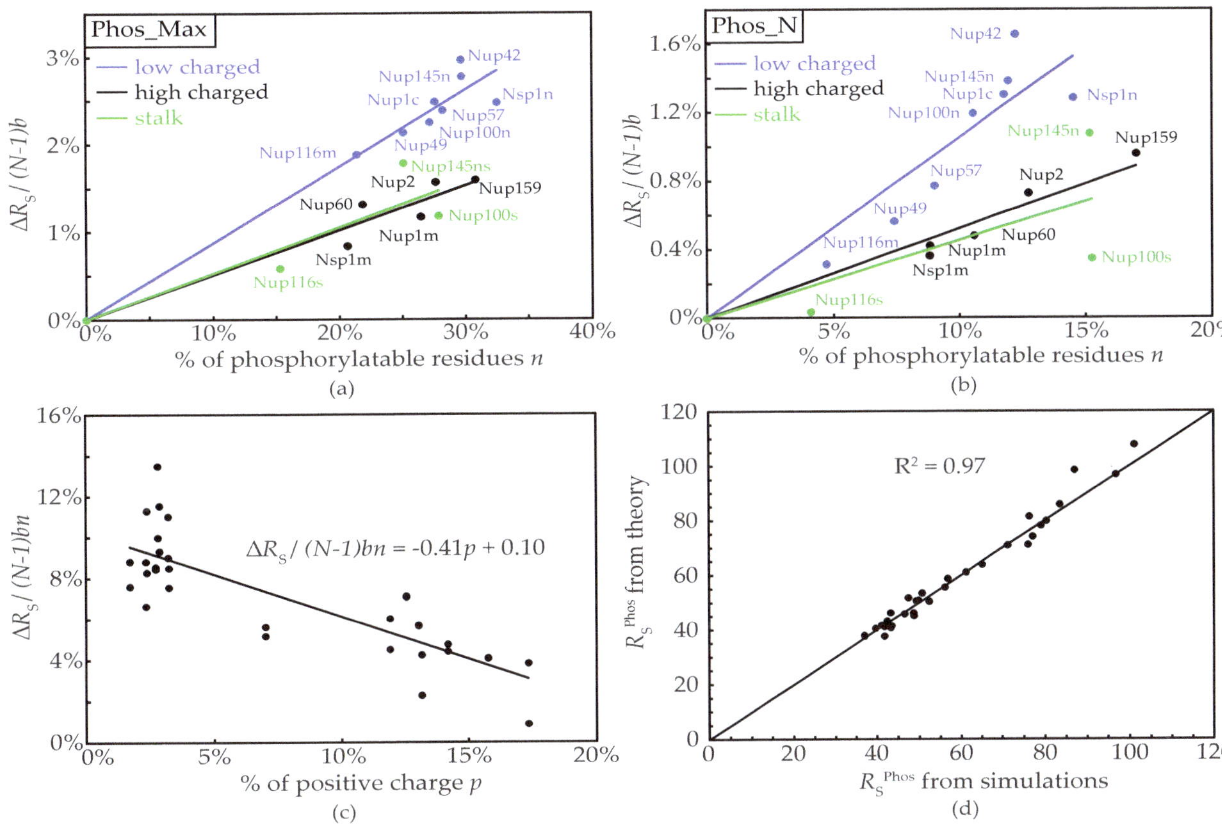

Figure 2. The normalized change in R_S (i.e., $\Delta R_S/(N-1)b$) due to phosphorylation as a function of the fraction of phosphorylatable residues (n) for (**a**) the Phos_Max and (**b**) the Phos_N scenarios [37]. The expression for the change in R_S (i.e., $\Delta R_S = R_S^{phos} - R_S^{native}$), with R_S^{phos} and R_S^{native} being the Stokes radii of the FG-Nup segments in the phosphorylated and native states, respectively, is normalized with $(N-1)b$ where N is the total number of residues of the FG-Nup segment, and b (= 3.8 Angstrom) is the coarse-grained bond length between neighboring amino acids [35,46]. The data for the FG-Nups from the high charged (*hc*), low charged (*lc*), and stalk (*s*) segments [35,39] are represented in red, blue, and green data points, respectively. The data points of each group are fitted to a straight line passing through the origin revealing different slopes for different groups. For Phos_Max we observe slopes of 0.087 for *lc* (R^2 = 0.94), 0.051 for *hc* (R^2 = 0.92), and 0.053 for *s* (R^2 = 0.80), respectively, whereas for Phos_N the slopes are 0.1 for *lc* (R^2 = 0.86), 0.052 for *hc* (R^2 = 0.95) and 0.045 for *s* (R^2 = 0.61). (**c**) The ratio of normalized change in R_S to the fraction of phosphorylatable residues (n) is plotted as a function of the fraction of positively charged residues (p) for all data points (black) from the Phos_Max and Phos_N scenarios. These data points are fitted to a linear equation, as shown in the figure (giving R^2 = 0.68). (**d**) The R_S^{phos} predicted from the theory in Equation (1) compared to R_S^{phos} computed from the MD simulations, both in Angstrom, show a good agreement with a fitness measure of R^2 = 0.97.

In order to investigate the varying response for the three groups, as shown in Figure 2a,b, we analyzed the change in hydrophobicity upon phosphorylation and found that it is roughly similar for the three groups, i.e., for FG-Nups from the *lc*, *hc*, and s groups, the hydrophobicity drops by 13-20%, 16–22%, and 12–21%, respectively, for Phos_Max (see Table S2). Similarly, for Phos_N, the reduction in hydrophobicity amounts to 3–10%, 7–10%, and 4–12% for the *lc*, *hc*, and *s* groups, respectively, showing no major difference across the three groups. Clearly, the effect of phosphorylation on hydrophobicity alone cannot account for the different R_S of the groups. It has been argued that the net proline content in IDPs plays an important role in determining the effective Stokes radius [47], as proline provides additional stiffness to the peptide chain because of its ring structure. This effect of proline is included in our 1BPA model in the form of the bonded potentials [46]. However, all the FG-Nup segments analyzed in this study (Figure 1) have a similar 3–8% proline content, and therefore cannot explain the different Stokes radii across the three groups (see Table S2). Next, we analyzed the effect of charge. Since the net charge of the three families is quite similar (i.e., 2–3%, 0–3%, and 0–1% for *lc*, *hc*, and *s*, respectively), we investigated the occurrence of positively charged residues R and K in the FG-segments and found that the *lc* group contains only 2–3% of positively charged residues in contrast to the *hc* and *s* groups, which have more positively charged residues (7–16% for *hc* and 13–17% for *s*, see Table S2). Thus, it seems that the larger amount of positive charge in *hc* and *s* is more efficient in screening the effect of the negative charge increase induced from phosphorylation than the small amount of positive charges in the *lc* group (see Table S2). In order to confirm this, we plotted the ratio of $\Delta R_S/(N-1)b$ normalized by the fraction of phosphorylatable residues (n) as a function of the percentage of positive charge content (p) in the FG-segments (see Figure 2c). The data points in Figure 2c can be fitted to a straight line with a slope of -0.41 and y-intercept of 0.1, with an R^2 value of 0.68. Using this observation, the Stokes radius for a phosphorylated FG-segment can be predicted using the following expression:

$$R_S^{\text{phos}} = R_S^{\text{native}} + bn\,(N-1)(-0.41p + 0.1) \qquad (1)$$

We show the predictive power of this formula in Figure 2d, where the R_S^{phos} predicted using Equation (1) is plotted against the computed R_S^{phos} from the MD simulations, showing a very good correlation (with $R^2 = 0.97$).

2.2. *Effect of Phosphorylation on NPC Structure*

For the next step, we analyzed the disordered protein distribution inside the yeast NPC upon phosphorylation. Due to phosphorylation according to the Phos_N scheme, 6432 (7.4%) of all residues (86520) are phosphorylated inside the NPC, whereas for Phos_Max, 20992 (24.3%) of the NPC residues are phosphorylated (see Table S5). We tethered the FG-Nups inside the yeast scaffold at the same anchoring points as in the wild type yeast NPC [35,48], and then switched on phosphorylation. In Figure 3, we show snapshots of the wild type and phosphorylated NPCs (for both the Phos_N and Phos_Max conditions). As expected from the conformational analysis of isolated FG-Nups (Section 2.1), the FG-Nups assumed extended conformations in the phosphorylated NPC compared to the wild type NPC. For the maximally phosphorylated (Phos_Max) state, the FG-Nups spilled out of the NPC over a distance of almost 100 nm from the NPC surface, which is considerably larger than that for the Phos_N state. In the case of isolated FG-Nups, only intra-molecular interactions are present, whereas inside the NPC, the residues of each FG-Nup also interact with residues from other FG-Nups (i.e., both intra- and inter-molecular interactions are present). Therefore, the enhanced electrostatic repulsion and reduced hydrophobic attraction due to phosphorylation is much more pronounced inside a confined space like the NPC, resulting in a strong reduction of protein density.

Figure 3. Snapshots of the coarse-grained MD models of (**a**) the wild type NPC, (**b**) the Phos_N NPC, and (**c**) the Phos_Max NPC. The FG-Nups (different color beads denote different amino acids) are attached at the same anchor positions, as in [35,48]. Phosphorylation-induced changes in the interaction leads to spilling of FG-Nups out of the NPC.

We analyzed the collective distribution of the FG Nups inside the NPC by calculating the time averaged radial density distribution (averaged over the axial and circumferential direction) inside the pore (i.e., for $|z| < 15.5$ nm) (note that the origin of the coordinate system coincides with the center of the NPC). The radial density profile for all residues and hydrophobic residues are plotted in Figure 4a,b, respectively, for the wild type and phosphorylated NPCs. In addition, the 2-dimensional (rz) density distribution (averaged over the circumferential direction) is plotted in Figure 4c for the wild type and phosphorylated NPCs. For technical details on calculating the density distributions, the reader is referred to the Materials and Methods section. Figure 4 clearly shows that phosphorylation significantly alters the density distribution of the FG-Nups inside the NPC. For a wild type NPC, the mean density at the center (0 nm $< r <$ 5 nm) is ~ 80 mg/mL (see Figure 4a), which gradually increases with the radial distance r from the center and attains a peak value (~180 mg/mL) at r ~ 15 nm, after which the density gradually decreases. This is consistent with the doughnut-like structure in Figure 4c (left panel). However, for the phosphorylated NPCs, the density drops drastically inside the pore, amounting to only ~50 mg/mL and ~20 mg/mL at the center (see Figure 4a) for the Phos_N and Phos_Max scenarios, respectively. In the case of Phos_N, the radial density profile follows a similar trend (but lower in magnitude) as the wild type. This results in a less dense doughnut-like structure, as shown in Figure 4c (middle panel). For Phos_Max, the density remains lower than 20 mg/mL throughout the full range of r-values, as shown in Figure 4a,c (right panel). The density distribution of hydrophobic residues (see Figure 4b) is highly correlated with the density distribution of the total amount of residues (see Figure 4a).

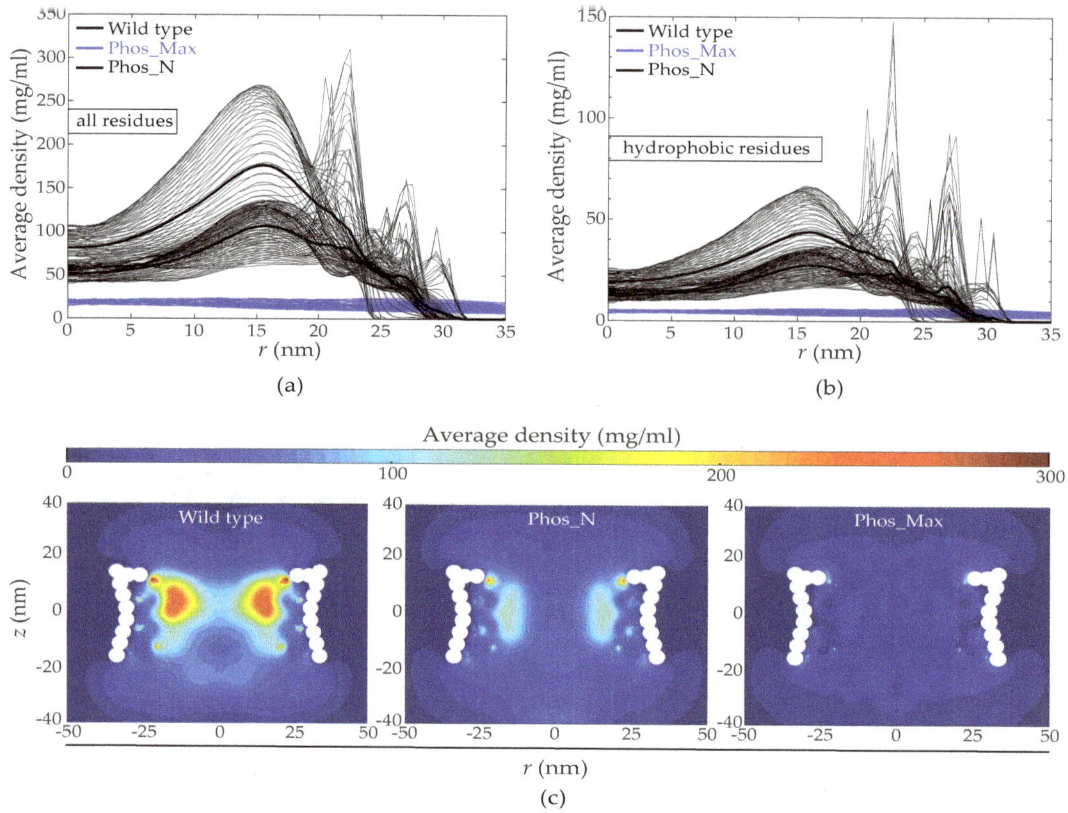

Figure 4. Disordered protein structure inside the wild-type and phosphorylated NPCs. (**a**,**b**) Time-averaged radial density distribution inside wild type (black), Phos_N (red) and Phos_Max (blue) NPCs for (**a**) all residues, and (**b**) hydrophobic residues. The thin lines represent the density at different positions along the *z*-axis separated by 1 nm, in the range of $|z| < 15.5$ nm (height of the NPC), with the mean density plotted as thick lines. (**c**) The *rz*-density map for the wild type (left panel), Phos_N (middle panel), and Phos_Max (right panel) NPCs. The wild type shows the characteristic highly dense doughnut-like structure [35]. The Phos_N NPC shows a density-depleted doughnut-like structure, whereas the Phos_Max NPC shows a significantly less dense and rather uniform density distribution.

To explore the main reason for the reduction in protein density for both phosphorylated NPCs, we computed the relative contribution of the hydrophobic and electrostatic energy to the total interaction energy inside the wild-type and phosphorylated NPCs [15,34,35], as shown in Figure 5a,b, respectively. In the wild type NPC, the time-averaged hydrophobic interaction energy amounts to approximately −76,300 kJ/mol, whereas for the Phos_N and Phos_Max NPCs, these values are about −49,000 kJ/mol and −6100 kJ/mol, respectively. Here, by far the largest reduction is in the Phos_Max NPC, with almost a twelve-fold decrease in hydrophobic interaction energy relative to the wild type. Note that this twelve-fold decrease cannot be explained by the reduction in net hydrophobicity alone (a reduction of 16%, see Table S5); also, the distance between the hydrophobic amino acids plays an important role, being much larger for Phos_Max than for the wildtype and Phos_N NPCs (see Figure 4). On the other hand, the Coulomb energy was measured to be two orders of magnitude smaller than the hydrophobic energy for the wild type and Phos_N NPCs (around −750 kJ/mol for wild type and −800 kJ/mol for Phos_N), while for Phos_Max, the total repulsive Coulomb energy (around 31,200 kJ/mol) is much larger than the hydrophobic energy. In summary, the wild type and Phos_N NPCs are highly hydrophobic, with only a small contribution from electrostatics. In sharp contrast to this, the energy in the Phos_Max NPC has a much more dominant (repulsive) Coulombic contribution corresponding to the large net negative charge, while the hydrophobic energy is much lower.

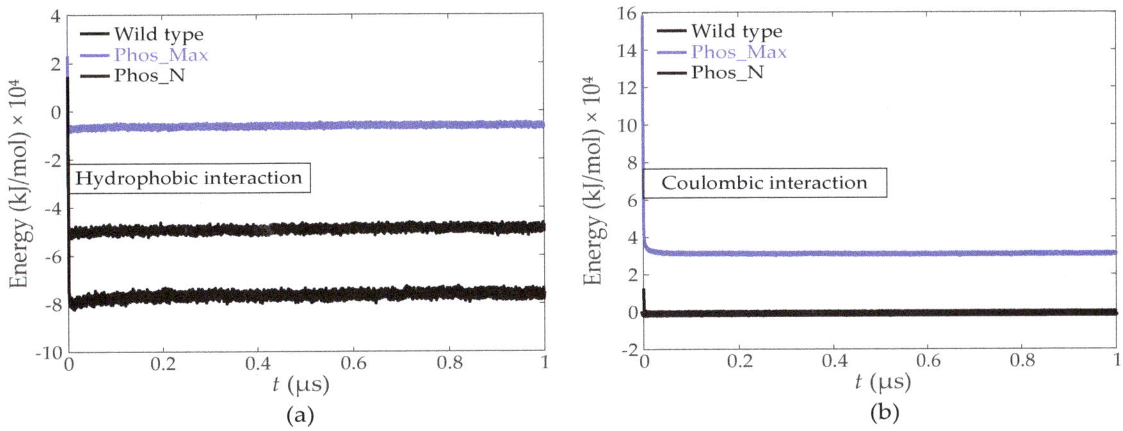

Figure 5. Time evolution of (**a**) the hydrophobic interaction energy, and (**b**) the coulombic interaction energy of the FG-Nups for the wild type (black), Phos_Max (blue), and Phos_N (red) NPCs.

2.3. Effect of Phosphorylation on Active and Passive Transport

In this section, we focus on the selective permeability barrier of the NPC and how phosphorylation affects this. The nuclear transport receptor Kap95 is known to interact with the FG-Nups via its hydrophobic binding sites and to translocate through the NPC by facilitated transport in the presence of RanGTP at the nucleoplasmic side, which dissociates the Kaps from the NPC [49,50]. Beside being hydrophobic, the Kaps are also negatively charged [51]. It is therefore expected that the interaction between the Kaps and FG-Nups is strongly affected by the phosphorylation-induced charge modification and reduction in hydrophobicity of the FG-Nups. To investigate this, we modelled a yeast NPC in the presence of ten Kap95 particles (with a diameter of 8.5 nm, 10 hydrophobic binding sites, and a uniformly-distributed surface charge of −43e, as used previously [15]) that are released at the cytoplasmic side to probe facilitated transport. After equilibrating the system (see Materials and Methods for details), the simulations were carried out for the same initial positions of the Kaps for the wild type and phosphorylated pores. Snapshots of the final state (at t = 2 μs) are shown in Figure 6, illustrating the inhibition of facilitated transport in phosphorylated NPCs. The bottom panels of Figure 6a–c depict a reduced binding affinity of the Kaps with the phosphorylated FG-Nups. To assess the propensity for translocation, we plot the initial (t = 0 μs) and final (t = 2 μs) z coordinate of the center of mass of the Kap95 particles in Figure 7a. We observed that for the wild type NPC, the Kap95-FG-Nup affinity is larger compared to the phosphorylated NPCs (see Figure 6), so that the Kap95 particles are able to enter the pore and translocate (Figure 7a). The large affinity is due to the fact that the pore is hydrophobic (see Figure 5 and Table S5) and has a weak positive charge [35]. In the course of 2 μs, a total of 9 Kaps translocated through the pore and the remaining Kap ended up inside the pore (Figures 6a and 7a). In the phosphorylated NPC, however, despite the lower FG-Nup density (see Figure 4), the Kaps are excluded from the pore (Figure 6b,c and Figure 7a). The results from Figure 5, Figure 6, and Figure 7a point towards a phosphorylation-induced decrease in hydrophobicity (for Phos_Max and Phos_N) and increase in coulombic repulsion (for the Phos_Max case) resulting in lowering of Kap95-FG-Nup binding affinity, which is instrumental for active transport.

Next, we study the effect of phosphorylation on passive transport by probing the transport of inert particles of the same size as the Kap95 particles (i.e., 8.5 nm in diameter) but without charge and hydrophobic binding spots. We used the same initial positions for the inert particles as for the Kap95 particles used in the case of active transport (Figures 6 and 7a). In Figure 7b, we plot the initial and final z location of the center of mass of the inert particles for the wild type and phosphorylated NPCs. For the wild type NPCs, it can be clearly observed that the inert particles stay at the cytoplasmic side and do not enter the NPC. On the other hand, in the Phos_Max NPC, two inert particles managed to translocate through the pore within 2 μs, showing that the permeability barrier of the Phos_Max NPC is jeopardized. For the Phos_N NPC, we did not observe any translocation. To further test the size-dependent permeability barrier of the Phos_N and wild type NPCs, we performed two additional transport simulations for ten spherical inert particles of diameter 4 nm. The results are shown in Figure S1 (see the Supplementary Information), revealing that in the wild type, the total number of translocations is 129, whereas in the Phos_N pore, there were 328 translocation events, indicating a 2.5-fold increase in passive transport rate compared to wild type. For the wild type, this is consistent with our previous work [33], where we found the energy barrier for inert particles of size 4 nm to be lower than $\kappa_B T$ (and thus likely to go through), while that for inert particles of 7 nm (and up) was found to be larger than 2 $\kappa_B T$ (and thus likely to not pass through). Our results for the wild type can be understood by recourse to the scaling relation of Timney et al. [8], which states that the characteristic time constant of passive transport scales with the third power of the molecular mass. Since the two inert particles used here are of size 4 and 8.5 nm, the mass dependence rule predicts that the translocation of the 8.5 nm particle should be 883 times slower compared to the smaller 4 nm particle. This is in qualitative agreement with our simulations, where for the 4 nm particles we see 129 translocation events within 2 μs of simulation time, whereas no translocation is observed for the 8.5 nm inert particle. Since Phos_Max NPCs have a lower protein density compared to Phos_N NPCs, we can expect the transport rates to be even higher for a 4 nm particle. In the case of passive transport, the inert particles interact with the FG-nups by means of steric repulsion only, and therefore the translocation events (Figure 7b and Figure S1) of inert molecules can be understood in terms of the density distribution of FG-Nups in the wild type and phosphorylated NPCs. Figure 4 shows that the FG-Nup density inside the NPC is significantly higher for the wild type NPC compared to the phosphorylated NPCs, which explains why in the wild type NPC no passive transport is observed (Figure 7b), whereas in the phosphorylated NPCs, passive transport occurs due to the lower permeability barrier.

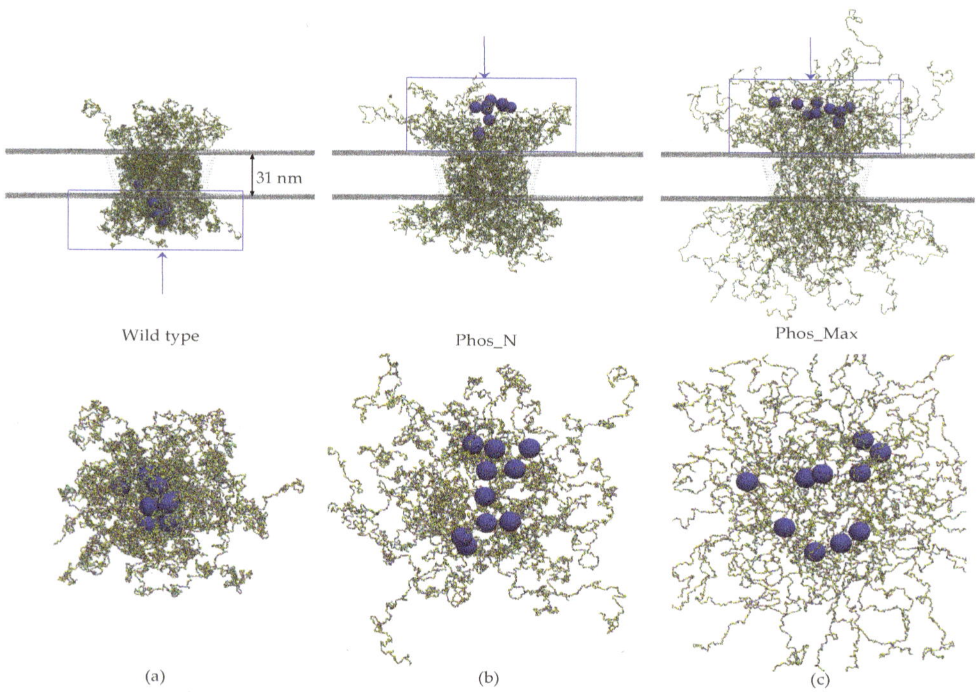

Figure 6. Snapshots at $t = 2$ μs of the FG-Nups and model Kap95 particles inside a wild type NPC (**a**), a phosphorylated NPC according to the Phos_N scheme (**b**), and a phosphorylated NPC according to the Phos_Max scheme (**c**). The Kap95 particles are shown in blue with the red hydrophobic binding spots on its surface. The 20 different amino acids of the FG-Nups are represented by different colors. The size of the scaffold beads (grey) is scaled down to make the Kap particles better visible. In the bottom panel we provide a magnified view from the bottom for (**a**) and from the top for (**b**) and (**c**), focusing on the region indicated by the blue boxes in the top panels. The Kaps are shown to traverse the wild type NPC, whereas the Kaps are not able to strongly partition into the FG-Nup meshwork for the phosphorylated NPCs.

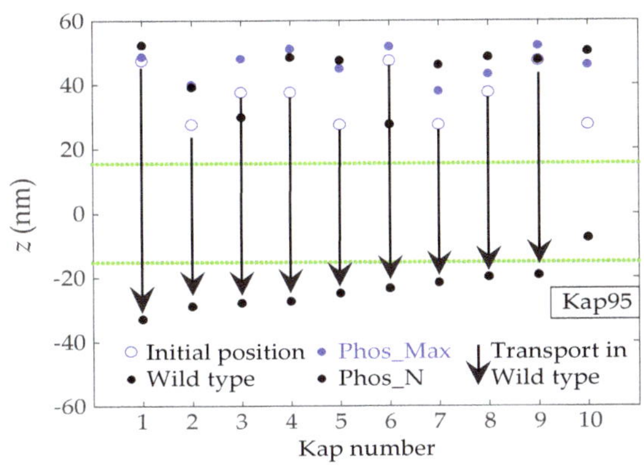

(a)

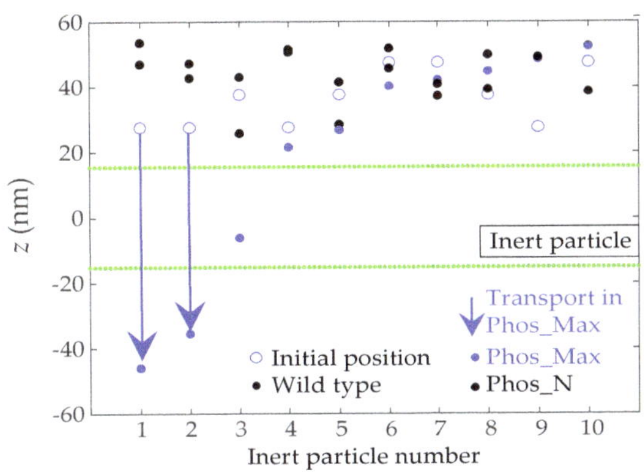

(b)

Figure 7. Effect of phosphorylation on active (**a**) and passive (**b**) transport. In both cases, the particles are released from the same position at the cytoplasmic side and are ordered from left to right based on the end position at $t = 2$ μs. (**a**) Initial ($t = 0$ μs) and final ($t = 2$ μs) axial (z) position of ten Kap95 particles, and (**b**) Initial ($t = 0$ μs) and final ($t = 2$ μs) axial (z) position of ten inert particles of the same size as the Kap95 particles (diameter = 8.5 nm, no charge and no hydrophobic binding spots). In both (**a**) and (**b**), the boundaries of the NPCs ($|z| = 15.5$ nm) are represented by green lines and the arrows represent translocations from the cytoplasm to the nucleoplasm.

Finally, we summarize our findings on active (from Figure 7a) and passive transport (Figure 7b and Figure S1) for the wild type, Phos_N, and Phos_Max NPCs in Figure 8. The wild type NPC is seen to have a selective permeability barrier, as it allows Kaps and small inert particles (diameter = 4 nm) to pass through, whereas larger inert particles (diameter = 8.5 nm) are excluded. The phosphorylated Phos_N NPC loses its selectivity, as transport of Kap95 is not observed, but its permeability barrier is still intact (8.5 nm particles are excluded). This indicates that the reduced steric hindrance due to the reduced amino acid density in the center (Figure 4a) is still sufficient to exclude large particles, but that the reduced hydrophobicity is no longer able to attract Kap95 particles into the FG-Nup mesh-work. Finally, the heavily phosphorylated Phos_Max NPC is observed to lose both its ability to facilitate active transport (due to the reduced hydrophobic attraction and increased electrostatic repulsion with respect to Kap95 particles), as well as its permeability barrier (due to the drastically reduced amino acid density allowing for transport of both inert particles). This, of course, is subject to the constraint of the limited time frame of our simulations (i.e., 2 μs).

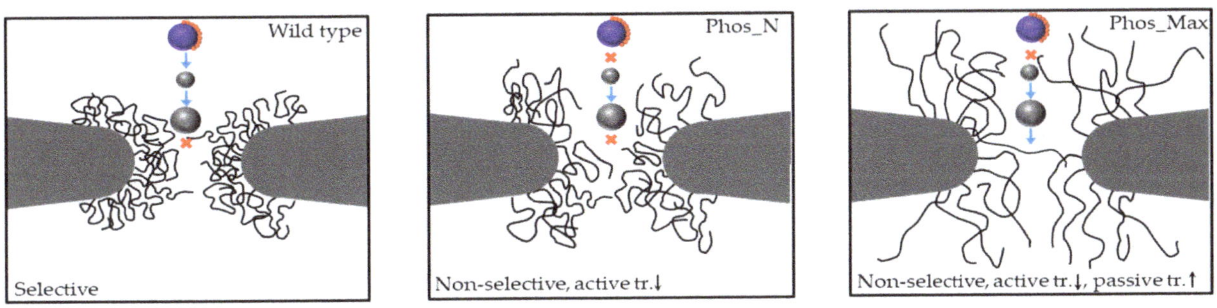

Figure 8. Summary of simulation results on wildtype and phosphorylated yeast NPCs, showing active transport of Kap95 (blue sphere with red dots representing the hydrophobic binding sites) and passive transport of inert particles (grey spheres of diameter 4 nm and 8.5 nm). The FG-Nups are represented by black filaments. The blue arrows indicate observed transport and the red crosses indicate prohibited transport within the 2 μs simulation time.

3. Discussion

In the current study, we incorporated phosphorylation-induced modifications of the hydrophobicity and charge of the four amino acids S, H, Y, and T (see Table 1) into our 1BPA model for IDPs. We addressed the effect of phosphorylation on the conformational changes of 16 different isolated FG-Nup segments of varying length and with varying numbers of charged and hydrophobic residues. We compared the predicted Stokes radius R_S of these FG-Nup segments in their phosphorylated state with the values in their native state (see Figure 1), and observed an increase in size due to phosphorylation. We found that R_S increases linearly with the fraction of phosphorylatable residues and decreases linearly with the percentage of positively-charged amino acids (see Equation (1) and Figure 2d. While the former dependence is straight-forward, the latter is subtler and points to the important role of positive charge in screening the effect of the phosphorylation-induced increase in negative charge.

Next, we investigated how the FG-Nups interact when confined inside the NPC and how phosphorylation alters these interactions and the resulting protein distribution. The density distributions demonstrate a considerable difference between the wild type and phosphorylated NPCs, with the total amino acid density and hydrophobic density dropping by almost a factor two and four for the Phos_N and Phos_Max NPCs, respectively, in comparison to wild type. Whereas the hydrophobicity changed both in terms of density (Figure 4b) and energy (Figure 5a) for the two phosphorylated NPCs, only the Phos_Max NPC showed a large increase in (repulsive) electrostatic energy, while for the Phos_N and wild type NPC, the electrostatic energy remained negligible compared to the hydrophobic energy (Figure 5a,b). All considered, we can conclude that the phosphorylated FG-nups resulted in a higher negative charge and lower hydrophobicity, resulting in a strong depletion of amino acid density in phosphorylated NPCs, with the effects (especially the electrostatic) much more pronounced in Phos_Max NPCs.

For those molecules that translocate through the NPC by means of active transport (for example Kap95 in this study) [15,33–35], the molecular interactions can be divided into three components: (i) steric repulsion by means of excluded volume; (ii) hydrophobic interactions; and (iii) Coulombic interactions. Firstly, as the density inside the phosphorylated pores (both Phos_N and Phos_Max NPCs) is significantly lower than in the wild type (see Figure 4), the steric repulsion component is lower. Secondly, as illustrated in Figure 5a, phosphorylation results in a serious reduction of the hydrophobic interaction energy, as the residues become more hydrophilic upon phosphorylation. Finally, as Kap95 carries negative charge, it will face electrostatic attraction when the pore is positively charged (wild type) and electrostatic repulsion when the pore is negatively charged (phosphorylated). All the energy components taken together indicate that the negatively-charged and hydrophobic Kap95 experiences a much more repulsive environment inside the phosphorylated NPCs due to the increased negative charge and reduced hydrophobicity compared to wild type. Therefore, the overall energy barrier for the translocation of Kap95 particles through a phosphorylated pore is much higher compared to the wild type pore. Our transport simulations (Figures 6 and 7a) for Kaps indeed reveal inhibition of facilitated transport upon phosphorylation, while wild type NPCs facilitate Kap translocation. We do not account for the presence of RanGTP in our model, which is known to play an important role in dissociating the Kaps from the FG-Nups. As a result, the model Kaps remain in the bound state towards the nuclear side of the pore thanks to a slightly higher affinity of the nuclear FG-Nups with the Kap95 [52]. In contrast to this, the phosphorylated pore inhibits the Kaps to enter the pore. Of course, we cannot completely rule out the fact that some of the Kaps might translocate through phosphorylated NPCs at longer simulation times. Nevertheless, the trend of a reduced probability for active transport through phosphorylated pores and an increased probability for passive transport upon phosphorylation is in accordance with experimental observations [27,29–32].

The results for wild type and Phos_N are comparable to our previous studies on biomimetic nanopores coated with Nsp1 and a more hydrophilic mutant, Nsp1-S, which illustrated that a lack of cohesion in the hydrophilic Nsp1-S pore can result in a depleted density (~twofold decrease) compared to the hydrophobic Nsp1 pore [15,53]. Despite the different nature of the modification (mutation versus phosphorylation), both cases result in a reduced protein density due to a depleted hydrophobicity, while the electrostatic interactions remain approximately unaffected (see Figure 5). However, the effect on the selective permeability was found to be different. Whereas the Nsp1-S pore lost its selectivity due to the fact that the permeability barrier was jeopardized (large inert particles were found to go through), the Phos_N NPC retained its permeability barrier, but lost its ability to actively transport Kap95 particles. This difference is most likely related to the different unfolded protein composition of the yeast NPC and the biomimetic nanopore, with the former consisting of 10 different FG-Nups, while the latter has only one.

It should be noted that in this study, we compared the wild type results with an NPC in which all S, H, T, and Y residues are maximally phosphorylated (Phos_Max), and a second variant (Phos_N) in which the phosphorylation sites are extracted from the NetPhosYeast 1.0 server [37], accounting for the phosphorylation of a subset of all S and T residues. Clearly, the Phos_Max scenario is not very relevant from a biology point of view, as phosphorylation of all S, T, H, and Y does not occur simultaneously in reality. The results of the Phos_Max scenario therefore serve as a theoretical limiting case of phosphorylated NPCs that feature a maximal phosphorylation-induced modification of charge and hydrophobicity. The Phos_N scenario predicts a higher number of phosphorylation sites compared to other phosphorylation databases, such as the fungi phosphorylation database (FPD) (see Tables S6 and S7 for FPD phosphosites). Despite this difference, for both scenarios we observe loss of selectivity (see Figure 7a and Figure S2), while the permeability barrier is retained (see the section "Sensitivity analysis" in the Supplementary Materials).

It is still not known what fraction of the phosphorylatable residues predicted in these databases actually undergo simultaneous phosphorylation inside the NPC in vivo. This will be an interesting aspect to be explored further, since our study indicates that the degree of phosphorylation can have a large impact on the structure of the NPC and the rate of transportation in passive and active pathways. Our work should therefore not be seen as an exact mimic of specific biological conditions, but as a qualitative study, in which NPC phosphorylation is explored in order to shed light on the fundamental mechanisms underlying in vitro experiments on the decreased kinetics for active import [25,27,29] and the increased kinetics of passive import [32] in phosphorylated NPCs.

4. Materials and Methods

4.1. Coarse-Grained Molecular Dynamics Simulations

The 1BPA Molecular Dynamics model used in this study accounts for the exact amino acid sequence of the FG-Nups, in which each bead is located at the C_α positions of the polypeptide chain [35,46]. We set the mass of each bead to the average amino acid mass (120 Da), and the distance between neighboring beads to ~0.38 nm through a stiff harmonic spring potential. The bending and torsion potentials are extracted from the Ramachandran data of the coiled regions of protein structures [46]. The solvent molecules are treated in an implicit manner. A distance-dependent dielectric constant is used to account for the solvent polarity, and ionic screening is incorporated through Debye screening with a screening constant $k = 1\ \text{nm}^{-1}$ corresponding to the physiological salt concentration inside the NPC [54]. The hydrophobic interactions between the amino acids are incorporated through a modified Lennard-Jones potential, which accounts for hydrophobicity scales of all 20 amino acids derived from normalized experimental partition energy data renormalized in a range from 0 to 1. For details of the method, the reader is referred to [35].

All MD simulations were carried out with a time step of 0.02 ps [35]. The simulations for the isolated disordered FG-Nup segments were carried out for 2.5×10^7 steps [35], which was found to be sufficiently long to reach convergence. For the NPC simulations with particles (Figure 6, Figure 7 and Figure S1) and without particles (Figure 3, Figure 4, and Figure 5), the systems were first energy minimized to remove any overlap of the amino acid beads. Then, all long-range forces were gradually switched on, and for the NPC with particle systems, the inert/Kap95 particles were kept at a fixed position on the cytoplasmic side. In the final production run for the NPC without particles, the simulations were carried out for 5×10^7 steps (with the first 5×10^6 steps ignored so that only the statistically meaningful results are extracted), which was found to be long enough to have converged results for the density distribution inside the pores. For the NPC with particles, we included one additional step before the production runs, in which we equilibrated the system for 5×10^6 steps with all long-range forces switched on while keeping the inert/Kap95 particles fixed at their position. In the final production runs for the NPC with transporting particles, the inert/Kap95 particles were allowed to move and the simulations were carried out for 10^8 steps. For the Kap95 simulations we modelled the hydrophobic binding sites on the Kaps as F beads [15,33].

The time-averaged density calculations presented in the main text (see Figure 4) were derived by using the "gmx densmap" tool in GROMACS. The nanopore is centered inside a box of size 100 nm $\times$ 100 nm $\times$ 200 nm, which was divided into discrete cells of size 0.5 nm $\times$ 0.5 nm $\times$ 0.5 nm. The trajectory files from the simulations were analyzed to compute the number density in each cell as a function of simulation time. A time averaged 3D mass density profile was obtained by multiplying the number density with the mass of each bead and then averaging over the simulation time. The 3D density was averaged in the circumferential direction to obtain two-dimensional (2D) *rz* density plots (as shown in Figure 4c). Finally, the radial density distribution was obtained by averaging these 2D density maps in the vertical direction (as shown in Figure 4a,b). To compute the Coulombic and hydrophobic interaction inside the NPC (see Figure 5), we used the "gmx energy" tool from GROMACS.

4.2. *Parametrization of Phosphorylated Amino Acids*

We used five different hydrophobicity-predictor programs to estimate the hydrophobicity of phosphorylated residues. These programs calculate the logarithmic value of the equilibrium partition coefficient *P*, i.e., the ratio of concentrations in a mixture of two immiscible phases, water, and 1-octanol, as a measure for hydrophobicity. They use experimental log *P* values of fragments (small groups of atoms) to calculate the log *P* for bigger molecules by adding the individual contributions from the constituting fragments, based on the structure additivity principle for hydrophobicity [55]. There are several challenges in incorporating the log *P* estimates for the complete molecules directly in our 1 BPA model, which are: (i) the error from the estimate of the log *P* values for the fragments accumulate while calculating the log *P* for the entire molecule; and (ii) each of the hydrophobicity-predictor programs are trained with different experimental data, and therefore generate different estimates of log *P* for a given molecular structure. In order to have the hydrophobicity values comparable to our 1BPA model, first we rescaled and then normalized the log *P* estimates for all amino acids obtained from a hydrophobicity-predictor program *k*, so that hydrophobicity of any amino acid *i* (i.e., $\varepsilon_{k,i}$) falls in the range from 0 to 1. Here, 0 and 1 corresponds to the hydrophobicity of the most hydrophilic and most hydrophobic amino acids, according to hydrophobicity-predictor program *k*. Next, to minimize the error we decided to incorporate the change in $\varepsilon_{k,i}$ values of the phosphorylatable residues (obtained from the five hydrophobicity-predictor programs) compared to the 1BPA model, for which the hydrophobicity values are extracted from three different partition coefficient measurements [35]. To account for the variation in the prediction of $\varepsilon_{k,i}$ by the different hydrophobicity-predictor programs, a weighted average approach is considered. The weights are assigned to individual predictor programs based on their accuracy in predicting the $\varepsilon_{k,i}$ values of the amino acids in their native state, as used in [35]. Thus, the assigned weight for hydrophobicity-predictor program *k* for amino acid *i* can be written as,

$$w_{k,i} = \frac{(1/\Delta\varepsilon_{k,i})^2}{\sum_{k=1}^{5}(1/\Delta\varepsilon_{k,i})^2}, \tag{2}$$

where $\Delta\varepsilon_{k,i} = \varepsilon_{k,i} - \varepsilon_{1\mathrm{BPA},i}$ represents the difference between the hydrophobicity for an amino acid in its native state used in our 1BPA model [35] and the hydrophobicity-predictor programs (see Tables S3 and S4 for the source data). Next, the change in hydrophobicity upon phosphorylation is calculated as

$\Delta\varepsilon_{k,i\text{-phos}} = \varepsilon_{k,i\text{-phos}} - \varepsilon_{1\text{BPA},i}$, where "$i$-phos" represents the amino acid i in its phosphorylated state. Finally, using the weights for the hydrophobicity-predictor programs (see Table S4) we computed the hydrophobicity for the phosphorylated amino acid as $\varepsilon_{\text{p},i} = \varepsilon_{1\text{BPA},i} + \sum_{k=1}^{5} w_{k,i}\, \Delta\varepsilon_{k,i\text{-phos}}$. The amino acids Serine (S), Histidine (H), Tyrosine (Y), and Threonine (T) undergo phosphorylation [18,19], and the introduction of a phosphate group results in the introduction of a $-2e$ charge, as shown in Table 1. The phosphorylation of these amino acids results in a more hydrophilic atomic composition, which can be seen in Table 1. As a reference, the prediction of the hydrophobicity of amino acid i in the native state, $\varepsilon_{\text{weighted},i} = \varepsilon_{1\text{BPA},i} + \sum_{k=1}^{5} w_{k,i}\, \Delta\varepsilon_{k,i}$, is also shown in Table 1. Note that the subscript i is dropped from $\varepsilon_{\text{p},i}$ and $\varepsilon_{\text{weighted},i}$ in Table 1 for clarity.

Author Contributions: Conceptualization, A.M. and P.R.O.; methodology, A.M., W.S.; software, A.M.; validation, A.M.; formal analysis, A.M.; investigation, A.M..; writing—original draft preparation, A.M.; writing—review and editing, A.M., W.S., L.M.V., E.V.G., and P.R.O.; visualization, A.M..; supervision, P.R.O. and E.V.G.; project administration, P.R.O.; funding acquisition, P.R.O.

Acknowledgments: We acknowledge the use of the Peregrine cluster (University of Groningen) and the Cartesius cluster (SURFsara, funding grant by NWO) for the large scale simulations carried out during this project.

Abbreviations

NPC	Nuclear pore complex
FG	phenylalanine-glycine
lc	Low charged
hc	High charged
s	Stalk
Kap	Karyopherin protein

References

1. Rout, M.P.; Blobel, G. Isolation of the yeast nuclear pore complex. *J. Cell Biol.* **1993**, *123*, 771–783. [CrossRef] [PubMed]
2. Yang, Q.; Rout, M.P.; Akey, C.W. Three-dimensional architecture of the isolated yeast nuclear pore complex: Functional and evolutionary implications. *Mol. Cell* **1998**, *1*, 223–234. [CrossRef]
3. Reichelt, R.; Holzenburg, A.; Buhle, E.L.; Jarnik, M.; Engel, A.; Aebi, U. Correlation between structure and mass distribution of the nuclear pore complex and of distinct pore complex components. *J. Cell Biol.* **1990**, *110*, 883–894. [CrossRef] [PubMed]
4. Hurt, E.; Beck, M. Towards understanding nuclear pore complex architecture and dynamics in the age of integrative structural analysis. *Curr. Opin. Cell Biol.* **2015**, *34*, 31–38. [CrossRef] [PubMed]
5. Hoelz, A.; Glavy, J.S.; Beck, M. Toward the atomic structure of the nuclear pore complex: When top down meets bottom up. *Nat. Struct. Mol. Biol.* **2016**, *23*, 624–630. [CrossRef] [PubMed]
6. Popken, P.; Ghavami, A.; Onck, P.R.; Poolman, B.; Veenhoff, L.M. Size-Dependent Leak of Soluble and Membrane Proteins Through the Yeast Nuclear Pore Complex. *Mol. Biol. Cell* **2015**, *26*, 1386–1394. [CrossRef] [PubMed]
7. Schmidt, H.B.; Görlich, D. Transport Selectivity of Nuclear Pores, Phase Separation, and Membraneless Organelles. *Trends Biochem. Sci.* **2016**, *41*, 46–61. [CrossRef]
8. Timney, B.L.; Raveh, B.; Mironska, R.; Trivedi, J.M.; Kim, S.J.; Russel, D.; Wente, S.R.; Sali, A.; Rout, M.P. Simple rules for passive diffusion through the nuclear pore complex. *J. Cell Biol.* **2016**, *215*. [CrossRef]
9. Iovine, M.K.; Watkins, J.L.; Wente, S.R. The GLFG repetitive region of the nucleoporin Nup116p interacts with Kap95p, an essential yeast nuclear import factor. *J. Cell Biol.* **1995**, *131*, 1699–1713. [CrossRef]

10. Bayliss, R.; Ribbeck, K.; Akin, D.; Kent, H.M.; Feldherr, C.M.; Görlich, D.; Stewart, M. Interaction between NTF2 and xFxFG-containing nucleoporins is required to mediate nuclear import of RanGDP. *J. Mol. Biol.* **1999**, *293*, 579–593. [CrossRef]
11. Rout, M.P.; Aitchison, J.D.; Magnasco, M.O.; Chait, B.T. Virtual gating and nuclear transport: The hole picture. *Trends Cell Biol.* **2003**, *13*, 622–628. [CrossRef] [PubMed]
12. Pante, N.; Kann, M. Nuclear Pore Complex Is Able to Transport Macromolecules with Diameters of 39 nm. *Mol. Biol. Cell* **2002**, *13*, 425–434. [CrossRef] [PubMed]
13. Lowe, A.R.; Siegel, J.J.; Kalab, P.; Siu, M.; Weis, K.; Liphardt, J.T. Selectivity mechanism of the nuclear pore complex characterized by single cargo tracking. *Nature* **2010**, *467*, 600–603. [CrossRef] [PubMed]
14. Frey, S.; Rees, R.; Schünemann, J.; Ng, S.C.; Fünfgeld, K.; Huyton, T.; Görlich, D. Surface Properties Determining Passage Rates of Proteins through Nuclear Pores. *Cell* **2018**, *174*, 202–217.e9. [CrossRef] [PubMed]
15. Ananth, A.N.; Mishra, A.; Frey, S.; Dwarkasing, A.; Versloot, R.; van der Giessen, E.; Görlich, D.; Onck, P.; Dekker, C. Spatial structure of disordered proteins dictates conductance and selectivity in nuclear pore complex mimics. *Elife* **2018**, *7*. [CrossRef] [PubMed]
16. Frey, S.; Richter, R.P.; Görlich, D. FG-rich repeats of nuclear pore proteins form a three-dimensional meshwork with hydrogel-like properties. *Science* **2006**, *314*, 815–817. [CrossRef] [PubMed]
17. Frey, S.; Görlich, D. A saturated FG-repeat hydrogel can reproduce the permeability properties of nuclear pore complexes. *Cell* **2007**, *130*, 512–523. [CrossRef]
18. Polyansky, A.A.; Zagrovic, B. Protein electrostatic properties predefining the level of surface hydrophobicity change upon phosphorylation. *J. Phys. Chem. Lett.* **2012**, *3*, 973–976. [CrossRef]
19. Petrov, D.; Margreitter, C.; Grandits, M.; Oostenbrink, C.; Zagrovic, B. A Systematic Framework for Molecular Dynamics Simulations of Protein Post-Translational Modifications. *PLoS Comput. Biol.* **2013**, *9*. [CrossRef]
20. Lee, T.; Hoofnagle, A.N.; Kabuyama, Y.; Stroud, J.; Min, X.; Goldsmith, E.J.; Chen, L.; Resing, K.A.; Ahn, N.G. Docking motif interactions in Map kinases revealed by hydrogen exchange mass spectrometry. *Mol. Cell* **2004**, *14*, 43–55. [CrossRef]
21. Vomastek, T.; Iwanicki, M.P.; Burack, W.R.; Tiwari, D.; Kumar, D.; Parsons, J.T.; Weber, M.J.; Nandicoori, V.K. Extracellular Signal-Regulated Kinase 2 (ERK2) Phosphorylation Sites and Docking Domain on the Nuclear Pore Complex Protein Tpr Cooperatively Regulate ERK2-Tpr Interaction. *Mol. Cell. Biol.* **2008**, *28*, 6954–6966. [CrossRef] [PubMed]
22. Kosako, H.; Imamoto, N. Phosphorylation of nucleoporins: Signal transduction-mediated regulation of their interaction with nuclear transport receptors. *Nucleus* **2010**. [CrossRef] [PubMed]
23. Lusk, C.P.; Waller, D.D.; Makhnevych, T.; Dienemann, A.; Whiteway, M.; Thomas, D.Y.; Wozniak, R.W. Nup53p is a target of two mitotic kinases, Cdk1p and Hrr25p. *Traffic* **2007**. [CrossRef] [PubMed]
24. Ficarro, S.B.; McCleland, M.L.; Stukenberg, P.T.; Burke, D.J.; Ross, M.M.; Shabanowitz, J.; Hunt, D.F.; White, F.M. Phosphoproteome analysis by mass spectrometry and its application to Saccharomyces cerevisiae. *Nat. Biotechnol.* **2002**. [CrossRef] [PubMed]
25. Ciomperlik, J.J.; Basta, H.A.; Palmenberg, A.C. Three cardiovirus Leader proteins equivalently inhibit four different nucleocytoplasmic trafficking pathways. *Virology* **2015**, *484*, 194–202. [CrossRef] [PubMed]
26. Porter, F.W.; Palmenberg, A.C. Leader-induced phosphorylation of nucleoporins correlates with nuclear trafficking inhibition by cardioviruses. *J. Virol.* **2009**. [CrossRef]
27. Kosako, H.; Yamaguchi, N.; Aranami, C.; Ushiyama, M.; Kose, S.; Imamoto, N.; Taniguchi, H.; Nishida, E.; Hattori, S. Phosphoproteomics reveals new ERK MAP kinase targets and links ERK to nucleoporin-mediated nuclear transport. *Nat. Struct. Mol. Biol.* **2009**, *16*, 1026–1035. [CrossRef]
28. Carlson, S.M.; Chouinard, C.R.; Labadorf, A.; Lam, C.J.; Schmelzle, K.; Fraenkel, E.; White, F.M. Large-scale discovery of ERK2 substrates identifies ERK-mediated transcriptional regulation by ETV3. *Sci. Signal.* **2011**, *4*. [CrossRef]
29. Kehlenbach, R.H.; Gerace, L. Phosphorylation of the nuclear transport machinery down-regulates nuclear protein import in vitro. *J. Biol. Chem.* **2000**, *275*, 17848–17856. [CrossRef]
30. Hazawa, M.; Lin, D.; Kobayashi, A.; Jiang, Y.; Xu, L.; Dewi, F.R.P.; Mohamed, M.S.; Hartono; Nakada, M.; Meguro-Horike, M.; Horike, S.; et al. ROCK-dependent phosphorylation of NUP62 regulates p63 nuclear transport and squamous cell carcinoma proliferation. *EMBO Rep.* **2018**. [CrossRef]

31. Borlido, J.; D'Angelo, M.A. Nup62-mediated nuclear import of p63 in squamous cell carcinoma. *EMBO Rep.* **2018**, *19*, 3–4. [CrossRef]
32. Shindo, Y.; Iwamoto, K.; Mouri, K.; Hibino, K.; Tomita, M.; Kosako, H.; Sako, Y.; Takahashi, K. Conversion of graded phosphorylation into switch-like nuclear translocation via autoregulatory mechanisms in ERK signalling. *Nat. Commun.* **2016**, *7*. [CrossRef]
33. Ghavami, A.; Van Der Giessen, E.; Onck, P.R. Energetics of transport through the nuclear pore complex. *PLoS ONE* **2016**, *11*. [CrossRef]
34. Tagliazucchi, M.; Peleg, O.; Kröger, M.; Rabin, Y.; Szleifer, I. Effect of charge, hydrophobicity, and sequence of nucleoporins on the translocation of model particles through the nuclear pore complex. *Proc. Natl. Acad. Sci. USA* **2013**, *110*, 3363–3368. [CrossRef]
35. Ghavami, A.; Veenhoff, L.M.; Van Der Giessen, E.; Onck, P.R. Probing the disordered domain of the nuclear pore complex through coarse-grained molecular dynamics simulations. *Biophys. J.* **2014**, *107*, 1393–1402. [CrossRef]
36. Hayama, R.; Rout, M.P.; Fernandez-Martinez, J. The nuclear pore complex core scaffold and permeability barrier: Variations of a common theme. *Curr. Opin. Cell Biol.* **2017**, *46*, 110–118. [CrossRef]
37. Ingrell, C.R.; Miller, M.L.; Jensen, O.N.; Blom, N. NetPhosYeast: Prediction of protein phosphorylation sites in yeast. *Bioinformatics* **2007**. [CrossRef]
38. Bai, Y.; Chen, B.; Li, M.; Zhou, Y.; Ren, S.; Xu, Q.; Chen, M.; Wang, S. FPD: A comprehensive phosphorylation database in fungi. *Fungal Biol.* **2017**. [CrossRef]
39. Yamada, J.; Phillips, J.L.; Patel, S.; Goldfien, G.; Calestagne-Morelli, A.; Huang, H.; Reza, R.; Acheson, J.; Krishnan, V.V.; Newsam, S.; et al. A bimodal distribution of two distinct categories of intrinsically disordered structures with separate functions in FG nucleoporins. *Mol. Cell. Proteomics* **2010**, *9*, 2205–2224. [CrossRef]
40. Leo, A.J. Calculating log Poct from Structures. *Chem. Rev.* **1993**, *93*, 1281–1306. [CrossRef]
41. Viswanadhan, V.N.; Ghose, A.K.; Revankar, G.R.; Robins, R.K. Atomic Physicochemical Parameters for Three Dimensional Structure Directed Quantitative Structure-Activity Relationships. 4. Additional Parameters for Hydrophobic and Dispersive Interactions and Their Application for an Automated Superposition of Certain. *J. Chem. Inf. Comput. Sci.* **1989**, *29*, 163–172. [CrossRef]
42. Meylan, W.M.; Howard, P.H. Atom/fragment contribution method for estimating octanol-water partition coefficients. *J. Pharm. Sci.* **1995**, *84*, 83–92. [CrossRef]
43. Tetko, I.V.; Gasteiger, J.; Todeschini, R.; Mauri, A.; Livingstone, D.; Ertl, P.; Palyulin, V.A.; Radchenko, E.V.; Zefirov, N.S.; Makarenko, A.S.; Tanchuk, V.Y.; Prokopenko, V.V. Virtual computational chemistry laboratory—Design and description. *J. Comput. Aided Mol. Des.* **2005**, *19*, 453–463. [CrossRef]
44. Carrasco, B.; De La Torre, J.G. Hydrodynamic properties of rigid particles: Comparison of different modeling and computational procedures. *Biophys. J.* **1999**. [CrossRef]
45. Garcia de la Torre, J.; Navarro, S.; Lopez Martinez, M.C.; Diaz, F.G.; Lopez Cascales, J.J. HYDRO: A computer program for the prediction of hydrodynamic properties of macromolecules. *Biophys. J.* **1994**. [CrossRef]
46. Ghavami, A.; van der Giessen, E.; Onck, P.R. Coarse-Grained Potentials for Local Interactions in Unfolded Proteins. *J. Chem. Theory Comput.* **2013**, *9*, 432–440. [CrossRef]
47. Marsh, J.A.; Forman-Kay, J.D. Sequence determinants of compaction in intrinsically disordered proteins. *Biophys. J.* **2010**, *98*, 2383–2390. [CrossRef]
48. Alber, F.; Dokudovskaya, S.; Veenhoff, L.M.; Zhang, W.; Kipper, J.; Devos, D.; Suprapto, A.; Karni-Schmidt, O.; Williams, R.; Chait, B.T.; et al. The molecular architecture of the nuclear pore complex. *Nature* **2007**, *450*, 695–701. [CrossRef]
49. Rexach, M.; Blobel, G. Protein import into nuclei: Association and dissociation reactions involving transport substrate, transport factors, and nucleoporins. *Cell* **1995**, *83*, 683–692. [CrossRef]
50. Görlich, D.; Panté, N.; Kutay, U.; Aebi, U.; Bischoff, F.R. Identification of different roles for RanGDP and RanGTP in nuclear protein import. *EMBO J.* **1996**, *15*, 5584–5594. [CrossRef]
51. Kersey, P.J.; Staines, D.M.; Lawson, D.; Kulesha, E.; Derwent, P.; Humphrey, J.C.; Hughes, D.S.T.; Keenan, S.; Kerhornou, A.; Koscielny, G.; et al. Ensembl Genomes: An integrative resource for genome-scale data from non-vertebrate species. *Nucleic Acids Res.* **2012**, *40*. [CrossRef]
52. Pyhtila, B.; Rexach, M. A Gradient of Affinity for the Karyopherin Kap95p along the Yeast Nuclear Pore Complex. *J. Biol. Chem.* **2003**, *278*, 42699–42709. [CrossRef]

53. Ketterer, P.; Ananth, A.N.; Laman Trip, D.S.; Mishra, A.; Bertosin, E.; Ganji, M.; Van Der Torre, J.; Onck, P.; Dietz, H.; Dekker, C. DNA origami scaffold for studying intrinsically disordered proteins of the nuclear pore complex. *Nat. Commun.* **2018**, *9*. [CrossRef]
54. Colwell, L.; Brenner, M.; Ribbeck, K. Charge as a selection criterion for translocation through the nuclear pore complex. *PLoS Comput. Biol.* **2010**. [CrossRef]
55. Fujita, T.; Iwasa, J.; Hansch, C. A New Substituent Constant, π, Derived from Partition Coefficients. *J. Am. Chem. Soc.* **1964**, *86*, 5175–5180. [CrossRef]

11

Acetylation Disfavors Tau Phase Separation

Josephine C. Ferreon [1,*], Antrix Jain [2], Kyoung-Jae Choi [1], Phoebe S. Tsoi [1], Kevin R. MacKenzie [1,3], Sung Yun Jung [4] and Allan Chris Ferreon [1,*]

1 Department of Pharmacology and Chemical Biology, Baylor College of Medicine, Houston, TX 77030, USA; Kyoungjae.Choi@bcm.edu (K.-J.C.); Phoebe.Tsoi@bcm.edu (P.S.T.); Kevin.MacKenzie@bcm.edu (K.R.M.)

2 Advanced Technology Cores, Baylor College of Medicine, Houston, TX 77030, USA; antrixj@bcm.edu

3 Department of Pathology and Immunology, Baylor College of Medicine, Houston, TX 77030, USA

4 Department of Biochemistry and Molecular Biology, Baylor College of Medicine, Houston, TX 77030, USA; syjung@bcm.edu

* Correspondence: josephine.ferreon@bcm.edu (J.C.F.); allan.ferreon@bcm.edu (A.C.F.);

Abstract: Neuropathological aggregates of the intrinsically disordered microtubule-associated protein Tau are hallmarks of Alzheimer's disease, with decades of research devoted to studying the protein's aggregation properties both in vitro and in vivo. Recent demonstrations that Tau is capable of undergoing liquid-liquid phase separation (LLPS) reveal the possibility that protein-enriched phase separated compartments could serve as initiation sites for Tau aggregation, as shown for other amyloidogenic proteins, such as the Fused in Sarcoma protein (FUS) and TAR DNA-binding protein-43 (TDP-43). Although truncation, mutation, and hyperphosphorylation have been shown to enhance Tau LLPS and aggregation, the effect of hyperacetylation on Tau aggregation remains unclear. Here, we investigate how the acetylation of Tau affects its potential to undergo phase separation and aggregation. Our data show that the hyperacetylation of Tau by p300 histone acetyltransferase (HAT) disfavors LLPS, inhibits heparin-induced aggregation, and impedes access to LLPS-initiated microtubule assembly. We propose that Tau acetylation prevents the toxic effects of LLPS-dependent aggregation but, nevertheless, contributes to Tau loss-of-function pathology by inhibiting Tau LLPS-mediated microtubule assembly.

Keywords: intrinsically disordered protein; membrane-less organelle; neurodegenerative disease; p300 HAT acetylation; post-translational modification; protein aggregation; Tau fibrillation

---

1. Introduction

Tau inclusions are key components of neurofibrillary tangles (NFTs) a recurring pathological feature for several neurodegenerative diseases including Alzheimer's disease (AD) [1–4]. There are two prevailing hypotheses on the mechanism of Tau pathology linked to protein misfolding and aggregation, both of which are not mutually exclusive [5–7]. One is that Tau has intrinsic aggregation motifs that enable fibrillation, leading to gain-in-toxic function(s) [8–11] exacerbated by the inability of the cellular degradation machinery to remove misfolded or aggregated Tau [12,13]. Another pathological mechanism is that aggregation-promoting Tau accumulation stems from loss-of-normal function(s). Tau is essential for microtubule dynamics and stability; impairment of this function linked to Tau sequestration into aggregates results in neuronal loss [14]. Alterations in protein sequence and structure (such as truncations, mutations, or post-translational modifications) contribute to both Tau loss-of-normal function and gain-in-toxic dysfunction by affecting the protein's ability to bind microtubules or propensity to misfold and aggregate [6,7,11].

Tau is an intrinsically disordered protein (IDP), a class of proteins characterized by a high degree of structural flexibility, conformational heterogeneity and binding promiscuity. Often, these properties not only allow for complex functions involving networks of interactions, but also facilitate dysfunctions as a result of misfolding or aggregation [15–17]. Tau is rich in serine/threonine (S/T) and lysine (K) residues, and is known to undergo post-translational modifications (PTMs), such as phosphorylation, acetylation, ubiquitination, and sumoylation. These PTMs are linked to both Tau function and pathology [18]. Phosphorylation is known to modulate Tau's ability to promote microtubule assembly. Abnormal Tau hyperphosphorylation, however, results in fibrillation, as evidenced by hyperphosphorylated Tau being the primary component of NFTs [1]. At least 20 phosphorylation sites [14] and 23 acetylation sites [19–21] have been reported for Tau.

Tau is a macromolecular polyampholyte consisting of negatively-charged N- and C-terminal domains, and a positively-charged central Proline-rich (P) domain with microtubule binding regions (MTBR, R1–R4; Figure 1A). An increase in negative charge via phosphorylation or removal of lysine positive charge by acetylation can have significant effects on Tau function (microtubule assembly/stabilization) and dysfunction (Tau aggregation). Tau acetylation can be mediated by p300/CREB-binding protein (CBP) HAT [19–23], and reports indicate that Tau, itself, has intrinsic acetyltransferase activity [19,20,22]. In fact, hyperacetylated Tau (Ac-Tau) has been used as a diagnostic marker for AD [20,21,24]. Although the role of hyperphosphorylation in facilitating Tau pathological aggregation is not debated, there are conflicting reports on the role of hyperacetylation in Tau pathology. Several groups have found acetylated Tau in pathological inclusions in vivo [20,21], as well as in co-deposits with hyperphosphorylated Tau [21]. However, Cook et al. report that acetylation at key Tau motifs ($K_{259/353}$IGS) can be protective through the inhibition of phosphorylation of a nearby serine that otherwise would promote aggregation [23]. In addition, observations from in vitro experiments are contradictory: Cohen et al. report that acetylation accelerates Tau heparin-induced fibrillation [20], whereas others indicate that acetylation inhibits Tau filament assembly [19,23].

Liquid-liquid phase separation (LLPS) has recently gained attention as a physical mechanism for proteins to self-assemble into compartments termed membrane-less organelles [25–28]. The LLPS-mediated enrichment of proteins into membrane-less organelles, such as stress granules [29], provides "hotbeds" or seeds for protein aggregation [30–32]. LLPS was shown to initiate the aggregation of several neurodegenerative disease-associated proteins, such as FUS [30], TDP-43 [31,33], and hnRNPA1 [34]. Recently, Tau LLPS has been implicated in both the functional role of Tau in promoting microtubule assembly [35] and dysfunction in initiating Tau self-interaction and fibrillation [36–38]. Although it has been shown that hyperphosphorylation accelerates Tau LLPS and aggregation, consistent with hyperphosphorylated Tau's abundance in pathological inclusions [36,37], there are no reports on the role of acetylation on Tau phase separation and LLPS-mediated aggregation. Since Tau LLPS is expected to be strongly influenced by electrostatics, here, we investigate the role of acetylation in driving LLPS and determine if this role is consistent with the current hypothesis that LLPS can initiate and mediate Tau aggregation.

2. Results and Discussion

2.1. p300-Mediated Acetylation of Tau

Hyperacetylated Tau (Ac-Tau) was prepared from wt Tau using p300 HAT (see Section 3). Tau acetylation was verified by Western blot against acetyl-lysines (Figure 1B) and mass spectrometry (Figure 1C–E). We identified 15 acetylation sites (99% sequence coverage) using tandem mass spectrometry (Figure 1C; Section 3), including K148 near the Tau N-terminal domain; K163, K174, K190, K224, K234, and K240 in the P1–2 regions; K254, K280, K281, K290, and K311 in the microtubule binding region (MTBR, R1–R4); and, K375, K385, and K395 in the P3 region (Figure 1A,C). Figure 1D,E shows the representative MS/MS spectra (VQIINK$_{280}$K$_{281}$ and VQIVYK$_{311}$, respectively). Fragmented b (red) and y (blue) ions from low energy collisions in mass spectrometer are marked. The b and y ions refer to the peaks corresponding to the prefix ions observed sequentially in the spectrum with each prefix offset from the previous by the mass of an amino acid. A 42.016 Da increase in mass difference due to lysine acetylation is included in the sequential mass difference to construct the peptide sequence. The acetylation sites that were identified are consistent with previous reports [19–21]. Notably, K280/K281 and K311, which are in the hexapeptide aggregation motifs VQIINK$_{280}$K$_{281}$ and VQIVYK$_{311}$, were found to be acetylated (Figure 1D,E). These motifs play critical roles in Tau interaction with negatively-charged microtubules and with polyanions such as heparin [9]. If we assume complete acetylation, we expect the theoretical pI for full-length Tau to change from 8.2 to 5.5 (Prot pi web tool; https://www.protpi.ch). Such modifications can significantly alter Tau electrostatic properties, with 50% acetylation already corresponding to a pI of 6.2. Interestingly, the observed sites of acetylation are concentrated in the positively-charged central region of Tau (Figure 1A).

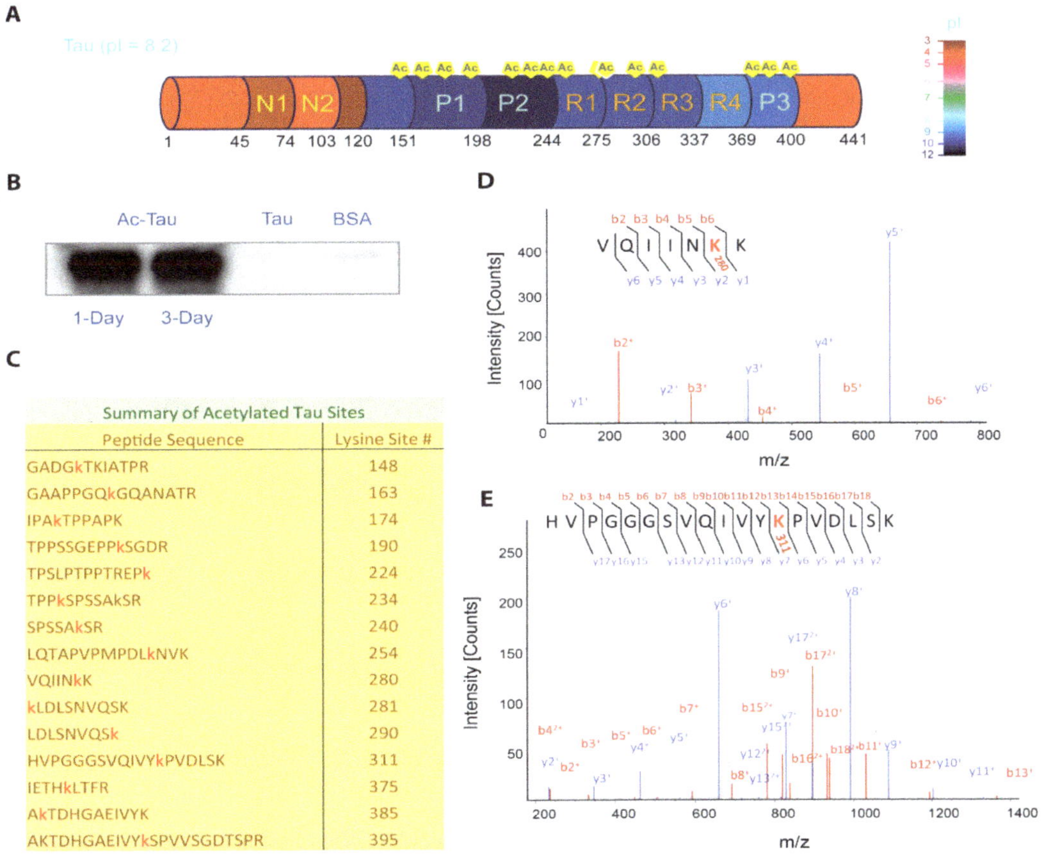

Figure 1. Tau hyperacetylation. (**A**) Domain organization of Tau. Protein segments are color-coded to reflect their respective pIs using the provided color palette. (**B**) In vitro Tau acetylation was verified by Western blot against acetyl-lysine (left to right lanes: acetylated Tau after one- and three-day reaction incubations, and negative controls using wild-type Tau and BSA, respectively). (**C**) Tau lysine acetylation sites (red) identified by mass spectrometry. (**D**,**E**) MS/MS spectra of peptides that contain the hexapeptide aggregation motifs VQIINK$_{280}$K$_{281}$ and VQIVYK$_{311}$, showing acetylation at K280 and K311, respectively (shown in red). The 'b' ions (shown in red) represent fragment peaks generated from the amino to carboxyl terminus. The 'y' ions (shown in blue) represent fragment peaks generated from the carboxyl to amino terminus. The suffix numbers represent the corresponding number of amino acids. See Section 3 for details.

2.2. *Acetylation Changes Tau Phase Behavior*

LLPS has recently been observed for wt and hyperphosphorylated Tau, and truncation mutant (K18) [35–38]. At low salt conditions (5 mM sodium phosphate, pH 7.8), we observed near-instantaneous formation of wt Tau droplets (Figure 2A). Subsequent fusions indicate the liquid nature of the wt Tau droplets. We characterized the protein concentration (2.5–20 μM) and salt concentration (0–250 mM NaCl) dependencies of wt Tau LLPS. Consistent with the literature [37,38], higher salt concentrations disfavor LLPS and higher protein concentrations favor LLPS (Figure 2E). For the case of Ac-Tau, we observed a dramatic reduction in droplet formation (Figure 2B,F). Similar results were also observed when LLPS experiments were performed with wt Tau or Ac-Tau in the presence of a crowding agent (10% PEG 8K, 200 mM NaCl, 10 mM acetate, 10 mM glycine, 10 mM sodium phosphate, pH 7.5; Figure 2C,D). Thus, independent of the presence or absence of crowding, the hyperacetylation of Tau disfavors LLPS.

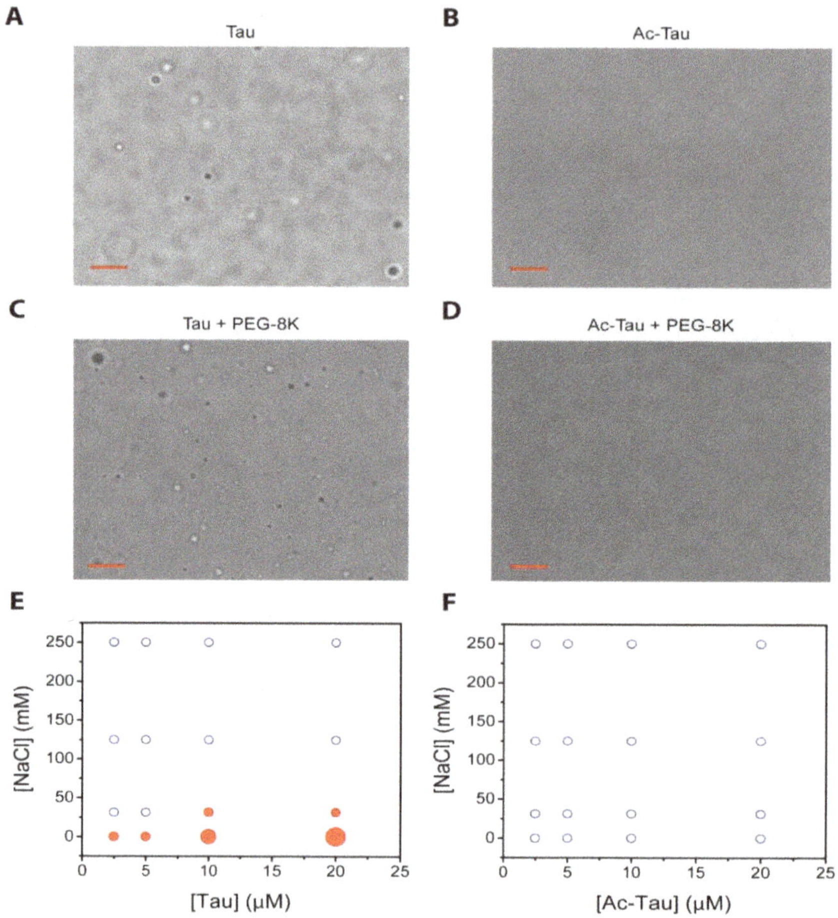

Figure 2. Acetylation disfavors Tau liquid-liquid phase separation (LLPS). Droplet formation of wt Tau versus hyperacetylated (Ac-Tau), respectively, in the absence (**A**,**B**) and presence of the crowding agent PEG-8K (**C**,**D**). Phase transition maps of Tau versus Ac-Tau (in the absence of crowders), respectively (**E**,**F**). The diagrams present protein and salt concentration dependencies of droplet formation. Blue open circles represent conditions of minimal droplet formation (<5 droplets/frame); small red solid circles represent 5–20 droplets, medium red solid circle 20–100 droplets, and large red circle >100 droplets. The bars in (**A**–**D**) represent 10 μm. See Section 3 for details.

Interestingly, even though both hyperphosphorylation and hyperacetylation decrease the overall pI of Tau, the two PTMs seem to have opposite effects on LLPS. In contrast to LLPS enhancement by hyperphosphorylation [36,37], hyperacetylation clearly disfavors Tau LLPS (Figure 2). Further experiments performed in identical or comparable conditions using the same Tau constructs, full-length or otherwise, are needed for a clear and direct comparison of LLPS behaviors of hyperphosphorylated, hyperacetylated, and wt Tau proteins. Nevertheless, we think that hyperacetylation disfavors full-length Tau LLPS by neutralizing the lysine positive charges, thereby affecting opposite-charge attractions that help support Tau self- and mesoscale interactions. Our data also give direct support that electrostatics plays a major role in Tau LLPS.

2.3. Acetylation of Tau Inhibits Heparin-Induced Aggregation

Heparin has been widely used to induce and accelerate Tau aggregation [11]. Utilizing a truncated Tau construct, Ambadipudi et al. demonstrated that heparin promotes Tau fibrillation via LLPS [37]. Similarly, we observed that heparin induces LLPS of full-length wt Tau and facilitates subsequent protein aggregation (Figure 3A–C,F). In contrast, Ac-Tau failed to undergo heparin-induced LLPS in the same experimental conditions (Figure 3D). Additionally, Ac-Tau (relative to wt Tau) exhibited a dramatic decrease in the fibrillation rate as reported by Th T fluorescence (Figure 3F). Residues in the VYINK$_{280}$K$_{281}$ and VQIVK$_{311}$ regions of the Tau microtubule binding repeats (R1–R4), which we identified as Tau acetylation sites (Figure 1), are also known interaction sites for heparin [39]. Thus, the observed effects of acetylation on Tau heparin-induced aggregation can be attributed to the loss of binding to heparin.

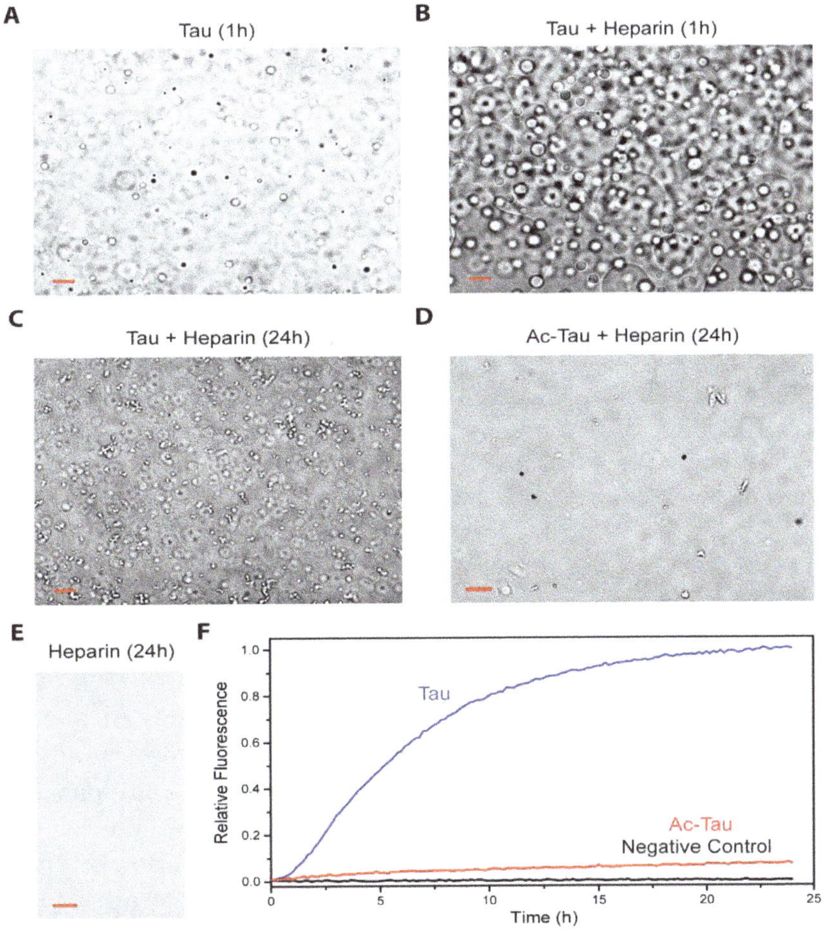

Figure 3. Acetylation disfavors heparin-induced Tau aggregation. (**A**,**B**) Heparin accelerates non-acetylated wt Tau LLPS. The presence of heparin (5 μM heparin: 20 μM Tau ratio) results in more droplets and larger fused droplets at the bottom of the dish. Droplet formation was not observed for Ac-Tau in the absence (Figure 2B,D) or presence of heparin (24 h, (**D**)). After 24 h incubation, an abundance of irregularly-shaped oligomers/aggregates was observed for wt Tau but not for Ac-Tau (**C**,**D**). No aggregation/LLPS was observed for the heparin control (**E**). Ac-Tau displayed minimal aggregation compared to wt Tau as reported by Th T fluorescence assay (**F**). Scale bars represent 10 μm. See Section 3 for details.

Although heparin accelerates wt Tau LLPS, it is unknown whether heparin is equally distributed in the Tau-rich and Tau-poor phases (which we think to be unlikely). LLPS, nevertheless, allows Tau to co-localize and thereby concentrate, with the Tau-rich condensed phase facilitating Tau aggregation nucleation and/or seeding.

2.4. Acetylation of Tau Prevents Access to LLPS-Mediated Microtubule Assembly

A recent report by Hernandez-Vega et al. suggests that Tau phase separated droplets (induced using the crowding agents PEG, Ficoll or dextran) can initiate microtubule assembly [35]. To assess LLPS-mediated microtubule assembly by wt Tau and Ac-Tau independent of crowding agents, we performed our phase separation experiments in low-salt conditions. After mixing rhodamine-labeled and unlabeled tubulin heterodimers with wt Tau, we observed an initial increase in solution turbidity. The ensuing dynamic microtubule assembly was visible by fluorescence microscopy within 1 h of incubation (Figure 4A). In contrast, Ac-Tau neither displayed turbidity nor detectable microtubule assembly up to 18 h of incubation (Figure 4B). Whereas previous studies have shown that acetylation reduces Tau's ability to bind to microtubules [20], our data clearly demonstrates that the failure of Ac-Tau to undergo LLPS affects its potential for microtubule assembly.

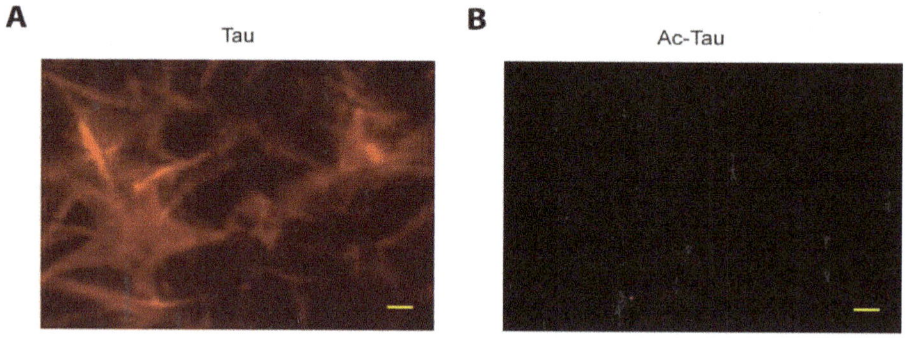

Figure 4. Tau acetylation prevents access to LLPS-mediated microtubule assembly. (**A**,**B**) Fluorescence microscopy images showing microtubule assembly from mixtures of rhodamine-labeled and unlabeled tubulin heterodimers with Tau or Ac-Tau, respectively. Scale bars represent 10 μm. See Section 3 for details.

Our in vitro data indicate that Ac-Tau is less prone to aggregation as compared to wt Tau. Cryo-EM structures of AD patient-derived filaments indicate that Tau residues 306–378 form the amyloid core [40]. The stable core is composed of several β-strands that pack intra- and inter-molecularly, with β1 ($_{306}$VYINK$_{311}$) in close proximity to β8 [40]. Our results show that in Ac-Tau, K311 (β1), and K375 (β8) are both acetylated; we speculate that this influences interactions within the amyloid core, and contributes to inhibition of Tau aggregation. Further experiments on the acetylation of the amyloid core residues will be needed to directly assess the effect of Tau acetylation on the amyloid structure.

Recent reports suggest that Tau aggregation is accelerated through LLPS [36,37]. Our data clearly show that acetylation decreases or abolishes Tau LLPS. Our findings are consistent with an LLPS-mediated model of aggregation (but do not prove whether such a mechanism is operative in vivo). Since acetylation reduces the propensity of Tau to undergo LLPS, we conclude that acetylation in vivo is unlikely to enhance or lead directly to condensation-mediated aggregation, in contrast to the demonstrated effect of hyperphosphorylation [36]. It is, however, possible that combinations of phosphorylations and acetylations can favor LLPS and/or aggregation; future experiments with Tau bearing homogeneous PTMs will be needed to address this conclusively.

Tau participates in microtubule formation and stabilization, and Tau LLPS has been shown as a mechanism by which a Tau-rich condensed phase can recruit tubulin dimers and facilitate their assembly [20]. Acetylation at key Tau sites that interfere with tubulin binding would affect this function, as would acetylation that disfavors partitioning of Tau into a Tau-rich phase. Thus, we speculate that the primary contribution of Tau acetylation to cellular dysfunction is not through a gain-of-function mechanism, such as toxic aggregation, but through a loss of physiologic function mechanism (i.e., reduced binding to tubulins/microtubules, and decreased LLPS-mediated initiation of microtubule assembly; Figure 5).

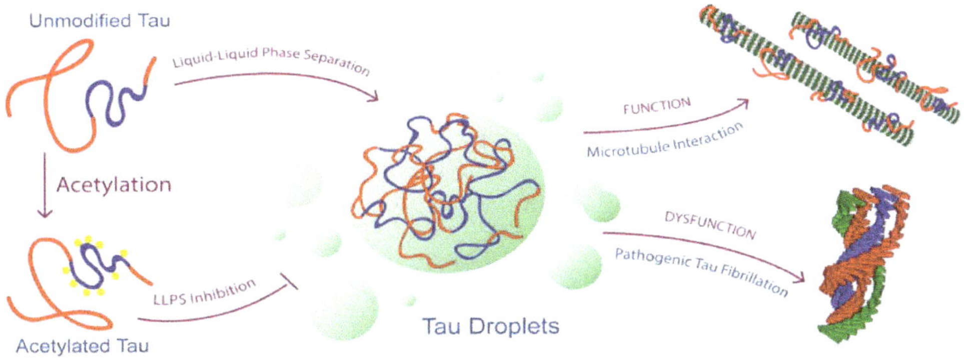

Figure 5. Model for Tau's loss of physiologic function and gain of pathologic dysfunction linked to its ability to undergo LLPS as modulated by acetylation.

Less direct effects on physiologic Tau function may also be important. Many of the same lysines (K254, K311, and K353) implicated as sites of ubiquitination [41] are also sites of acetylation and, thus, might be involved in evading the ubiquitin-lysosome proteasomal degradation machinery. Acetylation has been shown to inhibit Tau degradation by inhibiting its ubiquitination [21], and results in the accumulation of Tau, including hyperphosphorylated Tau. The presence of lysine deacetylase (SIRT1) has been shown to inhibit neuronal loss in an AD mouse model and deletion of SIRT1 results to pathologic levels of Tau in vivo [42]. Cross-talk between the different PTMs has also been reported. For example, hypoacetylation of Tau at key KIGS motifs in the *R1-4* regions increases vulnerability to hyperphosphorylation, which leads to filament aggregation [23]. Hyperphosphorylation of Tau has been reported to enhance Tau LLPS. However, other reports also show that hyperphosphorylation reduces microtubule assembly [14]. Thus, LLPS-mediated mechanisms by hyperphosphorylated Tau could be detrimental for both function and dysfunction pathways (Figure 5). We plan to carry out further experiments on hyperphosphorylated Tau to assess how this PTM of Tau can modulate microtubule assembly and protein aggregation, both in LLPS and non-LLPS conditions. Nevertheless, we speculate that the hyperacetylation of Tau is detrimental to Tau function, but not instrumental to LLPS-mediated Tau dysfunction (Figure 5). It would also be interesting to know the cross-talks between hyperphosphorylation and hyperacetylation in LLPS-mediated microtubule assembly and promotion of pathologic fibrils. Can hyperphosphorylated Tau also recruit hyperacetylated Tau into droplets? If so, this might explain the presence of hyperacetylated Tau in pathological inclusions of hyperphosphorylated Tau.

In conclusion, our data affirm the importance of electrostatics in Tau LLPS. Furthermore, we show that hyperacetylation disfavors Tau LLPS and, as a consequence, LLPS-facilitated aggregation. Finally, by preventing access to LLPS-mediated microtubule assembly and stabilization, hyperacetylation contributes to Tau dysfunction primarily through a loss-of-function mechanism.

3. Materials and Methods

3.1. Tau Expression and Purification

Wild-type (wt) Tau (2N4R isoform; 441 residues) plasmid (Addgene plasmid #16316, a gift from Peter Klein) was transformed into *Escherichia coli* BL21 star cells. Cells were grown at 37 °C in Terrific

Broth medium in the presence of kanamycin until the optical density at 600 nm (OD_{600}) reaches 0.8–1.0, then induced with 1 mM isopropyl β-D-1-thiogalactopyranoside (IPTG) and grown overnight at 18 °C.

wt Tau was purified using a similar procedure described by Barghorn et al. [43]. Briefly, wt Tau cell pellets were resuspended in 50 mM NaCl, 5 mM DTT, 50 mM sodium phosphate, pH 6.5, and supplemented with a protease inhibitor cocktail (GenDEPOT, Barker, TX, USA). The cells were lysed using a homogenizer (Avestin, Ottawa, ON, Canada). Additional salt was then added (for a final concentration of 450 mM NaCl) before the solution was incubated for 20 min in hot water (~80–90 °C). The supernatant was concentrated, diluted to a final salt concentration of 50 mM NaCl, and purified by FPLC (Bio-Rad, Hercules, CA, USA) using a salt gradient applied to a heparin sepharose HP column (GE, Marlborough, MA, USA). Fractions containing wt Tau were concentrated and further purified by reverse-phase HPLC (Agilent, Santa Clara, CA, USA), lyophilized, and stored at −80 °C until later use. Purified acetylated Tau (Ac-Tau) was prepared using reverse-phase HPLC after in vitro acetylation of wt Tau (see below).

3.2. p300 Histone Acetyltransferase (HAT) Domain Expression and Purification

Enzymatically-active p300 HAT was prepared as previously described [44]. Briefly, p300 HAT and Sir2 expression plasmids (generous gifts from Phillip Cole) were co-transformed into *E. coli* BL21 AI cells (Invitrogen, Carlsbad, CA, USA). Cells were grown at 37 °C in Terrific Broth medium until induction ($OD_{600} \approx 0.8$–1.0) with 1 mM IPTG, followed by overnight growth at 18 °C. Both proteins were purified using FPLC (Bio-Rad) with a Talon cobalt resin (GE) and a Q HP sepharose column (GE). Separate p300 HAT and Sir2 fractions were stored in −80 °C until later use. The final storage buffer for p300 HAT is ~150 mM NaCl, 125 mM TCEP, 25% (*v*/*v*) glycerol, 20 mM Tris, pH 8.

3.3. In Vitro p300 HAT-Mediated Acetylation Reactions

Acetylation of wt Tau by p300 HAT was performed by combining 500 µL of 86 µM purified wt Tau (dissolved in water), 200 µL of 15 µM p300 HAT, 25 µL of 10 mM acetyl-CoA (Sigma, Saint Louis, MO, USA), and 25 µL of 1 M Tris, pH 8. The acetylation reaction was allowed to proceed for three days at RT (unless stated otherwise).

3.4. Western Blot of Acetylated Tau

Tau acetylation was verified by western blot against acetyl-lysine. 100-ng samples of Ac-Tau, wt Tau and BSA were loaded on a 4–20% gradient SDS-PAGE gel (Mini-PROTEAN TGX Precast Gels, Bio-Rad). After electrophoresis, the gel was transferred to a polyvinylidene fluoride (PVDF) membrane using the Trans-Blot Turbo Transfer System, following the manufacturer's protocols (Bio-Rad). After incubation with 5% (*w*/*v*) nonfat milk in TBS-T (150 mM NaCl, 0.1% Tween-20, 20 mM Tris-HCl, pH 7.5) at RT for 2 h, the membrane was incubated with antibody against acetyl-lysine (1:100 in 1% nonfat milk/TBS-T; sc-32268, Santa Cruz Biotech, Dallas, TX, USA) overnight at 4 °C. The membrane was washed six times for 10 min with TBS-T and incubated with HRP-conjugated anti-mouse antibody (1:1000; #7076, Cell Signaling, Danvers, MA, USA) at RT for 30 min. The membrane was washed six times and developed with Clarity Western ECL Substrate according to the manufacturer's protocols (Bio-Rad). Chemiluminescent signals were measured using ChemiDoc MP Image System (Bio-Rad).

3.5. Mass Spectrometry of Acetylated Tau

Ac-Tau sample was boiled in 30 µL of 1× NuPAGE LDS sample buffer (Invitrogen) and subjected to SDS-PAGE (NuPAGE 10% Bis-Tris gel, Invitrogen) then visualized with Coomassie Brilliant blue-stain. The SDS-PAGE gel containing the band corresponding to Tau was excised, destained, and subjected to in-gel digestion using 100 ng trypsin (#T9600, GenDepot). The digested peptides were resuspended in 10 µL of 0.1% formic acid and subjected to a nanoHPLC-MS/MS system with an EASY-nLC 1200 coupled to Fusion Tribrid Orbitrap Lumos mass spectrometer (Thermo Fisher, Waltham, MA, USA). The peptides were loaded onto a Reprosil-Pur Basic C18 (1.9 µm, Dr. Maisch

GmbH, Ammerbuch-Entringen, Germany) pre-column of 2 cm × 100 μm size. The pre-column was switched in-line with an in-housed 50 mm × 150 μm analytical column packed with Reprosil-Pur Basic C18 equilibrated in 0.1% formic acid. The peptides were eluted using a 45-min discontinuous gradient of 4–28% acetonitrile/0.1% formic acid at a flow rate of 750 nL/min. The eluted peptides were directly electro-sprayed into mass spectrometer operated in the data-dependent acquisition mode acquiring fragmentation spectra of the top 30 strongest ions under direct control of Xcalibur software (4.0; Thermo Fisher). Parent MS spectrum was acquired in the Orbitrap with full MS range of 300–1400 m/z in the resolution of 120,000. CID fragmented MS/MS spectrum was acquired in ion-trap with rapid scan mode. Obtained MS/MS spectra were searched against the target-decoy human refseq database (June 2015 release, containing 73,637 entries) in Proteome Discoverer 1.4 interface (Thermo Fisher) with the Mascot algorithm (Mascot 2.4, Matrix Science, London, UK). Variable modifications of lysine and arginine acetylation, methionine oxidation, and N-terminal acetylation were allowed. The precursor mass tolerance was confined within 20 ppm with fragment mass tolerance of 0.5 Da and with a maximum of two missed cleavages allowed. The peptides identified in the Mascot results file were validated with a 5% false discover rate (FDR) and subjected to manual verification to confirm lysine acetylation.

3.6. Microscopy Imaging

Liquid-liquid phase separation (LLPS) experiments were performed using variable protein (wt Tau and Ac-Tau; 2.5–20 μM) and salt (0–250 mM NaCl) concentrations in 5 mM sodium phosphate buffer, pH 7.8. 10–20-μL drops were pipetted onto 35-mm glass bottom dishes (ibidi, Martinsried, Germany) and immediately monitored for droplet formation (with incubation time of 5 min). For heparin-induced LLPS experiments, heparin (8–25 kDa; Santa Cruz Biotech) was mixed with wt Tau or Ac-Tau at approximately 1:4 molar ratio (heparin:protein). LLPS experiments in the presence of crowding agent were carried out using 30 μM wt Tau or Ac-Tau in 10% PEG-8K, $\alpha\beta\gamma$ buffer (10 mM glycine, 10 mM acetate, 10 mM sodium phosphate, pH 7.5). Microscopy images were recorded at RT using an FL EVOS imaging system (Invitrogen).

3.7. Thioflavin T (Th T) Aggregation Assay

Heparin-induced wt Tau and Ac-Tau aggregation were detected following changes in Th T fluorescence using a Biotek Synergy H1 plate reader, employing 440 nm excitation and 480 nm emission wavelengths. Protein aggregation reactions were conducted using 20 μM wt Tau or Ac-Tau in 5 μM heparin, 10 μM Th T (GenDepot), 0.25 mM TCEP, 5 mM sodium phosphate, pH 7.8. Aggregation kinetics were monitored for ~24 h.

3.8. Microtubule Assembly Assay

The ability of wt Tau and Ac-Tau to promote microtubule assembly was investigated using fluorescence imaging. Rhodamine-labeled and unlabeled tubulin heterodimers (1:20 ratio; Cytoskeleton, Inc., Denver, CO, USA) were mixed with wt Tau or Ac-Tau (9 μM tubulin heterodimers and 27.5 μM Tau) in a final buffer condition of 0.2 mM $MgCl_2$, 0.1 mM GTP, 50 μM EDTA, 9 mM sodium PIPES, pH 6.9. Microtubule formation was visually monitored using an FL EVOS fluorescence microscope (Invitrogen) starting from 10 min up to 18 h.

Author Contributions: J.C.F. and A.C.F. designed the experiments. J.C.F., K.J.C., and P.S.T. performed the microscopy imaging and molecular biology experiments. A.J. and S.Y.J. acquired and analyzed the mass spectrometry data. J.C.F., K.R.M., and A.C.F. wrote the manuscript.

Acknowledgments: We thank Jin Wang for the use of the Biotek Synergy H1 instrument.

Abbreviations

Ac-Tau	Hyperacetylated Tau
AD	Alzheimer's disease
IDP	Intrinsically disordered protein
HAT	Histone acetyltransferase
LLPS	Liquid-liquid phase separation
PTMs	Post-translational modifications
wt	Wild-type

References

1. Buee, L.; Bussiere, T.; Buee-Scherrer, V.; Delacourte, A.; Hof, P.R. Tau protein isoforms, phosphorylation and role in neurodegenerative disorders. *Brain Res. Brain Res. Rev.* **2000**, *33*, 95–130. [CrossRef]
2. Dickson, D.W. Tau and synuclein and their role in neuropathology. *Brain Pathol.* **1999**, *9*, 657–661. [CrossRef] [PubMed]
3. Holtzman, D.M.; Morris, J.C.; Goate, A.M. Alzheimer's disease: The challenge of the second century. *Sci. Transl. Med.* **2011**, *3*, 77sr71. [CrossRef] [PubMed]
4. Ghetti, B.; Oblak, A.L.; Boeve, B.F.; Johnson, K.A.; Dickerson, B.C.; Goedert, M. Invited review: Frontotemporal dementia caused by microtubule-associated protein tau gene (MAPT) mutations: A chameleon for neuropathology and neuroimaging. *Neuropathol. Appl. Neurobiol.* **2015**, *41*, 24–46. [CrossRef] [PubMed]
5. Trojanowski, J.Q.; Lee, V.M. Pathological tau: A loss of normal function or a gain in toxicity? *Nat. Neurosci.* **2005**, *8*, 1136–1137. [CrossRef] [PubMed]
6. Feinstein, S.C.; Wilson, L. Inability of tau to properly regulate neuronal microtubule dynamics: A loss-of-function mechanism by which tau might mediate neuronal cell death. *Biochim. Biophys. Acta* **2005**, *1739*, 268–279. [CrossRef] [PubMed]
7. Iqbal, K.; Alonso Adel, C.; Chen, S.; Chohan, M.O.; El-Akkad, E.; Gong, C.X.; Khatoon, S.; Li, B.; Liu, F.; Rahman, A.; et al. Tau pathology in alzheimer disease and other tauopathies. *Biochim. Biophys. Acta* **2005**, *1739*, 198–210. [CrossRef] [PubMed]
8. Li, W.; Lee, V.M. Characterization of two VQIXXK motifs for tau fibrillization in vitro. *Biochemistry* **2006**, *45*, 15692–15701. [CrossRef] [PubMed]
9. Mukrasch, M.D.; Bibow, S.; Korukottu, J.; Jeganathan, S.; Biernat, J.; Griesinger, C.; Mandelkow, E.; Zweckstetter, M. Structural polymorphism of 441-residue tau at single residue resolution. *PLoS Biol.* **2009**, *7*, e34. [CrossRef] [PubMed]
10. Von Bergen, M.; Friedhoff, P.; Biernat, J.; Heberle, J.; Mandelkow, E.M.; Mandelkow, E. Assembly of tau protein into alzheimer paired helical filaments depends on a local sequence motif ((306)VQIVYK(311)) forming beta structure. *Proc. Natl. Acad. Sci. USA* **2000**, *97*, 5129–5134. [CrossRef] [PubMed]
11. Ramachandran, G.; Udgaonkar, J.B. Mechanistic studies unravel the complexity inherent in tau aggregation leading to alzheimer's disease and the tauopathies. *Biochemistry* **2013**, *52*, 4107–4126. [CrossRef] [PubMed]
12. Cripps, D.; Thomas, S.N.; Jeng, Y.; Yang, F.; Davies, P.; Yang, A.J. Alzheimer disease-specific conformation of hyperphosphorylated paired helical filament-tau is polyubiquitinated through Lys-48, Lys-11, and Lys-6 ubiquitin conjugation. *J. Biol. Chem.* **2006**, *281*, 10825–10838. [CrossRef] [PubMed]
13. Tai, H.C.; Serrano-Pozo, A.; Hashimoto, T.; Frosch, M.P.; Spires-Jones, T.L.; Hyman, B.T. The synaptic accumulation of hyperphosphorylated tau oligomers in alzheimer disease is associated with dysfunction of the ubiquitin-proteasome system. *Am. J. Pathol.* **2012**, *181*, 1426–1435. [CrossRef] [PubMed]
14. Tepper, K.; Biernat, J.; Kumar, S.; Wegmann, S.; Timm, T.; Hubschmann, S.; Redecke, L.; Mandelkow, E.M.; Muller, D.J.; Mandelkow, E. Oligomer formation of tau protein hyperphosphorylated in cells. *J. Biol. Chem.* **2014**, *289*, 34389–34407. [CrossRef] [PubMed]
15. Ferreon, A.C.; Moran, C.R.; Ferreon, J.C.; Deniz, A.A. Alteration of the alpha-synuclein folding landscape by a mutation related to parkinson's disease. *Angew. Chem. Int. Ed. Engl.* **2010**, *49*, 3469–3472. [CrossRef] [PubMed]
16. Ferreon, A.C.; Ferreon, J.C.; Wright, P.E.; Deniz, A.A. Modulation of allostery by protein intrinsic disorder. *Nature* **2013**, *498*, 390–394. [CrossRef] [PubMed]

17. Ferreon, J.C.; Lee, C.W.; Arai, M.; Martinez-Yamout, M.A.; Dyson, H.J.; Wright, P.E. Cooperative regulation of p53 by modulation of ternary complex formation with CBP/p300 and HDM2. *Proc. Natl. Acad. Sci. USA* **2009**, *106*, 6591–6596. [CrossRef] [PubMed]
18. Martin, L.; Latypova, X.; Terro, F. Post-translational modifications of tau protein: Implications for alzheimer's disease. *Neurochem. Int.* **2011**, *58*, 458–471. [CrossRef] [PubMed]
19. Kamah, A.; Huvent, I.; Cantrelle, F.X.; Qi, H.; Lippens, G.; Landrieu, I.; Smet-Nocca, C. Nuclear magnetic resonance analysis of the acetylation pattern of the neuronal tau protein. *Biochemistry* **2014**, *53*, 3020–3032. [CrossRef] [PubMed]
20. Cohen, T.J.; Guo, J.L.; Hurtado, D.E.; Kwong, L.K.; Mills, I.P.; Trojanowski, J.Q.; Lee, V.M. The acetylation of tau inhibits its function and promotes pathological tau aggregation. *Nat. Commun.* **2011**, *2*, 252. [CrossRef] [PubMed]
21. Min, S.W.; Cho, S.H.; Zhou, Y.; Schroeder, S.; Haroutunian, V.; Seeley, W.W.; Huang, E.J.; Shen, Y.; Masliah, E.; Mukherjee, C.; et al. Acetylation of tau inhibits its degradation and contributes to tauopathy. *Neuron* **2010**, *67*, 953–966. [CrossRef] [PubMed]
22. Cohen, T.J.; Friedmann, D.; Hwang, A.W.; Marmorstein, R.; Lee, V.M. The microtubule-associated tau protein has intrinsic acetyltransferase activity. *Nat. Struct. Mol. Biol.* **2013**, *20*, 756–762. [CrossRef] [PubMed]
23. Cook, C.; Carlomagno, Y.; Gendron, T.F.; Dunmore, J.; Scheffel, K.; Stetler, C.; Davis, M.; Dickson, D.; Jarpe, M.; DeTure, M.; et al. Acetylation of the kxgs motifs in tau is a critical determinant in modulation of tau aggregation and clearance. *Hum. Mol. Genet.* **2014**, *23*, 104–116. [CrossRef] [PubMed]
24. Min, S.W.; Chen, X.; Tracy, T.E.; Li, Y.; Zhou, Y.; Wang, C.; Shirakawa, K.; Minami, S.S.; Defensor, E.; Mok, S.A.; et al. Critical role of acetylation in tau-mediated neurodegeneration and cognitive deficits. *Nat. Med.* **2015**, *21*, 1154–1162. [CrossRef] [PubMed]
25. Brangwynne, C.P. Phase transitions and size scaling of membrane-less organelles. *J. Cell Biol.* **2013**, *203*, 875–881. [CrossRef] [PubMed]
26. Feric, M.; Vaidya, N.; Harmon, T.S.; Mitrea, D.M.; Zhu, L.; Richardson, T.M.; Kriwacki, R.W.; Pappu, R.V.; Brangwynne, C.P. Coexisting liquid phases underlie nucleolar subcompartments. *Cell* **2016**, *165*, 1686–1697. [CrossRef] [PubMed]
27. Kato, M.; Han, T.W.; Xie, S.; Shi, K.; Du, X.; Wu, L.C.; Mirzaei, H.; Goldsmith, E.J.; Longgood, J.; Pei, J.; et al. Cell-free formation of rna granules: Low complexity sequence domains form dynamic fibers within hydrogels. *Cell* **2012**, *149*, 753–767. [CrossRef] [PubMed]
28. Li, P.; Banjade, S.; Cheng, H.C.; Kim, S.; Chen, B.; Guo, L.; Llaguno, M.; Hollingsworth, J.V.; King, D.S.; Banani, S.F.; et al. Phase transitions in the assembly of multivalent signalling proteins. *Nature* **2012**, *483*, 336–340. [CrossRef] [PubMed]
29. Lin, Y.; Protter, D.S.; Rosen, M.K.; Parker, R. Formation and maturation of phase-separated liquid droplets by rna-binding proteins. *Mol. Cell* **2015**, *60*, 208–219. [CrossRef] [PubMed]
30. Patel, A.; Lee, H.O.; Jawerth, L.; Maharana, S.; Jahnel, M.; Hein, M.Y.; Stoynov, S.; Mahamid, J.; Saha, S.; Franzmann, T.M.; et al. A liquid-to-solid phase transition of the als protein fus accelerated by disease mutation. *Cell* **2015**, *162*, 1066–1077. [CrossRef] [PubMed]
31. Conicella, A.E.; Zerze, G.H.; Mittal, J.; Fawzi, N.L. Als mutations disrupt phase separation mediated by alpha-helical structure in the TDP-43 low-complexity C-terminal domain. *Structure* **2016**, *24*, 1537–1549. [CrossRef] [PubMed]
32. Li, Y.R.; King, O.D.; Shorter, J.; Gitler, A.D. Stress granules as crucibles of als pathogenesis. *J. Cell Biol.* **2013**, *201*, 361–372. [CrossRef] [PubMed]
33. Tsoi, P.S.; Choi, K.J.; Leonard, P.G.; Sizovs, A.; Moosa, M.M.; MacKenzie, K.R.; Ferreon, J.C.; Ferreon, A.C.M. The N-terminal domain of ALS-linked TDP-43 assembles without misfolding. *Angew. Chem. Int. Ed. Engl.* **2017**, *56*, 12590–12593. [CrossRef] [PubMed]
34. Molliex, A.; Temirov, J.; Lee, J.; Coughlin, M.; Kanagaraj, A.P.; Kim, H.J.; Mittag, T.; Taylor, J.P. Phase separation by low complexity domains promotes stress granule assembly and drives pathological fibrillization. *Cell* **2015**, *163*, 123–133. [CrossRef] [PubMed]
35. Hernandez-Vega, A.; Braun, M.; Scharrel, L.; Jahnel, M.; Wegmann, S.; Hyman, B.T.; Alberti, S.; Diez, S.; Hyman, A.A. Local nucleation of microtubule bundles through tubulin concentration into a condensed tau phase. *Cell Rep.* **2017**, *20*, 2304–2312. [CrossRef] [PubMed]

36. Wegmann, S.; Eftekharzadeh, B.; Tepper, K.; Zoltowska, K.M.; Bennett, R.E.; Dujardin, S.; Laskowski, P.R.; MacKenzie, D.; Kamath, T.; Commins, C.; et al. Tau protein liquid-liquid phase separation can initiate tau aggregation. *EMBO J.* **2018**, *37*, e98049. [CrossRef] [PubMed]
37. Ambadipudi, S.; Biernat, J.; Riedel, D.; Mandelkow, E.; Zweckstetter, M. Liquid-liquid phase separation of the microtubule-binding repeats of the alzheimer-related protein tau. *Nat. Commun.* **2017**, *8*, 275. [CrossRef] [PubMed]
38. Zhang, X.; Lin, Y.; Eschmann, N.A.; Zhou, H.; Rauch, J.N.; Hernandez, I.; Guzman, E.; Kosik, K.S.; Han, S. Rna stores tau reversibly in complex coacervates. *PLoS Biol.* **2017**, *15*, e2002183. [CrossRef] [PubMed]
39. Mukrasch, M.D.; Biernat, J.; von Bergen, M.; Griesinger, C.; Mandelkow, E.; Zweckstetter, M. Sites of tau important for aggregation populate {beta}-structure and bind to microtubules and polyanions. *J. Biol. Chem.* **2005**, *280*, 24978–24986. [CrossRef] [PubMed]
40. Fitzpatrick, A.W.P.; Falcon, B.; He, S.; Murzin, A.G.; Murshudov, G.; Garringer, H.J.; Crowther, R.A.; Ghetti, B.; Goedert, M.; Scheres, S.H.W. Cryo-em structures of tau filaments from alzheimer's disease. *Nature* **2017**, *547*, 185–190. [CrossRef] [PubMed]
41. Lee, M.J.; Lee, J.H.; Rubinsztein, D.C. Tau degradation: The ubiquitin-proteasome system versus the autophagy-lysosome system. *Prog. Neurobiol.* **2013**, *105*, 49–59. [CrossRef] [PubMed]
42. Kim, D.; Nguyen, M.D.; Dobbin, M.M.; Fischer, A.; Sananbenesi, F.; Rodgers, J.T.; Delalle, I.; Baur, J.A.; Sui, G.; Armour, S.M.; et al. Sirt1 deacetylase protects against neurodegeneration in models for alzheimer's disease and amyotrophic lateral sclerosis. *EMBO J.* **2007**, *26*, 3169–3179. [CrossRef] [PubMed]
43. Barghorn, S.; Davies, P.; Mandelkow, E. Tau paired helical filaments from alzheimer's disease brain and assembled in vitro are based on beta-structure in the core domain. *Biochemistry* **2004**, *43*, 1694–1703. [CrossRef] [PubMed]
44. Thompson, P.R.; Wang, D.; Wang, L.; Fulco, M.; Pediconi, N.; Zhang, D.; An, W.; Ge, Q.; Roeder, R.G.; Wong, J.; et al. Regulation of the p300 hat domain via a novel activation loop. *Nat. Struct. Mol. Biol.* **2004**, *11*, 308–315. [CrossRef] [PubMed]

12

Physical Background of the Disordered Nature of "Mutual Synergetic Folding" Proteins

Csaba Magyar [1,†], Anikó Mentes [1,†], Erzsébet Fichó [1], Miklós Cserző [1,2] and István Simon [1,*]

[1] Institute of Enzymology, Research Centre for Natural Sciences, Hungarian Academy of Sciences, Magyar Tudósok krt. 2, H-1117 Budapest, Hungary; magyar.csaba@ttk.mta.hu (C.M.); mentes.aniko@ttk.mta.hu (A.M.); ficho.erzsebet@ttk.mta.hu (E.F.); cserzo.miklos@ttk.mta.hu (M.C.)

[2] Department of Physiology, Faculty of Medicine, Semmelweis University, Tűzoltó u. 37-47, H-1094 Budapest, Hungary

* Correspondence: simon.istvan@ttk.mta.hu

† These authors contributed equally to the work.

Abstract: Intrinsically disordered proteins (IDPs) lack a well-defined 3D structure. Their disordered nature enables them to interact with several other proteins and to fulfil their vital biological roles, in most cases after coupled folding and binding. In this paper, we analyze IDPs involved in a new mechanism, mutual synergistic folding (MSF). These proteins define a new subset of IDPs. Recently we collected information on these complexes and created the Mutual Folding Induced by Binding (MFIB) database. These protein complexes exhibit considerable structural variation, and almost half of them are homodimers, but there is a significant amount of heterodimers and various kinds of oligomers. In order to understand the basic background of the disordered character of the monomers found in MSF complexes, the simplest part of the MFIB database, the homodimers are analyzed here. We conclude that MFIB homodimeric proteins have a larger solvent-accessible main-chain surface area on the contact surface of the subunits, when compared to globular homodimeric proteins. The main driving force of the dimerization is the mutual shielding of the water-accessible backbones and the formation of extra intermolecular interactions.

Keywords: dehydron; homodimer; hydrogen bond; inter-subunit interaction; intrinsically disordered protein; ion pair; mutual synergistic folding; solvent-accessible surface area; stabilization center

---

1. Introduction

Since the millennium it has been clear that Anfinsen's long-standing paradigm that was alleged to be valid for all proteins: "Protein structure is uniquely determined by its amino acid sequences" [1,2] is only valid for a specific subclass of proteins, while the rest of the proteins, termed intrinsically disordered proteins (IDPs), have no permanent 3D structures [3–6]. In our earlier effort to identify the physical background of protein disorder, the lack of sufficient pairwise interaction energy between the residues to ensure a stable 3D structure was pinpointed. When this energy is not enough to compensate the entropy-related free energy loss in the course of the formation of a unique structure, intrinsically disordered proteins are witnessed [7]. It has been shown that this pairwise energy can be calculated from the amino acid sequences without any structural information. On this basis we developed a widely used method, IUPred, to predict disordered proteins or protein segments from local composition data [8]. Another application of the estimation of the pairwise interaction energies led us to recognize the physical properties of the binding regions of disordered proteins, which can bind to ordered proteins [9]. When certain segments of a disordered protein interact with an ordered protein structure, part of their interactions will be manifested through elements of this stable globular protein having enough pairwise energy to stabilize their structures, i.e., to be folded, on the surface of ordered proteins. The contribution of a single residue depends only on the composition of the surrounding residues. Since ordered proteins have different amino acid compositions to disordered proteins, the resulting interaction energies of the residues at the contact surface can stabilize the structure (coupled folding and binding).

On the basis of this phenomenon, a binding site prediction method, termed ANCHOR, was developed [10]. These interacting segments generally appeared as short motifs of polypeptide chains (ELMs) [9,10]. More recently, the upgraded version of IUPred and ANCHOR were combined into a new server called IUPred2A [11].

While this phenomenon appeared to be general, over the years the number of "exceptions" increased, suggesting that the insufficient pairwise energy calculated by the IUPred algorithms was only valid for certain intrinsically disordered proteins and protein segments (IDSs), and that another kind of IDP and IDS also existed. Even in the early age of IDP studies, there was sporadic information that some IDPs exhibit mutual folding and binding together with other IDPs, without the help of already stable proteins or other stable macromolecules [12,13]. For example, NCDB segments of CBP form a complex with the ACTR domain of p160, see: protein data bank (PDB) entry 1kbh [14] or region C of WASP is I complex with the GBD segment of WASP [15]. In these examples, the interacting parts of the disordered proteins were not ELM sized, but rather have structural domain sizes [16]. In many cases the interacting disordered protein segments were alike, forming homodimer or homo-oligomers. Here the coupled folding and binding should not appear due to the difference in residue composition, as in the case of ELMs stabilized on the surface of an ordered protein. Therefore there should be another mechanism for coupled folding and binding than the one we can recognize by ANCHOR. Since macromolecular interactions are part of almost all the activity of disordered proteins, a new mechanism for coupled folding and binding, where there is no stable template to use, define a new subset of IDPs. Despite the sporadic information about these interactions, not too many of this kind of complexes were reported in the literature [17,18]. Therefore we performed a detailed analysis of several databases and on the scientific literature and collected information on these complexes and created the Mutual Folding Induced by Binding (MFIB) database [19]. These complexes exhibit large structural variations (see Figure 1).

Almost half of the MSF-complexes are homodimers, but there is a significant amount of heterodimers and other oligomeric states, including homo- and heterotetramers, as well as trimers, pentamers, and hexamers. To explore the unique features of the entries in the MFIB database and pinpoint those characters that differ between these entries and those of those disordered segments that can participate in coupled folding and binding with already stable proteins, we created the Disordered Binding Site (DIBS) database of the latter complexes [20]. Currently, a publication of the comparison of the structural differences of proteins of the MFIB and DIBS databases is in progress [21].

The elements of the pairwise interaction matrices used in the IUPred and ANCHOR algorithms were derived from the structure data of folded globular proteins, therefore this data includes the free energy from the average hydration of the residues in these proteins. We showed that this is similar for most globular protein, therefore a fair free energy contribution of a particular residue can be calculated from the composition of the rather large polypeptide segment centered by the particular residue, using the pairwise energy interaction matrix [7]. In the IUPred algorithm, when a particular residue is processed, whether it belongs to an ordered segment or a disordered one, the interaction of this residue in question and all other residues in a large surrounding region are considered. Therefore this calculated energy value has to be the same for all permutations of the residues of the segments located at both sides of the center residue, until or unless the compositions of the segments are changed. The amino acid sequences of proteins that have stable folded structures evolved in such a way that the side chains together shield the backbone from water, which minimizes the energetically unfavored water-accessible area on the polypeptide backbone. In this work we show that this statement is not valid for the disordered proteins listed in MFIB.

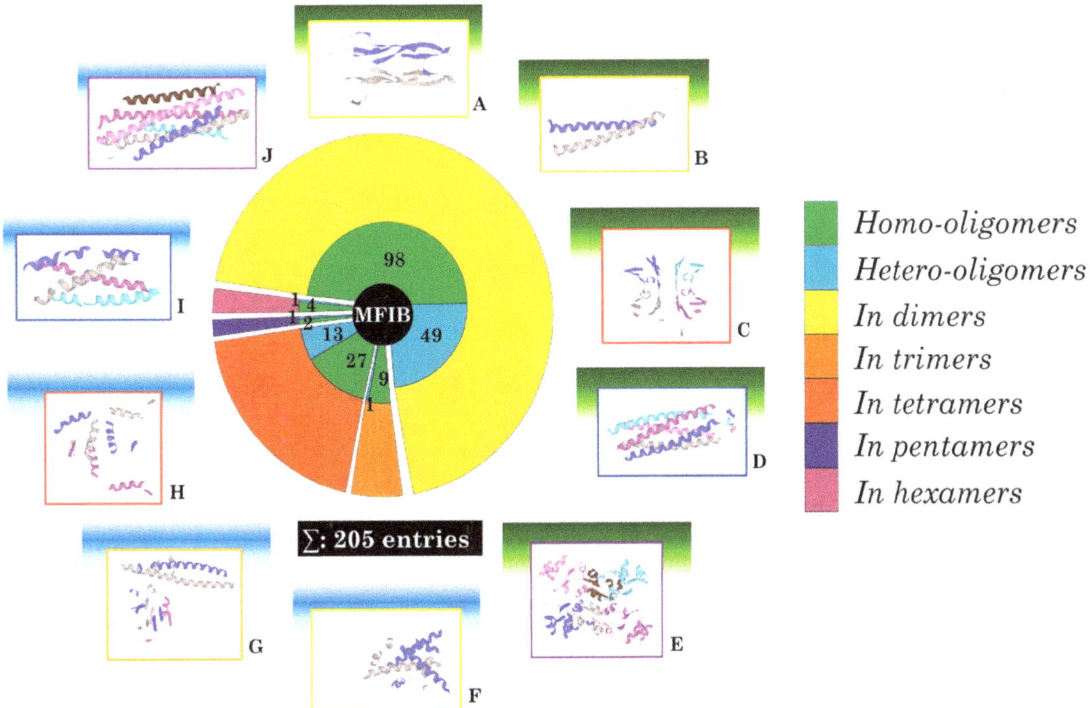

Figure 1. Oligomeric states in the MFIB database with example complexes. (**A**: 1BET, nerve growth factor (*Mus musculus*); **B**: 1AQ5, assembly domain of cartilage oligomeric matrix protein (*Gallus gallus*); **C**: 1GKE, Transthyretin (*Rattus norvegicus*); **D**: 1MZ9, assembly domain of cartilage oligomeric matrix protein (*Mus musculus*); **E**: 1NPK, nucleoside diphosphate kinase (*Dictyostelium discoideum*); **F**: 5GT0, H2A-H2B histone dimer, containing histone variants H2A type 1-A and H2B type 1-J (*Homo sapiens*); **G**: 2AZE, Rb C-terminal domain bound to an E2F1-DP1 heterodimer (*Homo sapiens*); **H**: 2NB1, p63/p73 hetero-tetramerization domain (*Homo sapiens*); **I**: 1VZJ, The synaptic acetylcholinesterase tetramer assembled around a polyproline-II helix (*Homo sapiens*); **J**: 1G2C, respiratory syncytial virus fusion protein core (*Homo sapiens*).

We investigated whether the interacting regions of these proteins can be identified based on their location in the whole polypeptide chain, on their biased amino acid composition or on specific physical properties. We discovered that their most unique characteristic is the high water accessibility of their peptide backbone, compared to the water accessibility of the folded proteins, which have similar amino acid compositions.

2. Results

2.1. Sequence-Based Analysis

In this study, homodimeric protein complexes from MFIB were analyzed regarding sequence and structural properties. First we checked the location of the MFIB homodimeric dataset (MFHD) PDB segments with a known 3D structure in the full UniProt protein sequences. In some cases, the MFHD PDB segments were located near the N-terminus, near the C-terminus, in the middle of the sequence or they were identical with the full sequence (Figure 2).

We examined the residue composition of the MFHD proteins (Figure 3, Table S2), which were compared with two reference datasets, the globular homodimeric dataset (GLHD) and the globular monomeric dataset (GLMD, see Section 4). To better understand the amino acid composition of the sequences, it was depicted by principal component analysis (PCA) (Figure 4, Table S3). PCA showed that the amino acid composition of the MFHD proteins did not differ significantly from the amino acid composition of the globular proteins (GLHD, GLMD). The PCA also demonstrated that MFHD formed a diverse group based on their amino acid composition.

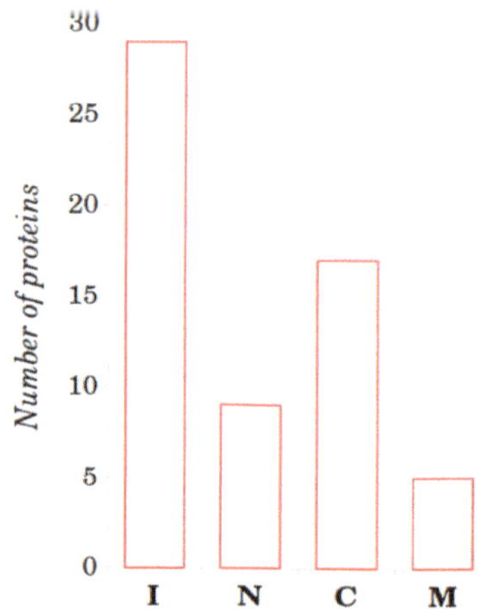

Figure 2. Distribution of MFHD PDB segments in the full UniProt sequences. (I: Amino acid sequence from UniProt is identical with amino acid sequences of MFHD PDB segment amino acid sequences; N: MFHD PDB segment is located in N-terminus of the amino acid sequences from UniProt; C: MFHD PDB segment is located in C-terminus of the amino acid sequences from UniProt; M: MFHD PDB segment is located in middle of the full amino acid sequence from UniProt).

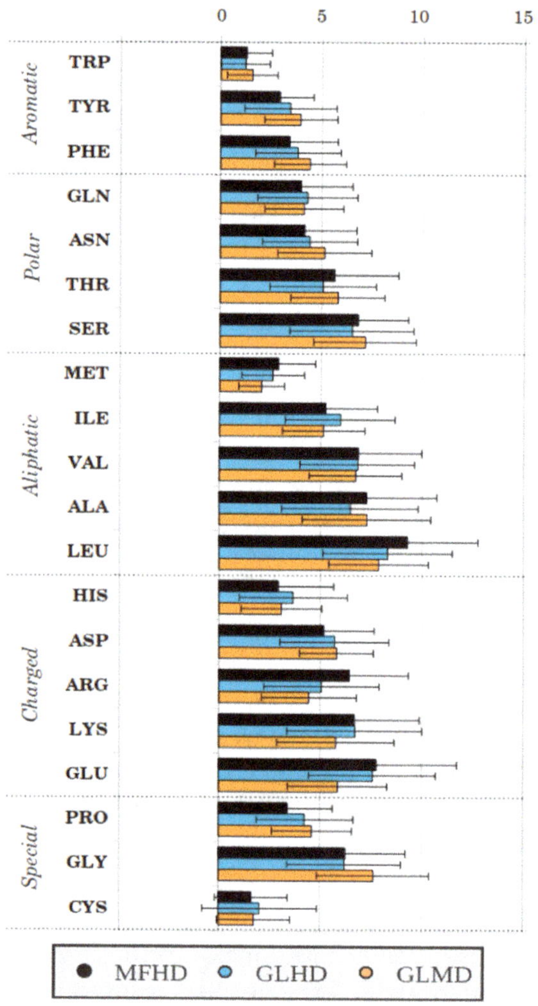

Figure 3. Sequence properties of MFHD, GLHD, and GLMD proteins (For values, see Table S2).

We investigated the MFHD with several protein disorder predictors (IUPred, ESpritz, GlobPlot, VSL2b, MobiDB Lite, MetaDisorder) [8,22–26], which worked well on the IDPs listed in DIBS, but did not recognize the polypeptide of MFHD complexes and other members of the MFIB database as disordered proteins. All methods predicted less than 30% of the protein residues as disordered, while the IUPred long/short methods, relying on a physical basis, predicted only 8 and 10% of the protein residues as disordered, respectively (for values, see Table S4). Other prediction methods based on amino acid composition bias also failed to detect MFHD PDB segments. Methods developed from the DAS and DAS-TMfilter [27,28] algorithms were tested on the dataset.

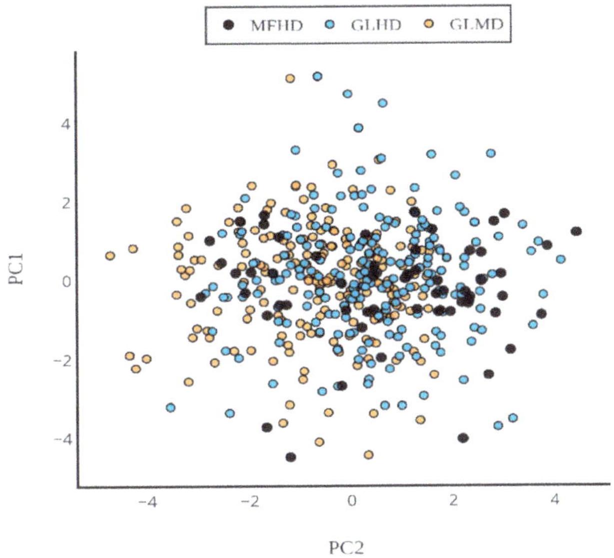

Figure 4. PCA ordination of the proteins from MFHD, GLHD, and GLMD based on their amino acid compositions (for values, see Table S3).

2.2. Structure-Based Analysis

We will use the term "interface" for the contact surface area of the two identical subunits in the dimeric structures. In cases where the term "monomeric structure" is used, calculations were carried out on structures from which the second chain was deleted since the PDB files contained dimer forms of the complexes. Residues belonging to the interface region were identified based on solvent accessible surface area (SASA) calculations. All-atom SASA values were calculated for the residues. Residues where the SASA value calculated from the dimer form were less than or equal to 20% of its counterpart from the monomeric structure defining the interface. We found that on average there were 26.4 interface residues per polypeptide-chain in the MFIB homodimeric dataset and 21.0 interface residues per polypeptide-chain in the reference globular homodimeric dataset. Considering the average size of the protein, this means that 27.13% of all residues in the MFHD and 22.34% of all residues in the GLHD belonged to the interface region. The higher value obtained for the MFIB homodimeric structures indicates that inter-subunit interactions may play an essential role in the stabilization of MFHD proteins.

We were looking for residues in the interface that have solvent-accessible spots in their main-chain in the monomeric structure, which become buried in the dimeric structures. We identified residues where the main-chain SASA in the dimeric form was less than 20% of the monomeric form value. Only residues with exposed main-chains, with a relative main-chain SASA larger than 0.2 in the monomeric structure, were taken into account. These residues with solvent-accessible main-chain patches (RSAMPs) were believed to be the main driving force of the dimerization of the disordered polypeptide chains collected in the MFIB database. We found a total of 183 such residues in the MFHD proteins; all structures contained at least one such residue. This was 3.14% of all residues. Considering that 27.13% of the residues were forming the interface, this means that 11.57% of the MFHD interface residues were RSAMPs. In the GLHD, 40.83% of the proteins did not contain such residues, on average 1.56% of all residues were RSAMPs. Since 22.34% of the residues form the interface, only 6.98% of the interface residues were RSAMPs. We calculated the average solvent-accessible surface area of the main chains. In the MFHD, the average solvent-accessible, main-chain area belonging to the interface region was 1154.56 Å^2 per polypeptide-chain, while in the GLHD this value was 790.54 Å^2. We can see that in the case of MFIB proteins a larger main-chain surface area is solvent accessible, which is energetically not favorable. The amino acid composition of the interface region and RSAMPs of the MFHD and GLHD complexes can be seen in Figure 5, Tables S5 and S6. Alanine and glycine were the most abundant residues under RSAMPs, which might be responsible for the higher solvent accessibility of the main chain in the MFHD. In the interface region, aliphatic residues are predominant. In the MFHD this was 50.6%, while in the GLHD 45.4% of the interface residues were aliphatic, making inter-subunit hydrophobic interactions even more prominent in MFIB proteins.

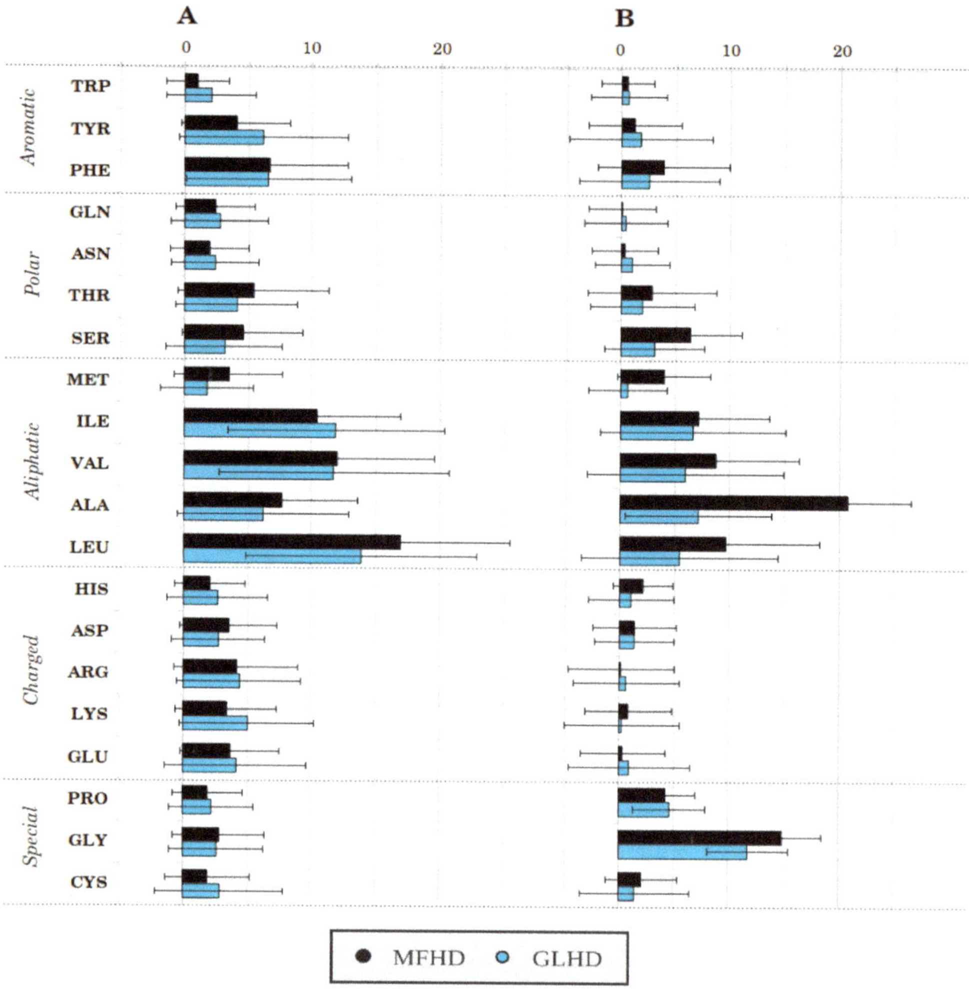

Figure 5. Amino acid composition of interface region (**A**) and RSAMPs (**B**) of the MFIB and globular homodimeric datasets (For values, see Tables S5 and S6).

We determined the secondary structural propensities in the MFHD, GLHD, and GLMD. We found that in the MFHD a significantly higher percentage of residues (39.4%) belonged to α-helices when compared to GLHD and GLMD (39.4% and 27.9%). In the MFHD, 21.2% of the residues belong to β-sheets, while in the GLHD and GLMD this value was 27.1% and 28.3%, respectively. The MFIB proteins show higher helical propensities than globular proteins.

We identified the hydrogen bonds formed between the two subunits. In the MFHD 6.97 inter-subunit H-bonds per structure were found, while in the GLHD this was only 4.58. Furthermore we identified underwrapped hydrogen bonds that are not well-enough shielded from the solvent, called dehydrons, in all structures [29]. In the MFHD we found 3.11 dehydrons per polypeptide chain under the inter-subunit H-bonds, while only 2.18 were found in the GLHD. Contrary to these results is the average wrapping of inter-subunit H-bonds, which was 16.0 for the MFHD and 13.6 for the GLHD. Although there were more dehydrons—i.e., underwrapped H-bonds—in the MFHD, the average wrapping value was still higher.

Due to the large difference found in the inter-subunit H-bonds, other inter-subunit interactions were also investigated. First we identified inter-subunit ion-pairs. We found that in the MFHD there were 1.17 inter-subunit ion-pairs on average, with only 0.66 in the GLHD. Charged residues tend to occur at the surface due to the desolvation of buried charges being energetically not favorable. Charged residues buried either in the interior of a protein or in the interface region of the dimeric structure should form ion pairs in order to compensate the desolvation penalty through favorable electrostatic interactions. Since the occurrence of charged residues is a bit higher in the interface region of the GLHD (16.2% vs. 14.7%), the lower number of inter-subunit ion-pairs was unexpected. We already noted in an earlier publication that inter-subunit ion pairs might contribute to the stabilization of proteins [30].

Stabilization centers (SCs) are pairs of residues involved in more than average long-range interactions [31]. These residue clusters are believed to contribute to the stabilization of protein structures through the cooperativity of the individual interactions [32,33]. The stabilization centers formed between different polypeptide chains can contribute to the stabilization of a protein complex [34]. We identified inter-subunit SCs in both the MFHD and GLHD. The two residues that form a stabilization center are called stabilization center elements (SCEs). We identified the SCEs belonging to the interface. In the MFHD, 3.86% of all residues form inter-subunit SCs, that is on average 14.22% of the interface residues form inter-subunit SCs. In the GLHD, only 1.83% of the residues belong to inter-subunit SCs. This means that only 8.19% of the interface residues form inter-subunit SCs. In MFIB dimers, the inter-subunit SCs were much more frequent than in the GLHD. We investigated whether SCEs overlap with RSAMPs or whether they are segregated. We found that there was a significant overlap, as 29.51% of the RSAMPs were SCEs in the MFHD. In the GLHD, we obtained a similar value of 29.19% for the overlap.

3. Discussion

In a recent study, we compared the residue composition of IDPs from the MFIB with complexes from the DIBS and other human protein databases [21] and we found that the composition of MFIB complexes was significantly different from that of the DIBS and only slightly different from that of human proteins. IDPs from the DIBS database are capable of coupled binding and folding on the surface of ordered proteins and can be predicted through bioinformatics methods like the ANCHOR algorithm, which is based on the different residue composition of the disordered monomer and the disordered–ordered protein complex. Therefore, in this work we studied MSF-homodimers to exclude this explanation for the case of mutual synergistic folding. We observed that in some cases the interacting segment of MFIB homodimers was the full polypeptide chain, while in other cases only a part of the chain was involved in the dimerization (Figure 2). We showed that they could be an order of magnitude longer than ELMs, which can be recognized by ANCHOR in other proteins.

In our current study the residue composition of the homodimeric complexes from the MFIB was determined and compared with that of homodimeric and monomeric globular proteins in similar amino acid sequence lengths (Figures 3 and 4). Our results showed that the IDPs listed in the MFIB had a similar amino acid composition to that of globular proteins. The PCA showed that the globular (GLHD, GLMD) and the MFHD proteins were not distinguishable. Although the points belonging to the complexes in the PCA figure were not certainly clustered, suggesting that MFHD is a distinct subgroup of IDPs. This was confirmed by the comparison of MFHD with the UniRef50 database, which showed that the main part of MFHD belongs to a distinct cluster and there is no significant similarity between their Pfam domains.

We investigated the MFHD with several protein disorder predictors (IUPred, ESpritz, GlobPlot, VSL2b, MobiDB Lite, MetaDisorder), which work well on the IDPs listed in DIBS. These methods did not recognize the full-length polypeptide chains of the MFHD complexes and other members of the MFIB database as disordered proteins. Since the disorder predictors IUPred and ANCHOR rely exclusively on solid physical principles, these methods were used to discover the physical principles behind the disordered character of the protein and the origin of the coupled folding and binding of the homodimers in the MFIB database. Our current study indicated that in the case of MFHD, the IUPred algorithm using its standard 20×20 pairwise free energy matrix overestimated the stabilizing energy because the energetically-unfavorable large solvent-accessible surface area of the peptide backbone in single protein chains resulted in less stabilizing energy. This can explain why these proteins were disordered in monomeric form. On the one hand, members of the MFIB dataset can be disordered for similar reason than other disordered proteins. That is, the sum of their pairwise interaction enthalpy did not compensate the free energy contribution of the entropy loss during folding. However, this is not the consequence of the amino acid composition of these polypeptides. Pairwise interactions of residue pairs, which have backbone parts not sufficiently shielded from the solvent, contribute less enthalpy to the stabilization than that found in globular proteins, from which the standard 20×20 pairwise free energy matrix was derived. Therefore, by using the free energy matrix in IUPred, we overestimated the stabilizing free energy of the proteins listed in the MFIB. This is why the IUPred algorithm predicted these monomers as structured proteins, while the experiments showed that they are disordered in their monomeric form [16].

We can conclude that the residue composition of MFHD is rather similar to that of the globular proteins (GLHD and GLMD), we were looking for structural differences among them. We found that the interface region had more residues in the MFHD than in the GLHD. MFIB homodimeric proteins had a larger solvent-accessible main-chain surface area in the interface when compared to globular homodimeric proteins. The polypeptide backbone of MFHD proteins was more accessible for water than in globular proteins. During dimerization, the solvent-accessible surface area of the backbone decreased and a high number of inter-subunit interactions (H-bonds, ion-pairs and stabilization centers) formed, leading to the stabilization of the of the disordered polypeptide-chains, enabling an ordered structure of MFIB proteins in the dimeric form. The driving force of the dimerization was the mutual shielding of the water-accessible backbones and the formation of extra intermolecular interactions.

4. Materials and Methods

Filters were applied to the homodimeric structures of the MFIB database. A reference dataset was created from homodimeric globular proteins, where the monomeric form was also globular in itself. Another reference dataset was created from monomeric globular proteins.

All homodimeric structures were collected from the MFIB database, and the modified PDB files were used. Entries belonging to the "coils and zippers" structure class were discarded since structures belonging to this class are both sequentially and structurally different from other homodimers. It is evident that a structure like a leucine-zipper cannot exist in monomeric form, thus no reference dataset can be created from "coils and zippers" where the monomer is not disordered in itself. A contact map matrix for all remaining structures was created. Entries with unusual contact maps were manually inspected. After inspection, the following entries were discarded: 2adl, 1r05, 4ath, 1aa0, 4w4k, 1ejp, resulting in a dataset of 60 homodimeric structures (Table S1). Heteroatoms were deleted from the structure. This dataset was referred to as MFIB homodimeric dataset (MFHD). We checked the secondary structure of the databases using the DSSP 2.0.4 program [35]. We found that 39.4% of the residues belonged to α-helices and 21.2% to β-sheets. The size distribution of the dataset was investigated. We counted the number of residues belonging to the N, N + 20 intervals. We found that the 140–240 interval was predominant, thus the reference datasets were created according to this size distribution.

A non-redundant reference dataset was created from homodimeric globular proteins. All homodimeric structures within the 140–240 amino acid size range were collected from the non-homologous PDB_Select database as of November, 2017 [36]. Structures containing coiled-coil structural elements identified with the Socket 3.0.3 program were excluded from the dataset [37]. The proper quaternary structure of the homodimers was created according to the BIOMT records of the PDB files. Entries with the following PDB ligand summary "ids" of large molecular sizes ligands and cofactors were discarded from the dataset because they could significantly alter the results of the solvent-accessible surface area calculations (017, 1BG, 1PE, 1PG, 5GP, C2E, FAD, HEC, KI1, MYA, MYR, MYS, NER, O8N, OLC, P33, P6G, PE5, UNL). Heteroatoms were deleted from the remaining structures. This procedure resulted in a list of 218 protein structures. This dataset was referred to as the globular homodimeric dataset (GLHD). For the PDB codes, see Table S1. According to DSSP, 27.9% of the residues belonged to α-helices and 27.1% to β-sheets.

An additional non-redundant reference dataset of the monomeric structures in the 140–240 amino acid size range containing only one structural domain was created from the PDB_SELECT database. The initial database was filtered by size and monomeric state criteria. All entries proved to be single domain according to the DDomain program using authors-trained parameters [38]. This dataset was referred to as the globular monomeric dataset (GLMD) and contained 191 entries (Table S1). According to DSSP, 24.9% of the residues belonged to α-helices and 28.3% to β-sheets.

Differences in the amino acid composition of the proteins sequences from the MFHD, GLHD, and GLMD datasets were revealed by principal component analysis (PCA) ordination using the plotly software according to Raska [39].

Hydrogen bonds were identified using the find_pairs command of PyMOL using 3.5 Å distance and 45 degree angle criteria between the donor and acceptor groups [40]. The calculation of the wrapping of hydrogen bonds and the identification of dehydrons was performed with the dehydron_ter.py program [41].

Stabilization centers (SCs) are pairs of residues, called stabilization center elements (SCEs), which are involved in several long-range interactions. These residues can be identified with our publicly available web server at http://scide.enzim.hu [42].

The solvent-accessible surface area (SASA) was calculated using the FreeSASA 2.03 program [43]. A residue was classified as buried when its relative SASA was below or equal to 0.2. Residues with a relative SASA value of over 0.2 were considered as exposed. A residue was classified as part of the interface region when its all-atom SASA calculated from the dimeric structure was less than 20% of the value calculated from the monomeric structure (created by deleting the second chain from the PDB file).

Ion-pairs were defined as pairs of negatively and positively charged residues, where the distance between the charged groups was equal to or less than 4 Å [44]. Ion pairs were identified using our own C++ program.

Supplementary Materials:
Table S1. List of PDB entries in the MFHD, GLHD, GLMD datasets. Table S2. Average amino acid sequence composition of proteins from MFHD, GLMD, and GLHD. Table S3. Amino acid sequence composition of proteins from MFHD, GLMD, and GLHD for PCA. Table S4. Disorder content by various predictors. Table S5. Average amino acid composition of interface region of the proteins from MFHD and GLMD. Table S6. Average amino acid composition of RSAMPs of the proteins from MFHD and GLMD.

Author Contributions: Conceptualization, I.S., C.M., M.C.; methodology, A.M., E.F., C. M.; software, A.M., E.F., M.C., C.M.; validation, A.M., C.M; formal analysis, C.M.; investigation, A.M., E.F.; resources, A.M, E.F.; data curation, A.M., E.F., C.M.; writing—original draft preparation, A.M., C.M., I.S.; writing—review and editing, E.F., A.M., C.M; visualization, A.M.; supervision, I.S., M.C.; project administration, I.S.; funding acquisition, I.S.

Abbreviations

IDPs	Intrinsically disordered proteins
MFIB	Mutual Folding Induced by Binding database
DIBS	Disordered Binding Site database
ELMs	Short motifs of polypeptide chains
MFHD	MFIB homodimeric dataset
MSF	Mutual synergistic folding
GLHD	Globular homodimeric dataset
GLMD	Globular monomeric dataset
PCA	Principal component analysis
PDB	Protein data bank
RSAMPs	Residues with solvent accessible main-chain patches
SC/SCE	Stabilization centers/stabilization center elements
SASA	Solvent accessible surface area

References

1. Anfinsen, C.B.; Haber, E.; Sela, M.; White, F.H., Jr. The kinetics of formation of native ribonuclease during oxidation of the reduced polypeptide chain. *Proc. Natl. Acad. Sci. USA* **1961**, *47*, 1309–1314. [CrossRef] [PubMed]
2. Anfinsen, C.B. The formation and stabilization of protein structure. *Biochem. J.* **1972**, *128*, 737–749. [CrossRef] [PubMed]
3. Wright, P.E.; Dyson, H.J. Intrinsically unstructured proteins: Re-assessing the protein structure-function paradigm. *J. Mol. Biol.* **1999**, *293*, 321–331. [CrossRef] [PubMed]
4. Uversky, V.N.; Dunker, A.K. Understanding protein non-folding. *Biochim. Biophys. Acta* **2010**, *1804*, 1231–1264. [CrossRef] [PubMed]
5. Tompa, P. Intrinsically unstructured proteins. *Trends Biochem. Sci.* **2002**, *27*, 527–533. [CrossRef]
6. Dyson, H.J.; Wright, P.E. Intrinsically unstructured proteins and their functions. *Nat. Rev. Mol. Cell Biol.* **2005**, *6*, 197–208. [CrossRef] [PubMed]
7. Dosztányi, Z.; Csizmók, V.; Tompa, P.; Simon, I. The pairwise energy content estimated from amino acid composition discriminates between folded and intrinsically unstructured proteins. *J. Mol. Biol.* **2005**, *347*, 827–839. [CrossRef] [PubMed]
8. Dosztányi, Z.; Csizmok, V.; Tompa, P.; Simon, I. IUPred: Web server for the prediction of intrinsically unstructured regions of proteins based on estimated energy content. *Bioinformatics* **2005**, *21*, 3433–3434. [CrossRef] [PubMed]
9. Mészáros, B.; Simon, I.; Dosztányi, Z. Prediction of Protein Binding Regions in Disordered Proteins. *PLoS Comput. Biol.* **2009**, *5*, e1000376. [CrossRef] [PubMed]
10. Dosztányi, Z.; Mészáros, B.; Simon, I. ANCHOR: Web server for predicting protein binding regions in disordered proteins. *Bioinformatics* **2009**, *25*, 2745–2746. [CrossRef] [PubMed]
11. Mészáros, B.; Erdos, G.; Dosztányi, Z. IUPred2A: Context-dependent prediction of protein disorder as a function of redox state and protein binding. *Nucleic Acids Res.* **2018**, *46*, W329–W337. [CrossRef] [PubMed]
12. Gunasekaran, K.; Tsai, C.-J.; Nussinov, R. Analysis of Ordered and Disordered Protein Complexes Reveals Structural Features Discriminating Between Stable and Unstable Monomers. *J. Mol. Biol.* **2004**, *341*, 1327–1341. [CrossRef] [PubMed]
13. Rumfeldt, J.A.O.; Galvagnion, C.; Vassall, K.A.; Meiering, E.M. Conformational stability and folding mechanisms of dimeric proteins. *Prog. Biophys. Mol. Biol.* **2008**, *98*, 61–84. [CrossRef] [PubMed]
14. Demarest, S.J.; Martinez-Yamout, M.; Chung, J.; Chen, H.; Xu, W.; Dyson, H.J.; Evans, R.M.; Wright, P.E. Mutual synergistic folding in recruitment of CBP/p300 by p160 nuclear receptor coactivators. *Nature* **2002**, *415*, 549–553. [CrossRef] [PubMed]
15. Garrard, S.M.; Capaldo, C.T.; Gao, L.; Rosen, M.K.; Macara, I.G.; Tomchick, D.R. Structure of Cdc42 in a complex with the GTPase-binding domain of the cell polarity protein, Par6. *EMBO J.* **2003**, *22*, 1125–1133. [CrossRef] [PubMed]

16. Tompa, P.; Fuxreiter, M.; Oldfield, C.J.; Simon, I.; Dunker, A.K.; Uversky, V.N. Close encounters of the third kind: Disordered domains and the interactions of proteins. *Bioessays* **2009**, *31*, 328–335. [CrossRef] [PubMed]
17. Zheng, Y.; Wu, Q.; Wang, C.; Xu, M.-Q.; Liu, Y. Mutual synergistic protein folding in split intein. *Biosci. Rep.* **2012**, *32*, 433–442. [CrossRef] [PubMed]
18. Ganguly, D.; Zhang, W.; Chen, J. Synergistic folding of two intrinsically disordered proteins: Searching for conformational selection. *Mol. Biosyst.* **2012**, *8*, 198–209. [CrossRef] [PubMed]
19. Fichó, E.; Reményi, I.; Simon, I.; Mészáros, B. MFIB: A repository of protein complexes with mutual folding induced by binding. *Bioinformatics* **2017**, *33*, 3682–3684. [CrossRef] [PubMed]
20. Schad, E.; Fichó, E.; Pancsa, R.; Simon, I.; Dosztányi, Z.; Mészáros, B. DIBS: A repository of disordered binding sites mediating interactions with ordered proteins. *Bioinformatics* **2017**, *34*, 535–537. [CrossRef] [PubMed]
21. Mészáros, B.; Dobson, L.; Fichó, E.; Tusnády, G.E.; Dosztányi, Z.; Simon, I. Interplay between folding and binding modulates protein sequences, structures, functions and regulation. *bioRxiv* **2017**, 211524. [CrossRef]
22. Walsh, I.; Martin, A.J.M.; Di Domenico, T.; Tosatto, S.C.E. ESpritz: Accurate and fast prediction of protein disorder. *Bioinformatics* **2012**, *28*, 503–509. [CrossRef] [PubMed]
23. Linding, R. GlobPlot: Exploring protein sequences for globularity and disorder. *Nucleic Acids Res.* **2003**, *31*, 3701–3708. [CrossRef] [PubMed]
24. Peng, K.; Radivojac, P.; Vucetic, S.; Dunker, A.K.; Obradovic, Z. Length-dependent prediction of protein intrinsic disorder. *BMC Bioinform.* **2006**, *7*, 208. [CrossRef] [PubMed]
25. Necci, M.; Piovesan, D.; Dosztányi, Z.; Tosatto, S.C.E. MobiDB-lite: Fast and highly specific consensus prediction of intrinsic disorder in proteins. *Bioinformatics* **2017**, *33*, 1402–1404. [CrossRef] [PubMed]
26. Kozlowski, L.P.; Bujnicki, J.M. MetaDisorder: A meta-server for the prediction of intrinsic disorder in proteins. *BMC Bioinform.* **2012**, *13*, 111. [CrossRef] [PubMed]
27. Cserzö, M.; Wallin, E.; Simon, I.; von Heijne, G.; Elofsson, A. Prediction of transmembrane alpha-helices in prokaryotic membrane proteins: The dense alignment surface method. *Protein Eng.* **1997**, *10*, 673–676. [CrossRef] [PubMed]
28. Cserzö, M.; Eisenhaber, F.; Eisenhaber, B.; Simon, I. On filtering false positive transmembrane protein predictions. *Protein Eng.* **2002**, *15*, 745–752. [CrossRef] [PubMed]
29. Fernández, A.; Scott, R. Dehydron: A structurally encoded signal for protein interaction. *Biophys. J.* **2003**, *85*, 1914–1928. [CrossRef]
30. Németh, A.; Svingor, A.; Pócsik, M.; Dobó, J.; Magyar, C.; Szilágyi, A.; Gál, P.; Závodszky, P. Mirror image mutations reveal the significance of an intersubunit ion cluster in the stability of 3-isopropylmalate dehydrogenase. *FEBS Lett.* **2000**, *468*, 48–52. [CrossRef]
31. Dosztányi, Z.; Fiser, A.; Simon, I. Stabilization centers in proteins: Identification, characterization and predictions. *J. Mol. Biol.* **1997**, *272*, 597–612. [CrossRef] [PubMed]
32. Magyar, C.; Gromiha, M.M.; Sávoly, Z.; Simon, I. The role of stabilization centers in protein thermal stability. *Biochem. Biophys. Res. Commun.* **2016**, *471*, 57–62. [CrossRef] [PubMed]
33. Simon, Á.; Dosztányi, Z.; Magyar, C.; Szirtes, G.; Rajnavölgyi, É.; Simon, I. Stabilization centers and protein stability. *Theor. Chem. Acc.* **2001**, *106*, 121–127. [CrossRef]
34. Simon, A.; Dosztányi, Z.; Rajnavölgyi, E.; Simon, I. Function-related regulation of the stability of MHC proteins. *Biophys. J.* **2000**, *79*, 2305–2313. [CrossRef]
35. Kabsch, W.; Sander, C. Dictionary of protein secondary structure: Pattern recognition of hydrogen-bonded and geometrical features. *Biopolymers* **1983**, *22*, 2577–2637. [CrossRef] [PubMed]
36. Griep, S.; Hobohm, U. PDBselect 1992–2009 and PDBfilter-select. *Nucleic Acids Res.* **2010**, *38*, D318–D319. [CrossRef] [PubMed]
37. Walshaw, J.; Woolfson, D.N. Socket: A program for identifying and analysing coiled-coil motifs within protein structures. *J. Mol. Biol.* **2001**, *307*, 1427–1450. [CrossRef] [PubMed]
38. Zhou, H.; Xue, B.; Zhou, Y. DDOMAIN: Dividing structures into domains using a normalized domain-domain interaction profile. *Protein Sci.* **2007**, *16*, 947–955. [CrossRef] [PubMed]
39. Raschka, S. *Python Machine Learning*; Packt Publishing Ltd.: Birmingham, UK, 2015; ISBN 9781783555147.
40. *The PyMOL Molecular Graphics System*; Version 1.6; Schrodinger, LLC: New York, NY, USA, 2011.

41. Martin, O.A. *Wrappy: A Dehydron Calculator Plugin for PyMOL*; IMASL-CONICET: San Luis, Argentina, 2012.
42. Dosztanyi, Z. Servers for sequence-structure relationship analysis and prediction. *Nucleic Acids Res.* **2003**, *31*, 3359–3363. [CrossRef] [PubMed]
43. Mitternacht, S. FreeSASA: An open source C library for solvent accessible surface area calculations. *F1000Research* **2016**, *5*, 189. [CrossRef] [PubMed]
44. Barlow, D.J.; Thornton, J.M. Ion-pairs in proteins. *J. Mol. Biol.* **1983**, *168*, 867–885. [CrossRef]

13

The C Terminus of the Ribosomal-Associated Protein LrtA is an Intrinsically Disordered Oligomer

José L. Neira [1,2,*,†], A. Marcela Giudici [1,†], Felipe Hornos [1], Arantxa Arbe [3] and Bruno Rizzuti [4,*]

1 Instituto de Biología Molecular y Celular, Edificio Torregaitán, Universidad Miguel Hernández, Avda. del Ferrocarril s/n, 03202 Elche (Alicante), Spain; marcela@umh.es (A.M.G.); fhornos@umh.es (F.H.)
2 Instituto de Biocomputación y Física de Sistemas Complejos, Joint Units IQFR-CSIC-BIFI, and GBsC-CSIC-BIFI, Universidad de Zaragoza, 50009 Zaragoza, Spain
3 Centro de Física de Materiales (CFM) (CSIC-UPV/EHU)—Materials Physics Center (MPC), 20018 San Sebastián, Spain; mariaaranzazu.arbe@ehu.eus
4 CNR-NANOTEC, Licryl-UOS Cosenza and CEMIF.Cal, Department of Physics, University of Calabria, Ponte P. Bucci, 87036 Rende, Italy
* Correspondence: jlneira@umh.es (J.L.N.); bruno.rizzuti@cnr.it (B.R.)

† These two authors contributed equally to this work.

Abstract: The 191-residue-long LrtA protein of *Synechocystis* sp. PCC 6803 is involved in post-stress survival and in stabilizing 70S ribosomal particles. It belongs to the hibernating promoting factor (HPF) family, intervening in protein synthesis. The protein consists of two domains: The N-terminal region (N-LrtA, residues 1–101), which is common to all the members of the HPF, and seems to be well-folded; and the C-terminal region (C-LrtA, residues 102–191), which is hypothesized to be disordered. In this work, we studied the conformational preferences of isolated C-LrtA in solution. The protein was disordered, as shown by computational modelling, 1D-^{1}H NMR, steady-state far-UV circular dichroism (CD) and chemical and thermal denaturations followed by fluorescence and far-UV CD. Moreover, at physiological conditions, as indicated by several biochemical and hydrodynamic techniques, isolated C-LrtA intervened in a self-association equilibrium, involving several oligomerization reactions. Thus, C-LrtA was an oligomeric disordered protein.

Keywords: disordered protein; folding; oligomer; ribosomal protein; protein stability

1. Introduction

The *lrtA* gene from *Synechococcus* sp. PCC702 is known to express a light-repressed protein [1,2]. Further investigations have shown that LrtA is involved in the stabilization the 70S ribosomal particles [3], as well as in cell survival during stress circumstances. LrtA is related with other proteins that take part in ribosome activity. Under environmental stress conditions, protein synthesis is stopped in a down-regulation process. Reduction of protein production involves: (i) Formation of the inactive 100S disome through dimerization of 70S particles [4], implicating the action of some proteins; or (ii) protein-ribosome interactions which involve the canonical ribosomal proteins [5,6]. The family of ribosomal proteins in *E. coli* includes YfiA (also known as PY or RaiA, ribosome associated inhibitor A); and YhbH (also known as HPF, hibernation promoting factor). YfiA likely inhibits translation indirectly, involving 70S particles [7,8]. Alternatively, HPF stops translation by stabilizing 100S dimers [8–10]. Most bacteria have one or more homologues related to HPF or YfiA [10]. These homologues can be classified in long HPF, short HPF and YFiA, on the basis of the length of their sequences and the presence of a specific domain. The conserved domain in all of them has the β-α-β-β-β-α fold [5,11], with a β-sheet formed by four strands and two α-helices packed against it. According to its sequence,

LrtA from *Synechocystis* sp. PCC 6803 belongs to the long HPF sub-family. We have previously shown that LrtA is involved in self-association equilibria [12], and has chameleonic structural properties. In particular, molecular dynamics (MD) simulations and experimental analyses suggest that the whole LrtA has a solvent-dependent conformation, where the N terminus adopts the β-α-β-β-β-α fold and the C terminus is disordered and compact [12].

In this work, we have studied the conformational preferences of the isolated C-terminal region of LrtA (residues 102-191), C-LrtA. We aimed to test whether: (i) C-LrtA was disordered and collapsed, as suggested by previous MD simulations of the whole LrtA; and (ii) isolated C-LrtA was oligomeric in solution. Characterizing the degree of disorder in proteins or protein domains, and whether this contributes to attaining a quaternary structure, is important to explain their functions; in fact, most of the intrinsically disordered proteins (IDPs) characterized so far are involved in protein-protein contacts [13], and it is essential to establish how specificity is achieved in those interactions. We show here that C-LrtA was disordered and with a strong self-association tendency, as shown by several biochemical, biophysical and hydrodynamic techniques: Blue-native gels, glutaraldehyde cross-linking, iodide quenching, small-angle X-ray scattering (SAXS), size exclusion chromatography (SEC) and isothermal titration calorimetry (ITC). MD simulations of isolated C-LrtA also predicted a disordered conformation, in reasonable agreement with the experiments. Therefore, we proved that: (i) former MD predictions on C-LrtA, based on the whole parental LrtA, were correct; (ii) the isolated domain has a tendency to self-associate; and (iii) the presence of quaternary interactions in C-LrtA did not induce any stable secondary nor tertiary structure, and therefore, C-LrtA was an oligomeric IDP.

2. Results

2.1. Isolated C-LrtA Was Intrinsically Disordered in Solution

To map the conformational features of C-LrtA in solution, we used NMR, fluorescence, far-UV (circular dichroism) CD and MD simulations. Fluorescence gives us information about the overall environment around the fluorescent residues (C-LrtA has 4 tyrosine residues). Far-UV CD provides information about the percentages of secondary structure. NMR gives further information about the presence of secondary and tertiary structures. Finally, MD simulations give indications on the conformation of isolated C-LrtA in solution and on the local propensity for secondary structure formation along the backbone.

(a) NMR: In the methyl (Figure 1A) and the amide (Figure 1B) regions of the NMR spectrum of C-LrtA, there was no dispersion, i.e., all the amide signals appeared clustered between 8.2 and 8.6 ppm and most of the alkyl chains appeared between 0.8 and 1.0 ppm. Only a small shoulder appeared up-field shifted at 0.7 ppm, indicative of a local conformation around the methyl group of a valine, leucine or isoleucine probably close to an aromatic residue; we hypothesize that this signal could correspond to the polypeptide patch VIYI (residues 173–176, in the numbering of the whole LrtA).

The NMR spectrum showed significant broadening in all the signals. On the basis of the results of the other techniques (see below in this section, and later in Section 2.2), this could be due to the presence of conformational exchange (equilibria) among protein species with different self-associated order. However, given the mobility of the protein (see MD simulations results, below in this section), we cannot exclude that the broadening observed could be due to conformational exchange in a single protein species.

Therefore, we can conclude from the NMR spectrum that C-LrtA was disordered.

(b) Far-UV CD: The far-UV CD spectrum of C-LrtA showed a minimum of around 200 nm and a wide shoulder at 222 nm (Figure 2A), which could be due to the presence of helix or turn-like structures, although the absorbance of aromatic residues (4 tyrosine and 3 phenylalanine ones) at the latter wavelength cannot be ruled out [14,15]. Although we cannot exclude protein adsorption to the cell at the lowest protein concentration used, the spectrum intensity was protein-concentration dependent (in the range of 10 to 20 μM of protein concentration, Figure S1) and its shape did not change. The fact that the other techniques used (see Section 2.2) indicate the presence of oligomeric species suggests that the variations observed in the far-UV CD spectra are due to the existence of self-associated species.

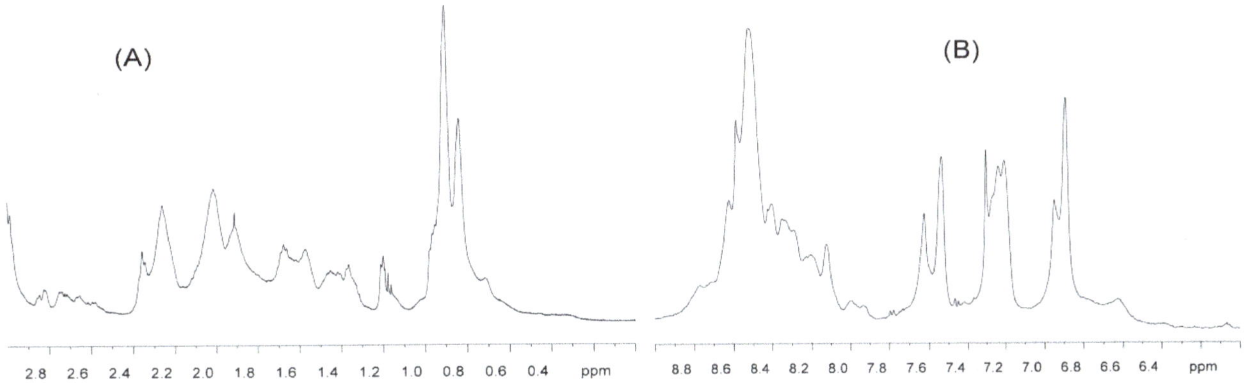

Figure 1. NMR characterization of C-LrtA: (**A**) Methyl and (**B**) amide regions of the 1D ^{1}H NMR spectrum of C-LrtA. Spectrum was acquired at 20 °C, and pH 7.2 (50 mM, Tris buffer).

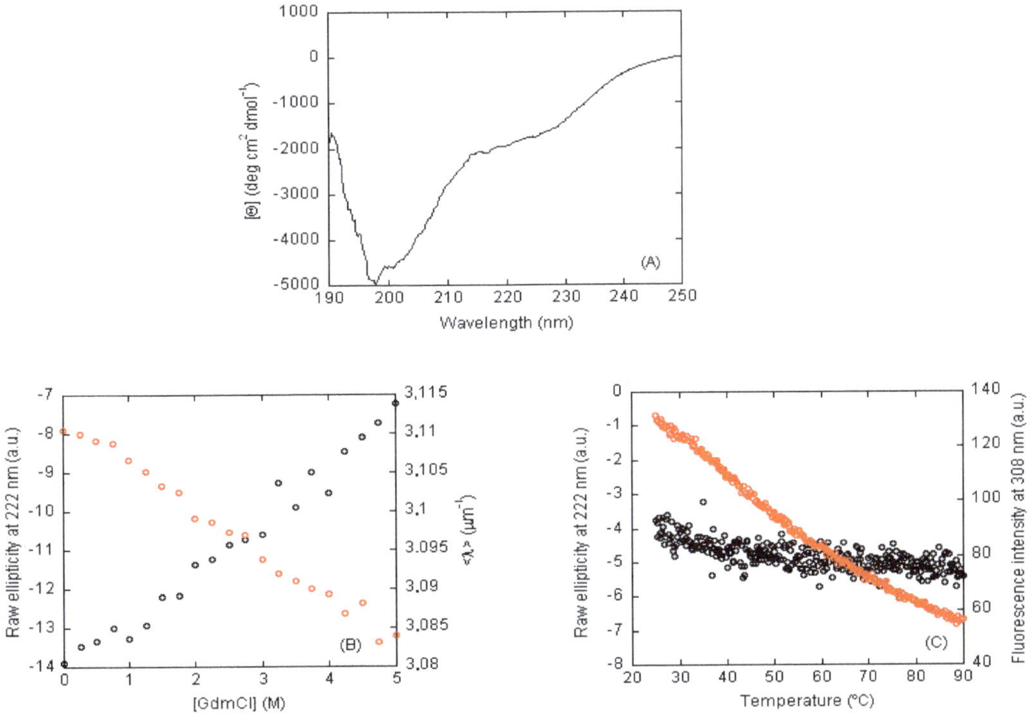

Figure 2. Spectroscopic characterization of C-LrtA: (**A**) Far-UV circular dichroism (CD) spectrum of C-LrtA. Spectrum was acquired at 20 °C, and pH 7.2 (50 mM, Tris buffer) with 20 μM (in protomer units) of protein concentration; (**B**) GdmCl-denaturations of C-LrtA followed by fluorescence (right axis, red circles) and CD (left axis, black circles), at 10 μM of protein concentration (in protomer units) and 20 °C; (**C**) Thermal denaturations of C-LrtA followed by fluorescence (right axis, red circles) and CD (left axis, black circles). Experiments were acquired at pH 7.2 (50 mM, Tris buffer) and 10 μM of protein concentration (in protomer units). The ellipticity (far-UV CD) units of thermal denaturations are arbitrary, because values are scaled up.

The shape of the spectrum of C-LrtA was characteristic of IDPs [16]. Its deconvolution, by using the algorithms available at the DICHROWEB site [17,18], yielded percentages of 7–8% for α-helix structure, 15–20% for β-turn, 28–44% for β-sheet and 45–48% for random-coil.

In the presence of increasing GdmCl concentrations, the shoulder at 222 nm of the far-UV CD spectrum decreased (Figure 2B, left axis, black circles) at the two concentrations explored (10 and 20 μM). These results suggest that the shoulder was not due to the presence of any well-fixed structure, but rather to flickering helix- or turn-like motifs, or even local conformations of the aromatic residues [14,15] Attempts to fit these data to the linear extrapolation model failed, as they led to thermodynamic parameters with non-physical meaning (i.e., negative values of m or large values of $[GdmCl]_{1/2}$). We further tested the disordered nature of C-LrtA by performing thermal denaturations. We observed a decrease in the ellipticity as the temperature was increased (Figure 2C, left axis, black circles) and therefore we did not observe a sigmoidal co-operative behaviour, as it should be expected for a well-folded globular domain [16,19].

Therefore, we can conclude from the far-UV CD data that C-LrtA was disordered.

(c) Fluorescence: Fluorescence spectra of C-LrtA showed a maximum at 307 nm, corresponding to its 4 tyrosine residues [20,21]. We carried out GdmCl denaturations by following the $\langle\lambda\rangle$ (at two different C-LrtA concentrations) after excitation at 280 nm. At both protein concentrations, we observed a linear decrease in the $\langle\lambda\rangle$ as the concentration of chemical denaturant was increased (Figure 2B, right axis, red circles). We could not fit these data to the linear extrapolation model, as fitting led to thermodynamic parameters (m- or $[GdmCl]_{1/2}$-values) with non-physical meaning (i.e., either negative values or values higher than the protein concentration explored). A similar linear tendency was observed in thermal denaturations (Figure 2C, right axis, red circles).

Therefore, we can conclude from the fluorescence data that C-LrtA was disordered.

(d) MD simulations: C-LrtA (the sequence present in the wild-type protein, i.e., without the His-tag) was simulated starting from an extended conformation with a radius of gyration R_g = 64 Å. The protein structure spontaneously collapsed in 15 ns, and for the subsequent 10 ns maintained a size of $R_g = 24 \pm 2$ Å, in excellent agreement with the value (R_g = 25 Å) predicted for a tag-free IDP of 90 residues with the sequence features of C-LrtA [22]. As reported in Figure 3, the secondary structure propensity of the protein in the time interval considered was reasonably consistent. In contrast, the conformations sampled at sufficiently large intervals (e.g., every 0.1 ns) were relatively different in terms of their three-dimensional arrangement, due to the dynamics of C-LrtA. Although not being long enough to obtain a complete statistical ensemble of conformations, the simulation was not prolonged to prevent well-known artifacts, due to over-compaction of the protein [23,24], because a small drift was observed in its size (decrease of R_g was ~ 0.3 Å/ns on average, Figure 3).

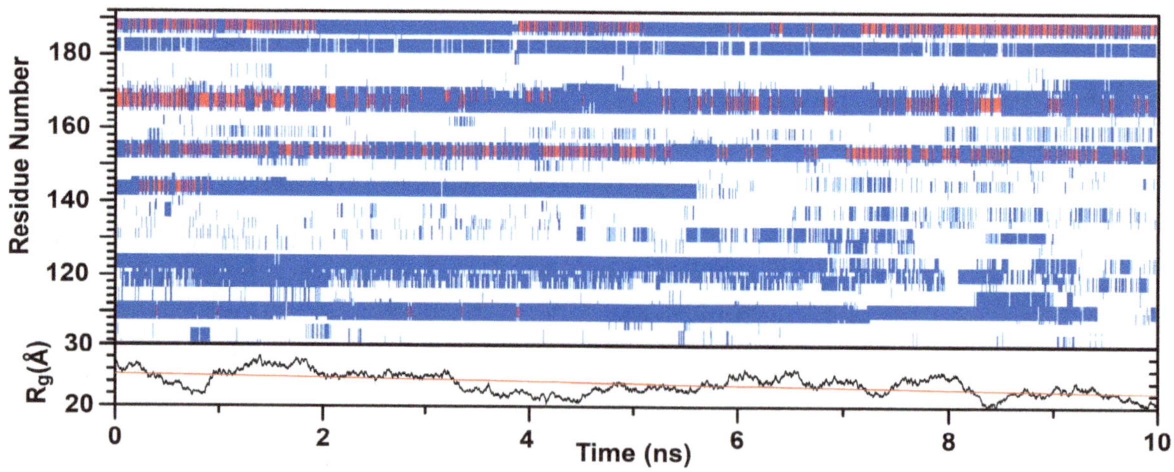

Figure 3. Simulated secondary structure and radius of gyration of C-LrtA: Backbone properties of the protein without the His-tag are calculated in a 10 ns time interval, following 15 ns of equilibration after starting from an elongated conformation. (**Up**) Secondary structure propensities calculated with VMD [25]: (Blue) β-structure, (red) helical structure, and (white) random coil; (**Down**) Radius of gyration of C-LrtA; the drift leading to a small decrease of R_g is also shown (red line).

The simulation results concurred to indicate that C-LrtA in solution was a very flexible protein with little secondary structure. The percentages of helical/β-structure were in good agreement with the range of those obtained from the deconvolution of far-UV CD spectra, but the corresponding backbone conformations were in all cases local and did not extend for more than a few residues. Interestingly, among the four tyrosine residues of C-LrtA, only Tyr182 (according to the numbering in intact LrtA) was in a region with β-structure propensity, whereas the other three were in random-coil regions and showed a large conformational freedom in the isolated domain.

In summary, taking into account all the data, as concluded from fluorescence, far- UV CD, NMR and MD simulations, C-LrtA appeared disordered in solution.

2.2. Isolated C-LrtA Was Involved in Self-Association Equilibria in Solution

To map the hydrodynamic properties of LrtA we used several biochemical, biophysical and hydrodynamic techniques: Blue-native gels; glutaraldehyde cross-linking; iodide quenching; DOSY-NMR (diffusion ordered spectroscopy NMR); SAXS (small-angle X-ray scattering); SEC (size exclusion chromatography) and ITC (isothermal titration calorimetry). We used such a plethora of different techniques to provide an unambiguous evidence of the presence of oligomerization in disordered C-LrtA. It is important to pinpoint, however, that with NMR we shall obtain information about the low-molecular weight species whose overall rotational tumbling is very fast, and then, we shall be able to obtain information only on the monomer and/or dimer species.

(a) DOSY-NMR: The DOSY-NMR measurements of C-LrtA yielded a translational diffusion coefficient, D, of $(5.0 \pm 0.2) \times 10^{-7}$ cm^2 s^{-1} (Figure 4A). By taking into account the hydrodynamic radius, R_S, of dioxane (2.12 Å), and its D under our conditions ($(8.53 \pm 0.02) \times 10^{-6}$ cm^2 s^{-1}), the estimated R_S for C-LrtA was 36 ± 4 Å. We can compare this value with that theoretically determined for a polypeptide with the sequence length of C-LrtA (including the N-terminal His-tag). The R value for an unsolvated, ideal, spherical molecule can be estimated from [21]: $R = \sqrt[3]{3M\overline{V}/4N_A\pi}$, where N_A is Avogadro's number, M is the molecular weight of the C-LrtA construct (12,449.89 Da), and $\overline{V}$ the specific volume of C-LrtA construct (0.721 mL/g). The calculated radius for C-LrtA is 15.3 Å, but taking into account the water shell [21,26], the hydration radius is 18.5 Å; this value is different from that obtained from experimental DOSY measurements. The R_s for a spherical, folded protein is given by [27]: $R_S = (4.75 \pm 1.11)N^{0.29}$, where N is the number of residues; in a 109-residue-long protein, such as C-LrtA (the His-tag and the 90-residue-long domain), this expression yields 18 ± 4 Å, in good agreement with the other theoretically calculated value. On the other hand, for an unfolded polypeptide chain, the R_S could be estimated from [27]: $R_S = (2.21 \pm 1.07)N^{0.57}$; for C-LrtA, the value is 32 ± 15 Å, which is closer to the values measured in the DOSY-NMR experiments; however, the use of that expression yields a value slightly higher (43 ± 15 Å) for a dimeric species. Therefore, by the DOSY-NMR experiments we are only detecting low-molecular weight species, which seemed to be unfolded.

(b) SEC: Different amounts of C-LrtA were loaded in an analytical Superose 12 10/300 GL column at pH 8.0 (50 mM Tris) and 0.250 M NaCl. The chromatograms did not show a sole peak (Figure S2), and the smaller the concentration of the protein the more were the peaks that appeared. This finding, given protein purity (Figure S3A), could be attributed to protein-column interactions of some species, which eluted at larger volumes than expected from their size. Similar delayed peaks, due to protein-column interactions, have been observed in the intact LrtA [12]. It is important to note that a small peak appearing at 16.48 mL was also present in the chromatogram of the most concentrated sample (500 μM) (Figure S2), as well as at any other protein concentration; we interpreted this peak as due to a monomeric species interacting with the column.

The elution volumes of one of the peaks showed a hyperbolic dependence as the concentration of protein was changed (Figure 4B). At very high concentrations (500 μM), the protein had elution volumes of 11.98 mL (obtained as the mean of three different measurements, although the elution peak was very broad). This value would correspond to a molecular weight of 100 kDa (for a comparison, a protein, such as ferritin, with a molecular weight of 400 kDa, elutes in the column at 10.11 mL, Figure S2); these results suggest that C-LrtA was probably an octamer under these conditions. On the other hand, at 30 μM, C-LrtA eluted at 13.88 mL, which would correspond to a molecular weight of 31.6 kDa, close to the expected molecular weight of a dimeric species. Therefore, in the column matrix the protein behaved as a self-associated species with several oligomerization orders, depending on the concentration used.

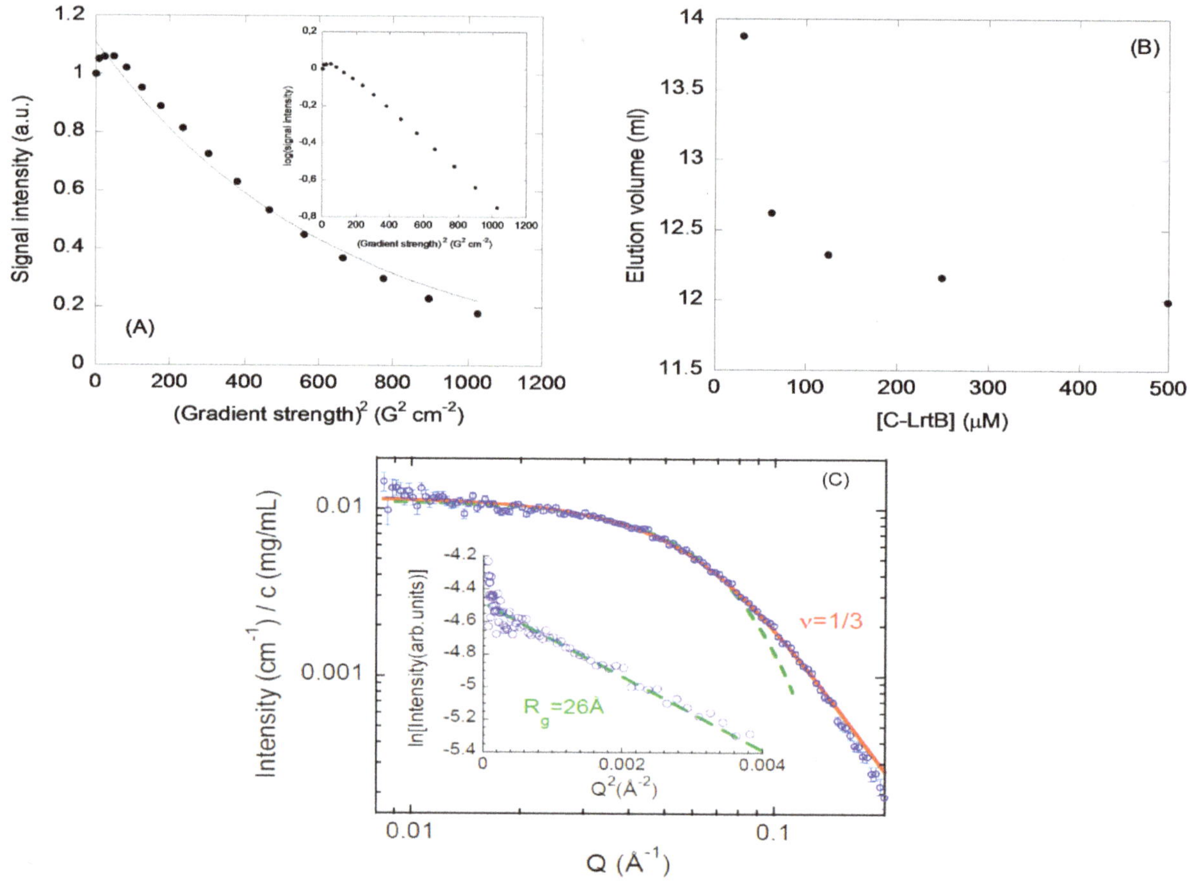

Figure 4. Hydrodynamic and biophysical measurements of C-LrtA: (**A**) DOSY measurements: Intensity decay (arbitrary units) of the methyl signals as the pulse field gradient strength was increased (x-axis). The line is the fitting to equation, as described in Section 4.6. The inset shows the linear relationship between the logarithm of the intensity and the square of the gradient strength; (**B**) size exclusion chromatography (SEC) measurements: Elution volumes of one of the peaks observed for C-LrtA in a Superose 12 10/300 GL at different protein concentrations in buffer pH 8.0 (50 mM Tris) and 0.250 M NaCl; the data have an error of 0.1 mL as obtained from three independent measurements at each particular C-LrtA concentration; (**C**) small-angle X-ray scattering (SAXS) results of C-LrtA are shown with the solid line representing a fit with a generalized Gaussian coil with a scaling exponent value of 1/3 (Section 4.11), and the dashed line is the Guinier description of the low-Q limit (see Guinier plot in the inset, and Section 4.11).

(c) BN-PAGE (blue native polyacrylamide gel electrophoresis): C-LrtA exhibited two species in these experiments, which corresponded to different self-associated species (Figure S3B). The protein species in the fastest migrating band corresponded to an apparent molecular weight of 33 kDa (close to the molecular weight of a dimer, and similar to that observed at the most diluted protein concentration in the SEC experiments). On the other hand, the other band corresponded to an apparent molecular weight of 66 kDa, denoting a pentamer. Our results also suggest that increasing the amount of SDS (well-below the concentration used in denaturing SDS-PAGE gels: 33 mM [28]) had significant effects on the population of self-associated C-LrtA species: the larger the proportion of SDS, the higher the amount of self-associated species detected (the critic micellar concentration of SDS is 1.33 mM). It is important to indicate, at this stage, that the BN-PAGE technique can lead to overestimation of the molecular weights, as some proteins can bind Coomassie dye [29].

(d) Glutaraldehyde cross-linking: To detect the presence of oligomeric species in C-LrtA, we also used the glutaraldehyde agent. We observed dimers (close to the band of the protein marker at 32 kDa) (Figure S3C) at shorter times after addition of the cross-linking agent, and other high-molecular-weight species at the top of the SDS-PAGE lanes. The population of these high-molecular weight self-associated species increased at the largest incubation times (Figure S3C).

(e) KI quenching: It is reasonable to assume that if the self-associated species form at low C-LrtA concentrations, and the tyrosine residues were involved in the association interfaces, then we should be able to follow protein self-association by KI quenching, and we should expect a decrease in the K_{sv} constant as the C-LrtA concentration was increased. We observed the following K_{sv} values in C-LrtA: 1.5 ± 0.3 M^{-1} (at 5 μM of protein); 1.1 ± 0.2 M^{-1} (at 20 μM of protein); and 1.04 ± 0.05 M^{-1} (at 40 μM of protein, all of them in protomer units). Then, there was a protein-concentration behaviour in the 5–40 μM concentration range for the K_{sv} and C-LrtA self-associates.

(f) ITC experiments: We also tried to test whether C-LrtA dissociated upon dilution, using the heat evolved in the reaction monitored by ITC. For experiments performed at a high protein concentration stock, the heat released upon dilution of the protein into the calorimetric cell was consistent with a dissociation reaction for all injections (Figure S4). We tried to fit the heat released to a simple dimer-monomer equilibrium, but the results of the fitting indicated that this assumption was not good enough, suggesting the presence of higher-order equilibria, as indicated by SEC results (see above in this section).

(g) SAXS experiments: The experiments with C-LrtA indicate a $R_g \approx 26$ Å with a $\upsilon \approx 0.33$, close to a compact species value, but with a value of R_g larger than that of a well-folded protein (Section 4, Figure 4C), which is within the range observed for unfolded polypeptide chains [27]. We obtained a good agreement with the expected Guinier regime at low Q values, indicating that, although the protein was self-associated, the size of the C-LrtA species was relatively small.

3. Discussion

The structural propensities and association features of IDPs and disordered protein domains are still poorly understood compared to those of well-folded protein regions. The difference is especially important for proteins, such as LrtA, which is formed by two distinct domains (i.e., N-LrtA and C-LrtA) with roughly the same sequence length, but with completely distinct conformational features. In a previous work [12], we have hypothesized that N-LrtA has a distinct folding topology that is in common with other members of the HPF protein family [5,11], whereas C-LrtA was predicted to be unfolded. In the present work, we have tested that hypothesis and we have found evidence that isolated C-LrtA is an IDP. The intrinsically disordered nature of C-LrtA was suggested by several pieces of evidence: (i) the lack of dispersion in NMR spectra (Figure 1); (ii) the shape (Figure 2A) and the deconvolution of far-UV CD spectrum; (iii) the absence of all-or-none co-operative transitions in the thermal and chemical denaturations (Figure 2B,C); and (iv) a model of the structure without the His-tag obtained with MD simulation (Figure 3).

It could be thought that the observed C-LrtA self-association is non-specific, that is, the protein has solvent-exposed hydrophobic patches which induce highly unspecific, self-association. However, there are two pieces of evidence that suggest that association was not the result of random solvent-exposed hydrophobic residues. First, the fact that Tyr residues were implicated in the oligomerization indicates that only regions containing those residues were involved. Second, if self-association was to be unspecific very large high-molecular species should be observed in some of the techniques used; in contrast, the highest molecular-weight species is observed to be an octamer (in SEC experiments) and glutaraldehyde cross-linking showed the presence of higher molecular-weight species only at very long incubation times (Figure S3C). Finally, it is interesting to note that we have shown that the intact protein, which also self-associates, did not have any large amount of close solvent-exposed hydrophobic patch at physiological pH [12].

Although it is devoid of secondary and tertiary structures, C-LrtA is involved in quaternary contacts, involving different orders of self-associated species, as indicated by several of the techniques used. In general, IDPs intervene in protein-protein contacts [13], but only a few are reported as self-associated species in solution, while keeping their disordered nature. In C-LrtA, some of the four tyrosine residues were involved in its self-association, as judged by the changes in the quenching parameters as the concentration of protein was increased. In the parental, whole protein, tyrosine residues were already found to participate in the oligomerization interface [12]. With the new results in hand, we suggest that some of the four tyrosine residues in C-LrtB were responsible for the self-oligomerization of the whole protein. Then, whereas tyrosine residues in N-LrtA (residues 1–101) seem to contribute to the rigidity of the β-sheet scaffold, tyrosine amino acids in C-LrtA (residues 102–191) seem to be responsible for the quaternary structure of the intact protein. It is important to note that the detected order of the self-associated species in C-LrtA varied among the different biochemical and hydrodynamic techniques used (as it happens for the whole LrtA [12]) indicating the presence of different oligomerization equilibria. That is, the self-association did not involve the simple dimer-monomer equilibrium, but rather a sequence of different order equilibria.

The exact biological significance of self-association in LrtA remains unknown, and it is still a matter of debate. In fact, in spite of similar structural organization and high sequence homology among the members of the HPF family, their functions during stress responses are very different in the organisms to which they belong to [4,7–10,30]. There are some examples of oligomeric HPFs reported in the literature: for instance, the short *Vibrio choleras* HPF is a dimer, whose dimerization occurs through Zn ions at one of the β-strands of the β-α-β-β-β-α fold; however, it is not known if such dimerization is due to the crystallization process [11]. On the other hand, the HPF of *Staphylococcus aureus* is also a member of the long HPF family, but its C-terminal region is shorter (60-residue long) and it is folded [5]. The protein is a dimer, and its C-terminal region is responsible for this dimerization; furthermore, interactions of the dimeric C-terminal region with ribosomes are responsible for ribosome dimerization. Thus, we hypothesize that in LrtA the oligomeric, disordered C-LrtA domain might be responsible for the dimerization of 70S particles (whose abundance in the cell has been associated with the presence of LrtA [3]) during stress conditions. The fact that C-LrtA is disordered (in contrast to that of *Staphylococcus aureus* HPF [5]) could provide the advantage that LrtA may be bound not only to the 70S particles, but also to other macromolecules, regulating several cyanobacterial processes triggered by stress conditions.

Evidence of the possibility of self-association and supra-molecular order in IDP are starting to be mounting. One of the first reported examples of oligomeric IDPs was the "fuzzy" dimer formed by the cytoplasmic domain of the T-cell receptor zeta subunit [31,32], although this putative dimer was later shown to be a monomer under a wide range of conditions, and its oligomerization was attributed to non-ideal protein-column-resin interactions [33]. However, other studies have used a plethora of biophysical and biochemical techniques to unambiguously show the existence of intact self-associated IDPs. For instance, there are reports describing oligomeric plant IDPs [34]; dimeric proteins in the disordered *umu*D gene products [35]; oligomeric mitochondrial IDPs [36]; oligomeric IDPs which bind to the Polycomb complex [37]; disordered, oligomeric acid-rich proteins of rod photoreceptors [38]; and disordered oligomeric oncogen products [39]. Many of these homoligomeric interactions are deposited in the MFIB database (http://mfib.enzim.ttk.mta.hu) [40], together with others involving hetero-oligomer assemblies, where in all cases mutual folding of the IDP chains occurs. The presence of such a quaternary (homo or hetero-oligomeric) structural organization for an IDP has been explained as due to multiple transient interactions or long-range contacts, which yield a fuzzy self-associated species [41–43], involving different degrees of organization in the association process [44].

4. Materials and Methods

4.1. Materials

Deuterium oxide, d_{11}-Tris acid and isopropyl-β-D-1-tiogalactopyranoside (IPTG) were purchased from Apollo Scientific (Stockport, UK). DNase, kanamycin, the Trizma base and its acid, sodium trimethylsilyl (2,2,3,3-$^2H_4$) propionate (TSP), imidazole, and the His-Select HF nickel resin were from Sigma-Aldrich (Madrid, Spain). The β-mercaptoethanol (β-ME) was provided by BioRad (Madrid, Spain). The protein marker, PAGEmark Tricolor, and Triton X-100 were from VWR (Barcelona, Spain). The protein gel-filtration-column calibration markers were from GE Healthcare (Barcelona, Spain). Dialysis tubing was from Spectrapor (Spectrum Laboratories, Shiga, Japan). Amicon centrifugal devices with a cut-off molecular weight of 3000 Da were from Millipore (Barcelona, Spain). The rest of the materials used were of analytical grade. Water was deionized and purified on a Millipore system.

4.2. Protein Expression and Purification

The C-LrtA region (residues 102-191 of LrtA) was cloned by NZytech (Lisbon, Portugal) in a pHTP1 *E. coli* expression vector (between XhoI and NcoI sites), with kanamycin resistance. The final construct contained an N-terminal His-tag to allow for purification (MGSSHHHHHHSSGPQQGLR), and had the overall sequence: MGSSHHHHHHSSGPQQGLRQHGNVKTSEIVEDKPVEENLIGDRA PELPSEVLRMKYFAMPPMAIEDALEQLQLVDHDFYMFRNKDTDEINVIYIRNHGGYGVIQPHQAS.

Expression of C-LrtA was carried out in the *E. coli* BL21 (DE3) strain (Novagen, VWR, Barcelona; Spain) strains with a final kanamycin concentration of 50 mg/mL. We used 1 L flasks to culture the cells. The expression of the protein was induced with a final concentration of 1.0 mM IPTG, when the absorbance observed for the cell culture was 0.4–0.9 at 600 nm, and the growth of the cells continued for 15–16 h at 25 °C 15–16 h at 25 °C (this temperature was chosen to decrease the possibility of aggregates in C-LrtA, as the parental LrtA had a tendency to form inclusion bodies). We harvested the cells in a JA-10 rotor (Beckman Coulter) for 15 min at 8000 rpm. We re-suspended the cellular pellet from 5 L of culture in 50 mL buffer A (500 mM NaCl, 5 mM imidazole, 20 mM Tris buffer (pH 8), 0.1% Triton X-100 and 1 mM β-ME) and adding a tablet of Sigma Protease Cocktail EDTA-free. In the first attempts to get the protocol of purification, we added 2 mg of DNase (per 1 L of culture). We incubated the mixture for 10 min, with gentle agitation in the fridge (4 °C). Next, we sonicated the mixture (by using a Branson sonicator, 750 W), with 10 cycles of 45 s each at 55% of maximal power output, with intervals of 15 s between the cycles, and always keeping the cells in ice. We separated the supernatant by centrifugation at 18,000 rpm for 40 min at 4 °C in a Beckman JSI30 centrifuge with a JA-20 rotor. C-LrtA was present in the supernatant, and we purified it by immobilized affinity chromatography (IMAC), by adding 5 mL of Ni-resin previously equilibrated in buffer A. The mixture was incubated for 20 min at 4 °C, and afterwards, the lysate was separated from the resin by gravity. The washing step was carried out with 20 mL of buffer B (20 mM Tris buffer (pH 8.0), 500 mM NaCl, 1 mM β-ME, and 20 mM imidazole); the protein was eluted by gravity from the column with buffer C (20 mM Tris buffer (pH 8.0), 500 mM NaCl, 1 mM β-ME, and 500 mM imidazole). The solution was dialyzed against 50 mM Tris buffer (pH 8.0). The protein was further purified in a Hi-Trap Mono Q (GE Healthcare) column by using a gradient step from 0 to 1 M NaCl (50 mM Tris buffer, pH 8.0) in 60 min and a flow of 1 mL/min, in an AKTA Basic system (GE Healthcare), while monitoring the absorbance at 280 nm. This column purification step was used based on the theoretical isoelectric point of C-LrtA (pI = 5.72).

In the initial purification attempts (carried out in the presence of DNase), we observed that the main peak coming from the Hi-Trap Mono Q column did not show an emission fluorescence spectrum expected for a protein containing only 4 tyrosine residues (i.e., with a maximum at 308 nm), but rather the spectrum of a protein containing tryptophan (a maximum at ~330 nm). Therefore, we thought of the possibility of contamination of the protein with the DNase used in the first steps of the purification, which also has a similar pI (5.2), and it has a tendency to self-associate. Then, in subsequent purifications we did not add DNase to the cell lysate. As a consequence, the protein coming from the Hi-Trap Mono Q showed absorbance at 260 nm, probably due to the presence of oligonucleotides resulting from the sonication step; these oligonucleotides must be present in a small amount, since no evidence of sharp peaks (contaminants of low-molecular weight) were observed in the 1D ^{1}H-NMR spectrum (Figure 1). The presence of traces of oligonucleotides is not infrequent in proteins of the same family as it has been also observed in the recombinant HPF from *S. aureus* after its purification [5]. After elution from the Hi-Trap Mono Q column, the sample was extensively dialyzed against water, and no precipitation was observed in the dialysis tubing. The concentration of the protein Pc (mg/mL) was calculated by [45]: $P_c = 1.55\ A_{280}–0.75\ A_{260}$, where A_{280} and A_{260} are the absorbances of the dialyzed solution of the protein at 280 and 260 nm, respectively.

4.3. Fluorescence

Fluorescence spectra were collected on a Cary Varian spectrofluorimeter (Agilent, Santa Clara, CA, USA), interfaced with a Peltier, at 25 °C. The C-LrtA concentrations used were 5 and 10 μM of protein (in protomer units) in the GdmCl denaturations carried out at pH 7.5 (50 mM Tris buffer). A 1-cm-pathlength quartz cell (Hellma; Sigma Aldrich, Madrid, Spain) was used. In the denaturation experiments with GdmCl, we prepared the samples the day before starting from a 7 M GdmCl concentrated stock that equilibrated overnight; samples were left at 25 °C for 1 h before performing the measurements.

The emission intensity weighted average of the inverse wavelengths (also called the spectrum mass centre, or the spectral average energy of emission), $<\lambda>$, was calculated as described [46].

(a) Steady-state spectra: The experimental set-up used in the case of the denaturation experiments with GdmCl was as previously described [46]. In brief, excitation of the protein samples was at 278 nm, and excitation and emission slits were 5 nm in all cases. The experiments were recorded between 300 and 400 nm. The signal was acquired for 1 s and the increment of wavelength was set to 1 nm.

(b) Thermal denaturations: Thermal denaturations of isolated C-LrtA were carried out with the same experimental set-up previously described [46]. These experiments were performed at constant heating rates of 60 °C/h, with an average time of 1 s. Thermal scans were collected at 308 nm after excitation at 278 nm from 25 to 90 °C and acquired every 0.2 °C. Protein concentration was 10 μM (in protomer units).

(c) Fluorescence quenching: Quenching by iodide was examined with concentrations ranging from 5 to 40 μM (in protomer units) at pH 7.0 (phosphate buffer, 50 mM). The experimental set-up for KI was the same described above for the intrinsic fluorescence experiments. The data were fitted to [47]: $F_0/F = 1 + K_{sv}[KI]$, where K_{sv} is the Stern-Volmer constant for collisional quenching; F_0 is the fluorescence intensity in the absence of KI; and F is that at any KI concentration. The range of KI concentrations explored was 0–0.7 M. Fittings to the above equation were carried out by using Kaleidagraph (Synergy software).

4.4. CD

The far-UV CD spectra were recorded at 25 °C on a Jasco J815 spectropolarimeter (Jasco, Easton, MD, USA City, Japan) equipped with a thermostated cell holder, and interfaced with a Peltier unit. A periodical calibration was performed with (+)-10-camphorsulphonic acid. We used two concentration values for C-LrtA (10 and 20 μM, in protomer units) to check for concentration-dependence of the shape and intensity of spectra, performed at pH 7.0 (50 mM, phosphate buffer). Molar ellipticity was determined as previously indicated [46].

(a) *Steady-state spectra*: Experiments were performed using the experimental set-up described previously [46]. Spectra were corrected by subtracting the baseline in all cases. Protein concentration was 10 and 20 μM (in protomer units) for GdmCl-denaturation experiments.

(b) Thermal denaturations: Experiments were carried out with the same experimental set-up described previously [46] and a protein concentration of 10 μM (in protomer units). Briefly, thermal denaturations were performed at a constant heating rate of 60 °C/h from 25 to 85 °C, a response time of 8 s, a band width of 1 nm, acquired every 0.2 °C, and following the ellipticity at 222 nm.

4.5. NMR Spectroscopy

The NMR experiments were acquired at 20 °C on a Bruker Avance DRX-500 spectrometer equipped with a triple resonance probe and z-pulse field gradients. All spectra were processed and analysed by using TopSpin 2.1 (Bruker GmbH, Karlsruhe, Germany). We used TSP as the external chemical shift reference [26].

(a) 1D ^{1}H-NMR experiments: The 1D-^{1}H-NMR spectrum were acquired with a C-LrtA concentration of 120 μM (in protomer units) in 0.5 mL, 50 mM d_{11}-Tris buffer (pH 7.2) in H_2O/D_2O (90%/10%, *v*/*v*), without any correction for deuterium isotope effects. The spectrum was acquired with 16 K data points. We acquired 2 K scans with a 6000 Hz spectral width (12 ppm), by using the WATERGATE sequence [48]. Before processing the data, baseline correction and zero-filling were applied.

(b) Translational diffusion measurements: The DOSY experiments of C-LrtA at pH 7.2 were performed with the pulse-field gradient (PFG) spin-echo sequence, as described previously [12,46], with sixteen gradient strengths ranging linearly from 2% to 95% of the total power of the gradient unit. The intensity of the methyl signals, *I*, was fit to: $\frac{I}{I_0} = -\exp\left(D\gamma_H^2\delta^2G^2\left(\Delta - \frac{\delta}{3} - \frac{\tau}{2}\right)\right)$, where I_0 is the maximum peak intensity of the methyl resonances at the smallest gradient strength; δ is the duration (in s) of the gradient (2.7 ms); *G* is the gradient strength (in T cm^{-1}); Δ is the time (in s) between the gradients (150 ms); γ_H is the gyromagnetic constant of the proton; and, τ is the recovery delay between the bipolar gradients (100 μs). Samples were exchanged in D_2O buffer (50 mM d_{11}-Tris buffer, pH 7.2, not corrected for isotope effects) by using Amicon centrifugal devices for 4 to 6 h.

4.6. Blue-Native PAGE (BN-PAGE)

BN-PAGE was performed in linear 4% to 16% (*w*/*v*) polyacrylamide-gradient gels [28,49,50]. Before running the BN-PAGE, sample aliquots (10 μL) containing 20 μg of C-LrtA were mixed with 1 μL of 5% Coomassie Brilliant blue G stock solution in 750 mM aminocaproic acid. Electrophoresis was initiated at 85 V for 30 min, and then continued at 200 V for 2.5 h, at 4 °C. After electrophoresis, the gels were stained overnight with colloidal Coomassie Blue G 250 [51]. In these gels, it is the negative charge from the Coomassie Blue, capable of binding to the hydrophobic protein surfaces, what determines the electrophoretic mobility. The technique is a method of choice to study the organization of protein complexes in their native state [28,49–51].

4.7. Glutaraldehyde Cross-Linking

Glutaraldehyde cross-linking was carried out at pH 7.2 (50 mM, Tris buffer). The volume of the sample was 200 μL. The C-LrtA concentration was 120 μM (in protomer units). The protocol used has been described previously [12], and experiments were repeated twice.

4.8. SEC

SEC experiments were performed at pH 8.0 (50 mM Tris) and 0.250 M NaCl in a Superose 12 10/300 GL column, which was connected to an AKTA FPLC (GE Healthcare, Barcelona, Spain); absorbance at 280 nm was monitored during the elution. Flow rate was 0.8 mL/min, and C-LrtA was loaded in a volume of 100 μL at several protein concentrations, in the range of 30–500 μM (in protomer concentration). The markers used to calibrate the column were ferritin, catalase, aldolase, albumin, bovine RNase A and blue dextran (GE Healthcare, Barcelona, Spain); the isolated protein markers were loaded in the column in the above described buffer, at the same flow rate, and three independent measurements were performed with each marker.

4.9. ITC

The protocol used during the ITC experiments was that previously described [46]. Briefly, the ITC experiments were carried out by using a VP-ITC instrument (Microcal, Northampton, MA, USA). Before the experiments, C-LrtA was dialysed at 4 °C against water at physiological pH. Dilution ITC experiments involved sequential injections of microliter amounts (10 μL) of a concentrated protein solution (498 μM, in protomer units) into the calorimetric cell (1.4 mL), which initially contained water alone.

4.10. SAXS

SAXS experiments were performed on a Rigaku 3-pinhole PSAXS-L instrument, at 45 kV and 0.88 mA. The MicroMax-002+ X-ray Generator Systems includes a module with a microfocus sealed tube source, and an X-ray generator unit producing Cu-Kα transition photons, with $\lambda = 1.54$ Å wavelength. Vacuum was maintained both in the flight path and sample chamber. A two-dimensional multiwire X-ray detector (Gabriel design, 2D-2000X) was used as a detector of the scattered X-rays. We obtained the azimuthally averaged scattered intensities as a function of the scattering vector Q (where $Q = 4\pi(\lambda)^{-1}\sin\theta$, and θ represents half the scattering angle). Silver behenate was used as standard for calibration (reciprocal space). The solutions were filling Boron-rich capillaries with an outside diameter of 2 mm and wall thickness of 0.01 mm. The contribution from the corresponding buffer (measured on the same capillary) was subtracted by applying the proper factors obtained from transmission measurements. The sample-detector distance was 2 m, allowing covering a Q-range from 0.008 to 0.2 Å^{-1}.

From the intensity scattered at low-Q values –in the so-called Guinier regime– we can determine the average gyration radius, R_g, of the protein under several conditions, by using the Guinier law: $I(Q) = A\exp(\frac{-R_g^2Q^2}{3})$. The pre-exponential factor, A, is determined by the molecule concentration, the scattering contrast and the mass of the macromolecules dispersed in the solution. On the other hand, we can estimate the compaction grade through the scaling exponent υ, which relates to Q as $I(Q) \approx Q^{-1/v}$, in the high Q-range here explored. The values of υ are 1/3 for a polymeric chain collapsed into a globule; 0.5 for a random-coil polymer (which is the conformation of a linear polymer chain in Θ-conditions); and 0.6 for a swollen chain in a good solvent (self-avoiding-walk conformation). The form factor of a coil, with scaling exponent υ, is described in terms of the so-called generalized Gaussian coil function, given by the expression [52]: $I(Q) \approx \frac{1}{\nu U^{1/2\nu}}\gamma\left(\frac{1}{2\nu}, U\right) - \frac{1}{\nu U^{1/\nu}}\gamma\left(\frac{1}{\nu}, U\right)$, where $U = (2\nu+1)(2\nu+2)Q^2R_g^2/6$ and $\gamma(a,x) = \int_0^x t^{a-1}\exp(-t)dt$. From the fits of this function to the experimental data, the value of the radius of gyration can also be obtained.

4.11. Molecular Modelling

A model of C-LrtA (without the His-tag) was obtained in MD simulations by using a protocol previously adopted for IDPs [23,24]. In brief, simulations were performed with the GROMACS package [53] starting from a protein model built by using VMD [25], and collapsing it in a brute-force run carried out in the isobaric-isothermal ensemble. C-LrtA was initially in an extended conformation, except for backbone turns in correspondence with proline residues. The protein was centered in a rhombic dodecahedron box with a minimum distance of 1 nm from the edge of the simulation box, and surrounded with explicit water molecules. Amino acid residues were adjusted to mimic neutral pH and Na^+ counterions were added to obtain an overall neutral molecular system. The AMBER ff99SB-ILDN force field [54] was used for the protein, and the TIP3P [55] model for water. Other simulation conditions, including modelling of the electrostatics and van der Waals interactions, and reference values and coupling times for the thermostat and barostat, were as previously described [56,57].

5. Conclusions

The C-terminal region of the cyanobacterial protein LrtA, a member of the HPF family, was a disordered domain that self-associated through a mechanism that involved some of its tyrosine residues. These findings clarify a number of important features of intact LrtA, including demonstrating, in an unambiguous way, the presence of two distinct and structurally different domains in this protein. Moreover, in the absence of studies of the N-terminal region (residues 1-101) our results suggest that the unstructured C-terminal one is the region driving the supramolecular self-organization of the whole protein. This study contributes to clarify the structural and stability features of proteins that interact and modulate the ribosome activity, and especially those belonging to the almost unexplored subfamily of long HPF proteins.

Author Contributions: Conceptualization, J.L.N. and B.R.; Methodology, J.L.N., A.M.G., F.H., A.A. and B.R.; Software, B.R.; Investigation, J.L.N., A.M.G., F.H., A.A. and B.R.; Writing-Original Draft Preparation, J.L.N., A.M.G., F.H., A.A. and B.R.; Writing-Review and Editing, J.L.N. and B.R.

Acknowledgments: B.R. acknowledges kind hospitality and use of computational resources in the European Magnetic Resonance Center (CERM), Sesto Fiorentino (Florence), Italy. J.L.N. thanks F. J. Florencio and M. I. Muro-Pastor for introducing to him the LrtA system. J.L.N. and F.H. thank Javier Gómez for helpful discussions. All the authors thank the three anonymous reviewers for their encouragement to formulate some hypotheses on the biological significance of the oligomerization in LrtA.

Abbreviations

β-ME	β-mercaptoethanol
BN	Blue native
CD	Circular dichroism
C-LrtA	C-terminal half of LrtA protein (comprising residues 102–191)
DOSY	Diffusion ordered spectroscopy
GdmCl	Guanidine hydrochloride
HPF	Hibernating promoting factor
IDP	Intrinsically disordered protein
IMAC	Immobilized affinity chromatography
IPTG	Isopropyl-β-D-1-tiogalactopyranoside
ITC	Isothermal titration calorimetry
MD	Molecular dynamics
N-LrtA	N-terminal half (residues 1–101) of LrtA protein
NMR	Nuclear Magnetic Resonance
PAGE	Polyacrylamide gel electrophoresis
PFG	Pulse field gradient
RaiA	Ribosome associate inhibitor A
RNase	Ribonuclease
SAXS	Small-angle X-ray scattering
SDS	Sodium dodecyl sulphate
SEC	Size exclusion chromatography
UV	Ultraviolet

References

1. Tan, X.; Varughese, M.; Widger, W.R. A light-repressed transcript found in *Synechococcus* sp. PCC 7002 is similar to a chloroplast-specific small subunit ribosomal protein and to a transcription modulator protein associated with sigma 54. *J. Biol. Chem.* **1994**, *269*, 20905–20912. [PubMed]

2. Samartzidou, H.; Widger, W.R. Transcriptional and post-transcriptional control of mRNA from lrtA, a light-repressed transcript in *Synechococcus* sp. PC 7002. *Plant Physiol.* **1998**, *117*, 225–234. [CrossRef] [PubMed]
3. Galmozzi, C.V.; Florencio, F.J.; Muro-Pastor, M.I. The cyanobacterial ribosomal-associated protein LrtA is involved in post-stress survival in *Synechocystis* sp. PCC 6803. *PLoS ONE* **2016**. [CrossRef] [PubMed]
4. Yoshida, H.; Wada, A. The 100S ribosome: Ribosomal hibernation induced by stress. *Wiley Interdiscipl. Rev. RNA* **2014**, *5*, 723–732. [CrossRef] [PubMed]
5. Khusainov, I.; Vicens, Q.; Ayupov, R.; Usachev, K.; Myasnikov, A.; Simonetti, A.; Validov, S.; Kieffer, B.; Yuspova, G.; Yusupov, M.; et al. Structures and dynamics of hibernating ribosomes from *Staphylococcus aureus* mediated by intermolecular interactions of HPF. *EMBO J.* **2017**, *36*, 2073–2087. [CrossRef] [PubMed]
6. Starosta, A.L.; Lasak, J.; Jung, K.; Wilson, D.N. The bacterial translation stress response. *FEMS Microbiol. Rev.* **2014**, *38*, 1172–1201. [CrossRef] [PubMed]
7. Agafonov, D.E.; Spirin, A.S. The ribosome-associated inhibitor A reduces translation errors. *Biochem. Biophys. Res. Commun.* **2004**, *320*, 354–358. [CrossRef]
8. Polikanov, Y.S.; Blaha, G.M.; Steitz, T.A. How hibernation factors RMF, HPF and YfiA turn off protein synthesis. *Science* **2012**, *336*, 915–918. [CrossRef]
9. Ueta, M.; Yoshida, H.; Wada, C.; Baba, T.; Mori, H.; Wada, A. Ribosome binding proteins YHbH and YfiA have opposite functions during 100S formation in the stationary phase of *Escherichia coli*. *Genes Cells* **2005**, *10*, 1103–1112. [CrossRef]
10. Ueta, M.; Ohniwa, R.L.; Yoshida, H.; Maki, Y.; Wada, C.; Wada, A. Role of HPF (hibernation promoting factor) in translational activity in *Escherichia coli*. *J. Biochem.* **2008**, *143*, 425–433. [CrossRef]
11. De Bari, H.; Berry, E.A. Structure of *Vibrio cholerae* ribosome hibernation factor. *Acta Cryst. Sect. F* **2013**, *69*, 228–236. [CrossRef] [PubMed]
12. Contreras, L.M.; Sevilla, P.; Cámara-Artigas, A.; Hernández-Cifre, J.G.; Rizzuti, B.; Florencio, F.J.; Muro-Pastor, M.I.; García de la Torre, J.; Neira, J.L. The Cyanobacterial ribosomal-associated protein LrtA from *Synechocystis* sp. PCC 6803 is an oligomeric protein in solution with chameleonic sequence properties. *Int. J. Mol. Sci.* **2018**. [CrossRef] [PubMed]
13. Sickmeier, M.; Hamilton, J.A.; LeGall, T.; Vacic, V.; Cortese, M.S.; Tantos, A.; Szabo, B.; Tompa, P.; Chen, J.; Uversky, V.N.; et al. DisProt: The database of disordered proteins. *Nucleic Acids Res.* **2007**, *35*, D786–D793. [CrossRef] [PubMed]
14. Woody, R.W. Circular dichroism. *Methods Enzymol.* **1995**, *246*, 34–71. [PubMed]
15. Kelly, S.M.; Jess, T.J.; Price, N.C. How to study proteins by circular dichroism. *Biochim. Biophys. Acta* **2005**, *1751*, 119–139. [CrossRef] [PubMed]
16. Chemes, L.B.; Alonso, L.G.; Noval, M.G.; de Prat-Gay, G. Circular dichroism techniques for the analysis of intrinsically disordered proteins and domains. *Methods Mol. Biol.* **2012**, *895*, 387–404. [PubMed]
17. Whitmore, L.; Wallace, B.A. Protein secondary structure analysis from circular dichroism spectroscopy: Methods and reference databases. *Biopolymers* **2008**, *89*, 392–400. [CrossRef]
18. Whitmore, L.; Wallace, B.A. DICHROWEB, an online server for protein secondary structure analyses from circular dichroism spectroscopic data. *Nucleic Acids Res.* **2004**, *32*, W668–w673. [CrossRef]
19. Receveur-Bréchot, V.; Bourhis, J.M.; Uversky, V.N.; Canard, B.; Longhi, S. Assessing protein disorder and induced folding. *Proteins* **2006**, *62*, 24–45.
20. Neira, J.L. Fluorescence, circular dichroism and mass spectrometry as tools to study virus structure. *Subcell. Biochem.* **2013**, *68*, 177–202.
21. Cantor, C.R.; Schimmel, P.R. *Biophysical Chemistry*; W. H. Freeman: New York, NY, USA, 1980.
22. Marsh, J.A.; Forman-Kay, J.D. Sequence determinants of compaction in intrinsically disordered proteins. *Biophys. J.* **2010**, *98*, 2383–2390. [CrossRef] [PubMed]
23. Neira, J.L.; Rizzuti, B.; Iovanna, J.L. Determinants of the pK_a values of ionizable residues in an intrinsically disordered protein. *Arch. Biochem. Biophys.* **2016**, *598*, 18–27. [CrossRef] [PubMed]
24. Cozza, C.; Neira, J.L.; Florencio, F.J.; Muro-Pastor, M.I.; Rizzuti, B. Intrinsically disordered inhibitor of glutamine synthetase is a functional protein with random-coil-like pK_a values. *Protein Sci.* **2017**, *26*, 1105–1115. [CrossRef] [PubMed]
25. Humphrey, W.; Dalke, A.; Schulten, K. VMD: Visual molecular dynamics. *J. Mol. Graph. Model* **1996**, *14*, 33–38. [CrossRef]

26. Cavanagh, J.F.; Wayne, J.; Palmer, A.G., III; Skelton, N.J. *Protein NMR Spectroscopy: Principles and Practice*, 1st ed.; Academic Press: San Diego, CA, USA, 1996.
27. Wilkins, D.K.; Grimshaw, S.B.; Receveur, V.; Dobson, C.M.; Jones, J.A.; Smith, L.J. Hydrodynamic radii of native and denatured proteins measured by pulse field gradient NMR techniques. *Biochemistry* **1999**, *38*, 16424–16431. [CrossRef] [PubMed]
28. Giudici, A.M.; Molina, M.L.; Ayala, J.L.; Montoya, E.; Renart, M.L.; Fernandez, A.M.; Encinar, J.A.; Ferrer-Montiel, A.V.; Poveda, J.A.; Gonzalez-Ros, J.M. Detergent-labile, supramolecular assemblies of KcsA: Relative abundance and interactions involved. *Biochim. Biophys. Acta* **2013**, *1828*, 193–200. [CrossRef] [PubMed]
29. Wittig, I.; Beckhaus, T.; Wumaier, Z.; Karas, M.; Schägger, H. Mass estimation of native proteins by blue native electrophoresis: Principles and practical hints. *Mol. Cell Proteom.* **2010**, *9*, 2149–2161. [CrossRef] [PubMed]
30. Basu, A.; Yap, M.F.N. Ribosome hibernation factor promotes Staphylococcal survival and diferentially represses translation. *Nucleic Acids Res.* **2016**, *44*, 4881–4893. [CrossRef]
31. Sigalov, A.B.; Aivazian, D.; Stern, L. Homo-oligomerization of the cytoplasmic domain of the T-cell receptor zeta chain and of other proteins containing the immunoreceptor tyrosine-based activation motif. *Biochemistry* **2004**, *43*, 2049–2061. [CrossRef]
32. Sigalov, A.B.; Zhurauleva, A.V.; Orekhov, V.Y. Binding of intrinsically disordered proteins is not necessarily accompanied by a structural transition to a folded form. *Biochimie* **2007**, *89*, 419–421. [CrossRef]
33. Nourse, A.; Mittag, T. The cytoplasmic domain of the T-cell receptor zeta subunit does not form disordered dimers. *J. Mol. Biol.* **2014**, *426*, 62–70. [CrossRef]
34. Rivera-Nájera, L.Y.; Saab-Rincón, G.; Battaglia, M.; Amero, C.; Pulido, N.O.; García-Hernández, E.; Solórzano, R.M.; Reyes, J.L.; Covarrubias, A.A. A group 6 late embryogenesis abundant protein from common bean is a disordered protein with extended helical structure and oligomer-forming properties. *J. Biol. Chem.* **2014**, *289*, 31995–32009. [CrossRef] [PubMed]
35. Simon, S.M.; Sousa, F.J.R.; Mohana-Borges, R.; Walker, G.C. Regulation of *Escherichia coli* SOS mutagenesis by dimeric intrinsically disordered *umu*D gene products. *Proc. Natl. Acad. Sci USA* **2008**, *105*, 1152–1157. [CrossRef] [PubMed]
36. Neira, J.L.; Martínez-Rodríguez, S.; Hernández-Cifre, J.G.; Cámara-Artigas, A.; Clemente, P.; Peralta, S.; Fernández-Moreno, M.Á.; Garesse, R.; García de la Torre, J.; Rizzuti, B. Human COA3 is an oligomeric highly flexible protein in solution. *Biochemistry* **2016**, *55*, 6209–6220. [CrossRef] [PubMed]
37. Neira, J.L.; Román-Trufero, M.; Contreras, L.M.; Prieto, J.; Singh, G.; Barrera, F.N.; Renart, M.L.; Vidal, M. The transcriptional repressor RYBP is a natively unfolded protein which folds upon binding to DNA. *Biochemistry* **2009**, *48*, 1348–1360. [CrossRef] [PubMed]
38. Batra-Safferling, R.; Abarca-Heidermann, K.; Körscehn, H.G.; Tziatzios, C.; Stoldt, M.; Budyak, I.; Willbold, D.; Schwalbe, H.; Klein-Seetharaman, J.; Kaupp, U.B. Glutamic acid-rich proteins of rod photoreceptors are natively unfolded. *J. Biol. Chem.* **2006**, *281*, 1449–1460. [CrossRef] [PubMed]
39. Neira, J.L.; López, M.B.; Sevilla, P.; Rizzuti, B.; Cámara-Artigas, A.; Vidal, M.; Iovanna, J.L. The chromatin nuclear protein NUPR1L is intrinsically disordered and binds to the same proteins as its paralogue. *Biochem. J.* **2018**, *475*, 2271–2291. [CrossRef] [PubMed]
40. Fichó, E.; Reményi, I.; Simon, I.; Mészáros, B. MFIB: A repository of protein complexes with mutual folding induced by binding. *Bioinformatics* **2017**, *33*, 3682–3684. [CrossRef] [PubMed]
41. Uversky, V.N. Intrinsically disordered proteins in overcrowded milieu: Membrane-less organelles, phase separation and intrinsic disorder. *Cur. Opin. Struct. Biol.* **2017**, *44*, 17–30. [CrossRef] [PubMed]
42. Tompa, P.; Fuxreiter, M. Fuzzy complexes: Polymorphism and structural disorder in protein-protein interactions. *Trends. Biochem. Sci.* **2008**, *33*, 2–8. [CrossRef] [PubMed]
43. Fuxreiter, M. Fuzziness in protein interactions: A historical perspective. *J. Mol. Biol.* **2018**, *430*, 2278–2287. [CrossRef] [PubMed]
44. Borgia, A.; Borgia, M.B.; Bugge, K.; Kissling, V.M.; Heidarsson, P.O.; Fernandes, C.B.; Sottini, A.; Soranno, A.; Buholzer, K.J.; Nettels, D.; et al. Extreme disorder in an ultrahigh-affinity protein complex. *Nature* **2018**, *555*, 61–66. [CrossRef] [PubMed]
45. Dunn, M.J. Initial planning: Determination of total protein concentration. In *Protein Purification Methods*; Harris, E.L.V., Angal, S., Eds.; Oxford University Press: Oxford, UK, 1995; pp. 10–20.

46. Neira, J.L.; Hornos, F.; Bacarizo, J.; Cámara-Artigas, A.; Gómez, J. The monomeric species of the regulatory domain of Tyrosine Hydroxylase has a low conformational stability. *Biochemistry* **2016**, *55*, 6209–6220. [CrossRef] [PubMed]
47. Lakowicz, J.R. *Principles of Fluorescence Spectroscopy*, 2nd ed.; Plenum Press: New York, NY, USA, 1999.
48. Piotto, M.; Saudek, V.; Sklenar, V. Gradient-tailored excitation for single-quantum NMR spectroscopy of aqueous solutions. *J. Biomol. NMR* **1993**, *2*, 661–665. [CrossRef]
49. Schagger, H.; von Jagow, G. Blue native electrophoresis for isolation of membrane protein complexes in enzymatically active form. *Anal. Biochem.* **1991**, *199*, 223–231. [CrossRef]
50. Schagger, H.; Cramer, W.A.; von Jagow, G. Analysis of molecular masses and oligomeric states of protein complexes by blue native electrophoresis and isolation of membrane protein complexes by two-dimensional native electrophoresis. *Anal. Biochem.* **1994**, *217*, 220–230. [CrossRef] [PubMed]
51. Neuhoff, V.; Stamm, R.; Pardowitz, I.; Arold, N.; Ehrhardt, W.; Taube, D. Essential problems in quantification of proteins following colloidal staining with coomassie brilliant blue dyes in polyacrylamide gels and their solution. *Electrophoresis* **1990**, *11*, 101–117. [CrossRef] [PubMed]
52. Hammouda, B. Small angle scattering from branched polymers. *Macromol. Theory Simul.* **2012**, *21*, 372–381. [CrossRef]
53. Abraham, M.J.; Murtola, T.; Schulz, R.; Páll, S.; Smith, J.C.; Hess, B.; Lindahl, E. GROMACS: High performance molecular simulations through multi-level parallelism from laptops to supercomputers. *SoftwareX* **2015**, *1–2*, 19–25. [CrossRef]
54. Lindorff-Larsen, K.; Piana, S.; Palmo, K.; Maragakis, P.; Klepeis, J.L.; Dror, R.O.; Shaw, D.E. Improved side-chain torsion potentials for the Amber ff99SB protein force field. *Proteins* **2010**, *78*, 1950–1958. [CrossRef] [PubMed]
55. Jorgensen, W.L.; Chandrasekhar, J.; Madura, J.D.; Impey, R.W.; Klein, M.L. Comparison of simple potential functions for simulating liquid water. *J. Chem. Phys.* **1983**, *79*, 926–935. [CrossRef]
56. Rizzuti, B.; Bartucci, R.; Sportelli, L.; Guzzi, R. Fatty acid binding into the highest affinity site of human serum albumin observed in molecular dynamics simulation. *Arch. Biochem. Biophys.* **2015**, *579*, 18–25. [CrossRef] [PubMed]
57. Evoli, S.; Mobley, D.L.; Guzzi, R.; Rizzuti, B. Multiple binding modes of ibuprofen in human serum albumin identified by absolute binding free energy calculations. *Phys. Chem. Chem. Phys.* **2016**, *18*, 32358–32368. [CrossRef] [PubMed]

14

A Protein Intrinsic Disorder Approach for Characterising Differentially Expressed Genes in Transcriptome Data: Analysis of Cell-Adhesion Regulated Gene Expression in Lymphoma Cells

Gustav Arvidsson and Anthony P. H. Wright *

Clinical Research Center, Department of Laboratory Medicine, Karolinska Institutet, Huddinge SE 141 57, Sweden; gustav.arvidsson@ki.se

* Correspondence: anthony.wright@ki.se

Abstract: Conformational protein properties are coupled to protein functionality and could provide a useful parameter for functional annotation of differentially expressed genes in transcriptome studies. The aim was to determine whether predicted intrinsic protein disorder was differentially associated with proteins encoded by genes that are differentially regulated in lymphoma cells upon interaction with stromal cells, an interaction that occurs in microenvironments, such as lymph nodes that are protective for lymphoma cells during chemotherapy. Intrinsic disorder protein properties were extracted from the Database of Disordered Protein Prediction (D^2P^2), which contains data from nine intrinsic disorder predictors. Proteins encoded by differentially regulated cell-adhesion regulated genes were enriched in intrinsically disordered regions (IDRs) compared to other genes both with regard to IDR number and length. The enrichment was further ascribed to down-regulated genes. Consistently, a higher proportion of proteins encoded by down-regulated genes contained at least one IDR or were completely disordered. We conclude that down-regulated genes in stromal cell-adherent lymphoma cells encode proteins that are characterized by elevated levels of intrinsically disordered conformation, indicating the importance of down-regulating functional mechanisms associated with intrinsically disordered proteins in these cells. Further, the approach provides a generally applicable and complementary alternative to classification of differentially regulated genes using gene ontology or pathway enrichment analysis.

Keywords: intrinsic disorder; intrinsic disorder prediction; intrinsically disordered region; protein conformation; transcriptome; RNA sequencing; Microarray; differentially regulated genes; gene ontology analysis; functional analysis

---

1. Introduction

Genome-wide approaches to identify genes that are differentially expressed under different conditions of interest have become a standard approach to investigating mechanisms involved in biological processes. The analysis pipeline used in such studies generally leads quickly to some form of gene ontology analysis, in order to identify biological functions that are associated with the differentially regulated genes. A complementary approach would be to analyse differentially regulated genes in relation to predicted conformational properties of the proteins they encode but such an approach has not been reported.

Characterisation of differentially regulated genes in relation to predicted or known conformational properties of the proteins they encode would be of interest in the light of recent discoveries showing overall relationships between conformational properties and different types of protein functionality or mechanism of action [1]. For example, the catalytic domains of enzymes are generally ordered globular conformations while transcription factors are characterised by a preponderance of intrinsic disorder leading to ensembles of many alternative conformational forms [2]. It is now clear that about half the proteins in eukaryotes contain at least one extended (>30 amino acid residues) intrinsically disordered region (IDR) and some proteins are completely disordered [3].

Interestingly, IDRs occur more frequently in regulatory proteins and disease-related proteins [4,5]. We recently identified genes that are differentially expressed in mantle cell lymphoma (MCL) cells that adhere to stromal cells with which they are co-cultured compared to non-adherent MCL cells in the same culture [6]. The differentially regulated gene set defined in this in vitro model system showed substantial overlap with genes that are differentially regulated in the lymph node microenvironment of MCL and chronic lymphoblastic leukaemia (CLL) patients. Retention of lymphoma cells in microenvironments is thought to lead to minimal residual disease, in which a subpopulation of cancer cells receives survival signals from normal cells in microenvironments, thus allowing them to survive during treatment and to subsequently cause disease relapse. In vitro, minimal residual disease is mimicked by cell adhesion mediated drug resistance whereby, for example, lymphoma cells residing in close proximity to stromal cells manifest an enhanced level of resistance to cytostatic drugs [7]. Thus, differentially regulated genes in co-cultured adherent lymphoma cells are likely to represent processes important for cell adhesion mediated drug resistance and minimal residual disease.

In our recent study, we identified 1050 genes that were differentially regulated in MCL cells adhered to stromal cells compared to non-adherent MCL cells in the same co-culture. The four main functional themes characterised by the differentially regulated gene set were cell adhesion, anti-apoptosis and B-cell signalling/immune-modulation, associated with up-regulated genes in adherent cells, as well as early mitotic processes, associated with down-regulated genes [6]. Here we test whether the differentially regulated gene set or its subsets encode proteins that differ in IDR properties compared to non-regulated genes.

2. Results

To determine whether there might be a difference in the frequency of IDRs (defined as predicted IDRs $\geq$ 30 amino acid residues in length) in proteins encoded by adhesion-regulated genes (adsu, n = 1009) compared to other proteins (nadsu, n = 17,612), we calculated the percentage of IDRs in adhesion-related proteins for each IDR predictor (Figure 1, blue line) and compared it to the proportion of genes in the adhesion-regulated gene set (5.4%, Figure 1, red line). For all predictors, the proportion of predicted IDRs associated with the adhesion gene set exceeded the frequency expected based on the proportion of proteins in the set. For many predictors, Figure 1 also shows a tendency towards a larger number of longer IDRs in proteins encoded by the adhesion gene set at the expense of shorter IDRs.

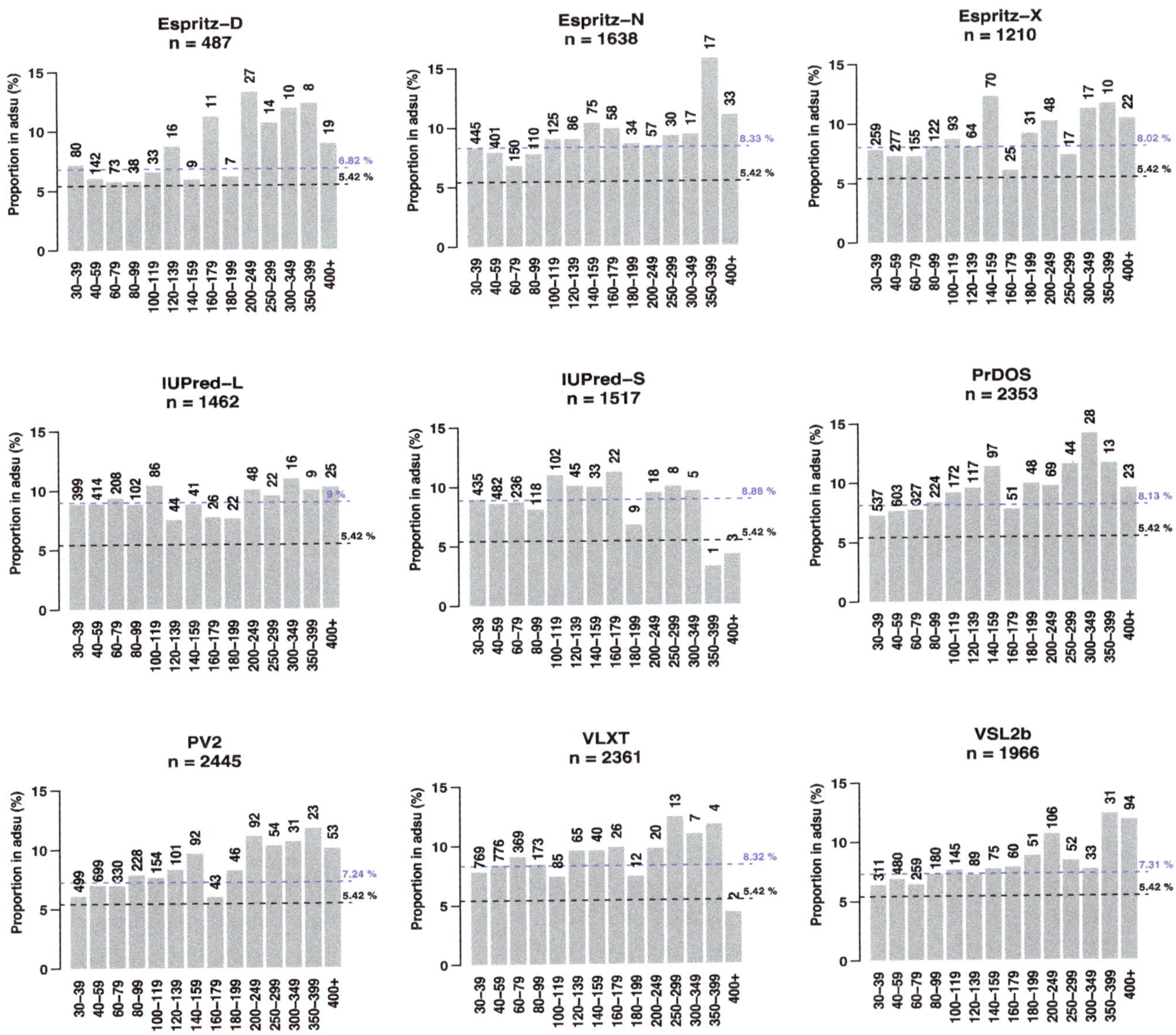

Figure 1. Enrichment of intrinsically disordered regions (IDRs) in proteins encoded by genes that are differentially expressed in lymphoma cells upon adhering to stromal cells. The number (n) of IDRs ($\geq$30 residues) for each predictor is shown as well as how the detected IDRs are distributed in relation to length. The number of IDRs in each size category is shown. The blue line shows the percentage of all IDRs encoded by adhesion-related genes (adsu) and non-adhesion-related genes (nadsu) that are associated with the adsu set, while the red line shows the percentage expected if IDRs are equally distributed between the adsu and nadsu sets.

To determine whether the enhanced frequency of IDRs in proteins encoded by adhesion-regulated genes was significant, we used a resampling approach to test whether the IDR frequency associated with the 1009 adhesion-regulated genes lay outside the distribution of frequencies generated by 1000-fold resampling of 1009 genes from the control gene set (n = 17,612). A z-score and associated p-value was generated for data from each predictor. As shown in Table 1 (adsu vs. nadsu), the enrichment of IDRs in proteins encoded by adhesion-regulated genes (adsu) was significant for all predictors.

Table 1. Intrinsically disordered regions are enriched in proteins encoded by down-regulated genes in lymphoma cells upon adherence to stromal cells.

Gene Set Comparison	Espritz-D #	Espritz-N #	Espritz-X #	IUPred-L #	IUPred-S #	PrDOS #	PV2 #	VLXT #	VSL2b #
Adsu vs. Nadsu									
IDR number in adsu	487	1638	1210	1462	1517	2353	2445	2361	1966
IDR number in nadsu *	382	1036	796	847	892	1520	1794	1492	1431
Adjusted p-value	2.38×10^{-8}	1.32×10^{-27}	2.81×10^{-23}	4.34×10^{-27}	1.32×10^{-27}	4.24×10^{-34}	2.85×10^{-19}	2.75×10^{-25}	8.29×10^{-19}
Adsu_Down vs. Adsu_Up									
IDR number in adsu_down	276	1072	760	983	1025	1507	1508	1572	1216
IDR number in adsu_up *	171	458	363	387	397	758	683	639	606
Adjusted p-value	3.51×10^{-78}	$<1.00 \times 10^{-99}$	$<1.00 \times 10^{-99}$	$<1.00 \times 10^{-99}$	$<1.00 \times 10^{-99}$	$<1.00 \times 10^{-99}$	$<1.00 \times 10^{-99}$	$<1.00 \times 10^{-99}$	$<1.00 \times 10^{-99}$

* mean of 1000 resamples of n proteins encoded by genes in nadsu or adsu_up, where n = the number of genes in adsu or adsu_down, respectively. Abbreviations: adsu (adhesion-regulated genes); nadsu (non-adhesion-regulated genes); adsu_down (down-regulated adsu); adsu_up (up-regulated adsu). # Predictors of intrinsic disorder that appear in the D^2P^2 database.

Next, we tested whether the enrichment of IDRs associated with the adhesion-regulated gene set could be ascribed to subsets of the adhesion-regulated genes. Comparison of proteins encoded by genes manifesting a greater degree of regulation (fold change ≥ 1.3) relative to the remaining regulated genes showed fewer IDRs in more highly regulated genes compared to less highly regulated genes for all predictors and with lower levels of significance compared to the comparison of regulated and non-regulated genes (data not shown). Thus, there is an enrichment of IDRs in adhesion-regulated genes but the enrichment is not related to the extent of their regulation. Comparison of the up-regulated subset (adsu_up, change >1) relative to the down-regulated subset (adsu_down, change <1), on the other hand, showed an enhanced enrichment of IDRs in proteins encoded by the adsu_down subset compared to the enhancement levels in Figure 1, with high levels of significance (Table 1, adsu_down vs. adsu_up). Thus, the enrichment in IDRs in proteins encoded by adhesion-regulated genes is mainly associated with proteins encoded by down-regulated genes.

We next investigated whether the length of IDRs in proteins encoded by adsu genes tends to be longer than in other proteins (nadsu). As expected, IDR length is not normally distributed, as indicated by the consistently higher value of the mean compared to the median (Table 2), as well as tests of normality (data not shown). Thus, a Mann–Whitney test was used to test the significance of differences in IDR length between groups. Table 2 shows that some predictors (notably PV2, PrDOS and VSL2b) predict longer IDRs in proteins encoded by adsu genes that in other proteins (nadsu), but for other predictors the difference is less significant or lacking in statistical support. IDRs encoded by adsu_down genes were significantly longer than IDRs encoded by adsu_up genes for all predictors. Thus, IDRs in proteins encoded by genes that are down-regulated in adherent cells tend to be both more frequent and longer than IDRs in other proteins.

Table 2. Intrinsically disordered regions tend to be longer in proteins encoded by down-regulated genes in lymphoma cells upon adherence to stromal cells.

Gene Set Comparison	Espritz-D	Espritz-N	Espritz-X	IUPred-L	IUPred-S	PrDOS	PV2	VLXT	VSL2b
				Adsu vs. Nadsu					
Median (mean) IDR length (adsu)	64 (125)	57 (97)	66 (98)	54 (87)	52 (67)	61 (98)	61 (90)	48 (62)	74 (128)
Median (mean) IDR length (nadsu)	61 (98)	56 (91)	62 (93)	54 (86)	51 (68)	55 (86)	56 (82)	47 (61)	66 (107)
Adjusted *p*-value *	6.04×10^{-2}	7.33×10^{-2}	**2.50×10^{-2}**	6.02×10^{-1}	5.10×10^{-1}	**3.75×10^{-11}**	**4.97×10^{-9}**	**2.42×10^{-2}**	**4.97×10^{-9}**
				Adsu_Down vs. Adsu_Up					
Median (mean) IDR length (adsu_down)	68.5 (154)	60 (104)	76 (108)	56 (93)	53 (70)	64 (107)	65 (96)	49 (65)	79.5 (142)
Median (mean) IDR length (adsu_up)	59 (89)	54 (86)	58 (83)	50 (75)	50 (62)	58 (85)	57 (80)	46 (57)	67 (104)
Adjusted *p*-value *	**1.24×10^{-3}**	**5.26×10^{-3}**	**2.14×10^{-7}**	**5.08×10^{-3}**	**1.66×10^{-2}**	**3.38×10^{-4}**	**1.42×10^{-5}**	**1.73×10^{-3}**	**7.05×10^{-5}**

* Mann–Whitney test; bold text = $p < 0.05$. Abbreviations: adsu (adhesion-regulated genes); nadsu (non-adhesion-regulated genes); adsu_down (down-regulated adsu); adsu_up (up-regulated adsu).

We next addressed how IDRs are distributed among the proteins encoded by the adsu_down gene set in relation to proteins associated with the adsu_up and nadsu gene sets (Table 3).

Table 3. Distribution of IDRs predicted by VSL2b in proteins encoded by genes that are differentially regulated in lymphoma cells upon interaction with stromal cells.

Gene Set Comparison	Number of Proteins	Number (%) of Completely Disordered Proteins	Number (%) of Proteins with IDR	Median Percent IDR Per Protein (All Proteins)	Median Percent IDR Per Protein (IDR-Containing Proteins)
adsu_down	445	19 (4.3)	367 (82.5)	38.1	48.1
adsu_up	556	11 (2)	370 (66.5)	18.1	37.6
nadsu	17,459	476 (2.7)	11,248 (64.4)	17.8	37.6

adsu_down (down-regulated adhesion-regulated genes); adsu_up (up-regulated adhesion-regulated genes); nadsu (non-adhesion-regulated genes).

The proportion of completely disordered proteins was higher for the adsu_down sets than for proteins encoded by the other gene sets, as was the proportion of proteins containing at least one IDR. The median proportion of the protein sequences that were predicted as IDR was higher for the adsu_down group, irrespective of whether all proteins were considered or only proteins containing IDRs. Table 3 shows data for the VSL2b predictor but other predictors generally produced a similar result, especially PrDOS and PV2. The frequency of IDR-containing proteins with different IDR proportions for the different gene sets is compared graphically in Figure 2A.

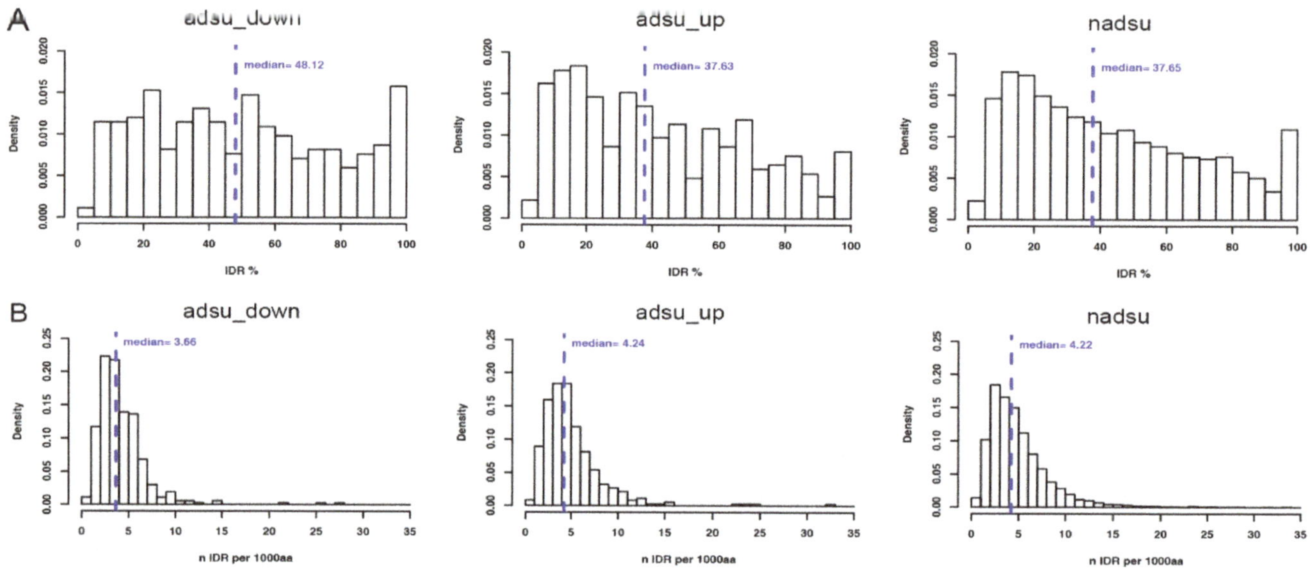

Figure 2. Relative frequency distributions of proportion of IDR per protein and length-normalized number of IDRs per protein for proteins encoded by adsu_down genes in relation to adsu_up and nadsu genes. IDR predictions were made using VSL2b. (**A**) Relative frequency distributions (Density) of IDR-containing proteins with different percent IDR content. The median position and value are shown in blue. (**B**) Relative frequency distributions (Density) of numbers of IDRs per IDR-containing protein, normalized for differences in protein length (IDR number per 1000 amino acid residues). The median position and value are shown in blue.

For proteins encoded by nadsu and adsu_up, the relative frequency declines progressively as the proportion of IDR per protein increases. Contrastingly, a more even distribution of relative frequencies is seen for adsu_down proteins, with relatively fewer low-IDR content proteins and an increased proportion of high-IDR content proteins. Interestingly, the protein length-normalized number of IDRs per protein is somewhat lower for proteins encoded by adsu_down genes, compared to adsu_up and nadsu genes (Figure 2B). Thus, the greater IDR content of adsu_down encoded genes tends to be associated with fewer and longer IDRs when only IDR-containing proteins are analyzed.

To further investigate differences in IDR lengths between groups, we plotted the length of the longest IDR in each protein as a function of protein length to compare adsu_down and adsu_up encoded proteins (Figure 3).

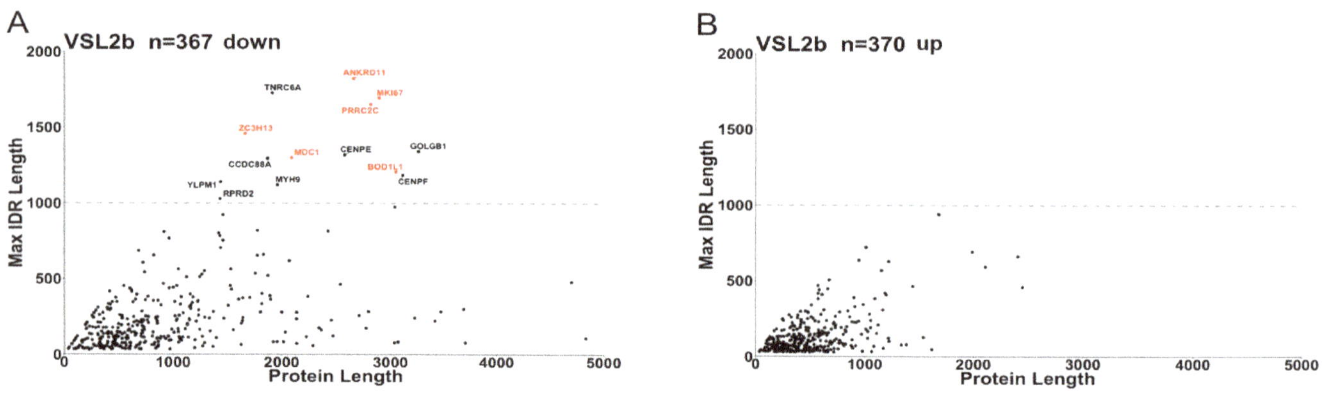

Figure 3. Comparison of proteins encoded by down- or up-regulated adhesion-regulated genes with regard to longest IDR length per protein and protein length. IDR-containing proteins encoded by (**A**) down-regulated adhesion-regulated genes (adsu_down) and (**B**) up-regulated adhesion-regulated genes (adsu_up) are shown. Of the 14 proteins in (**A**) for which the maximum IDR length is greater than 1000 residues (above dotted line), 6 proteins (red text) were also found in the sets of 14 proteins with the longest IDRs predicted by the PV2 and PrDOS predictors.

adsu_down encoded proteins are characterized by both longer protein length and longer length of the longest IDR (VSL2b). There are 14 adsu_down encoded proteins with IDRs longer than 1000 residues and these are also among proteins with the longest IDRs for most other predictors (notably PrDOS and PV2). The IDR score profiles for the 6 proteins that are reproducibly found in the top 14 proteins with longest IDRs by the VSL2b, PV2 and PrDOS predictors (red text in Figure 3A) are shown in Figure 4.

Consistent with Figure 3A, most of the proteins are predicted to be disordered throughout most of their length. Some contain extended regions with close to maximal intrinsic disorder scores (e.g., ZC3H13), while others are characterized by fluctuating levels of intrinsic disorder (e.g., MKI67). Some proteins contain both patterns in different regions of the protein (e.g., BOD1L1). Many of the proteins have short regions that are predicted to be ordered and that could correspond to folded protein domains. The different types of predicted conformation could inform about mechanisms involved in the function of proteins encoded by down-regulated genes in relation to up-regulated genes (see Discussion).

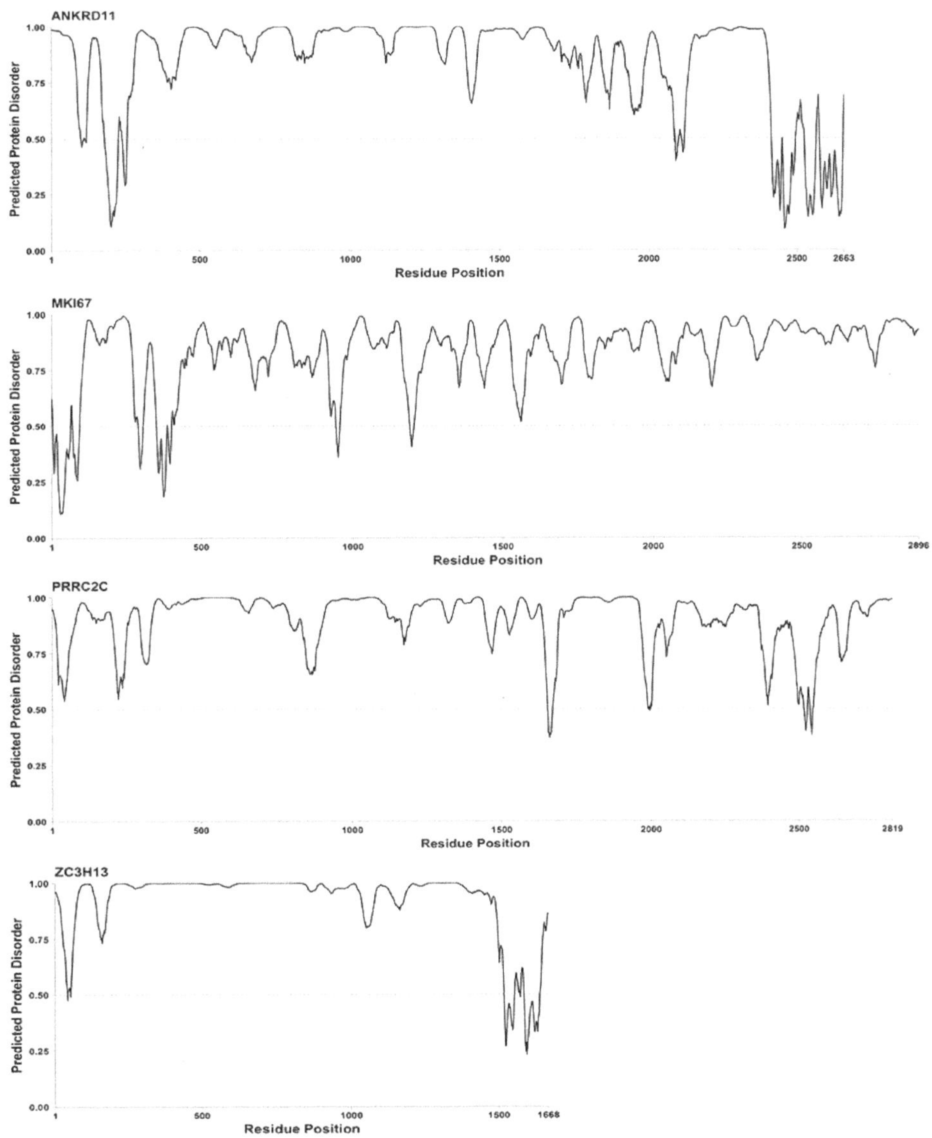

Figure 4. *Cont.*

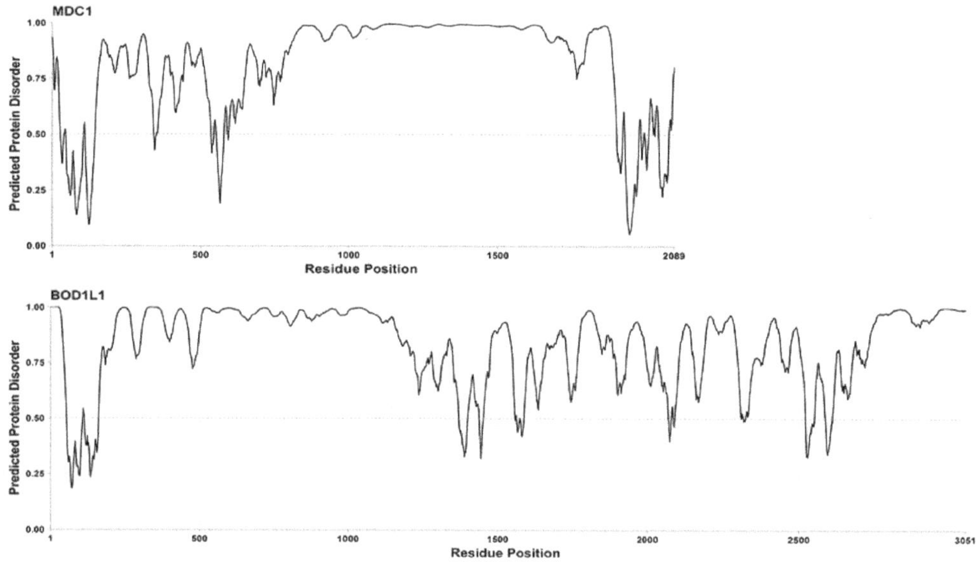

Figure 4. Examples of proteins with long IDRs. Proteins that are reproducibly found by the VSL2b, PV2 and PrDOS predictors in the set of 14 proteins with the longest predicted IDRs (red text in Figure 3A) are shown. The residue-by-residue intrinsic disorder score (VSL2b) is plotted as a function of residue number throughout the length of the respective proteins. The horizontal gridline at a score of 0.5 distinguishes regions predicted to be ordered (<0.5) or intrinsically disordered (>0.5).

3. Discussion

The main finding of this work is that proteins encoded by genes that are down-regulated in lymphoma cells upon adhering to stromal cells, typically found in microenvironments that increase cancer-cell survival, tend to have more frequent and longer regions of predicted intrinsically disordered conformation than proteins encoded by up-regulated genes or other expressed genes in the same cells. Our previous work has shown that many proteins encoded by down-regulated genes in adherent cells are involved in early stages of mitosis [6]. The present results complement this observation by suggesting that proteins encoded by the down-regulated gene set tend to function by mechanisms that are associated with intrinsically disordered regions. A secondary finding is that many of the proteins encoded by down-regulated genes are larger than proteins encoded by up-regulated genes.

Intrinsically disordered protein regions can be broadly divided into regions that are always disordered and disordered regions that form one or more ordered conformations in particular molecular environments, such as during coupled binding and folding interactions with partner proteins [8]. Some IDRs have been shown to bind partners in the disordered state via multi-valent interactions, mediated by short linear motifs that are distributed along the length of the IDR [5,9–11]. However, IDRs have other functions in addition to interaction with partners. One such function is mediation of phase transitions in cells that allow for compartmentalization of cellular regions in so-called "membrane-less organelles" that include nucleoli, nuclear speckles, P-bodies and chromatin [12–16]. These kinds of functional mechanisms might be associated with the IDRs that have consistently close-to-maximal prediction scores over extended regions of proteins encoded by down-regulated genes, as exemplified by some of the proteins in Figures 3 and 4.

The clearest example of a protein that is predicted to be maximally disordered throughout most of the protein sequence is ZC3H13. Interestingly, ZC3H13 is part of the WTAP complex, which is involved in RNA splicing and processing and is localized in nuclear speckles [17]. It is likely that such speckles result from phase transition processes and it is possible that the disordered region of ZC3H13 is important for speckle formation or ZC3H13 localization to the speckle. In fact, many documented types of so-called proteinaceous membrane-less organelles are located in the nucleus and include chromatin in addition to nuclear speckles, nucleoli and many other bodies [14]. The MDC1 protein

(Figures 3 and 4) contains a central region predicted to be completely disordered, flanked by less disordered/structured regions, which are known to mediate binding to several partner proteins at chromatin regions containing double-stranded DNA breaks [18]. Thus, MDC1 has been regarded as a "scaffold" protein responsible for spreading of DNA-repair factors over the damaged chromatin region and it is tempting to speculate that the central disordered region could play a role in phase-transitions. Other proteins in Figures 3 and 4 that have extensive regions predicted to be completely disordered and that work in a chromatin environment are YLPM1, involved in regulating telomerase activity, and BOD1L1, a protein that protects stalled DNA replication forks.

MKI67 is predicted to be disordered (with varying score) throughout almost its entire length (see Figure 4). Interestingly, MKI67 orchestrates formation of the perichromosomal layer, which coats the condensed chromosomes during mitosis in order to prevent chromosome aggregation [19]. In mitotic mammalian cells, the nuclear membrane and nucleolus are broken down and nucleolar proteins including the known phase-transition proteins, Nucleophosmin and Fibrillarin, that drive nucleolus formation in interphase cells [20], are also found in the mitotic perichromosomal layer. This fact, taken together with the RNA-binding activity associated with MKI67, suggests that the perichromosomal layer may be formed by phase transition phenomena. Interestingly, higher expression of MKI67 is a negative prognostic marker for MCL patients [21].

In the IDR class that conditionally adopts ordered conformations in some molecular contexts, the ordered conformations are characterized by varying degrees of "fuzziness", defined as the existence of a heterogeneous range of ordered conformations in the context of, for example, interaction with a single partner [22]. Many proteins that conditionally adopt ordered conformations contain pre-structure motifs (PreSMos), defined as short protein regions within IDRs that have a weak propensity for secondary structure formation leading to formation of unstable secondary structure elements in a minority sub-population of IDR-containing proteins [23]. PreSMos become stabilized during coupled binding and folding, and form part of the folded protein conformation that is seen in complexes with partner proteins. Protein regions encoded by down-regulated genes that show alternating sub-regions of higher and lower intrinsic disorder scores might correspond to these kinds of IDR since the short regions with lower intrinsic disorder scores may represent PreSMos. The CENPE and CENPF proteins are characterized by disordered regions interspersed with regions with lower disorder scores that could represent regions containing PreSMos. This would be consistent with the multiple interactions made by these proteins within the kinetochore structure that binds to the centromeric chromatin of chromosomes during mitosis. TNRC6A is a member of the GW182 family of scaffold proteins that are important for organization of proteins needed for RNA-mediated gene silencing and are found in P-bodies that are formed by a phase transition process [24].

Although somewhat speculative, the preceding sections suggest mechanisms by which some of the large proteins with large amounts of intrinsic disorder might contribute the propagation of lymphoma cells in suspension as well as how their down-regulation could lead to reduced proliferation of lymphoma cells adhered to stromal cells. Reduced proliferation is known to increase the survival of cancer cells during chemotherapy, which primarily targets proliferating cells [7,25]. Further, the cell cycle arrest that occurs in adherent MCL cells [26] would be expected to reduce the need for apoptotic responses and we previously showed that adherence to stromal cells is associated with up-regulation of anti-apoptotic genes [6].

We have shown that predicted intrinsic disorder can be used to interrogate proteins encoded by transcriptome data and that identification of gene sets encoding proteins with characteristic predicted disorder properties can provide information relevant for understanding the mechanisms underlying the functionality of groups of proteins. This approach complements the commonly used gene ontology analysis approach, which primarily gives information about the cellular components or processes that are characteristic for the function of protein sets. Both approaches provide information that can be used for hypothesis building and the design of further experiments.

In this work, we have only analyzed predicted protein disorder as a conformational characteristic. There are other predictors that could be used to expand the approach in the future and new predictors are continuously being developed as more is learned about how protein functionality is coupled to the conformational flexibility of proteins. Examples are the s2D predictor [27], which predicts secondary structure elements in relation to random coil regions, and Dynamine [28], which predicts the rigidity of the peptide backbone throughout protein sequences, as well as the ANCHOR [29] and MoRFpred [30] predictors, which predict protein interaction sites. More recently developed predictors include prediction of protein regions involved in phase transitions [31], prediction of decomposed residue-by-residue solvation free energy [32] and prediction of residue-by-residue compactness/secondary structure [33]. Thus, it is easy to see that a battery of predictors could be used to reveal many different conformational aspects of protein sets encoded by groups of differentially regulated genes identified in transcriptome data. Databases like the Database of Disordered Protein Prediction (D^2P^2) [34] or the more recently developed MobiDB [35], which contain collections of prediction data from different sources, will be useful tools for this purpose.

4. Materials and Methods

4.1. Data

Human protein regions predicted to be disordered and related data were downloaded from the publically available D^2P^2 database (available online: http://d2p2.pro/search/build) on 11 September 2017. Default options were used for the download except that "Genome" was set to "Homo sapiens 63_37" and the "Limit to" option was set to "all". The downloaded data contained all predicted IDRs detected in a total of 917,132 features for each of 9 different intrinsic disorder predictors (Espritz_Disprot, Espritz_NMR, Espritz_Xray, IUPred_long, IUPred_short, PV2, PrDOS, VL-XT, and VSL2b). See the D^2P^2 website (available online: http://d2p2.pro) or [34] for details. Mean fold-change transcriptome values for 1050 genes that show significantly altered transcript levels when Jeko-1 mantle lymphoma cells adhere to MS-5 stromal cells were taken from a recently published study from our group [6].

4.2. Data Analysis

Data were imported into and analysed using the *R* statistical programming platform (version 3.4.3, https://cran.r-project.org) [36] using packages shipped with the standard version, together with the following additional packages: data.table [37], nortest [38], formattable [39], org.Hs.eg.db [40].

To match gene expression data to IDR data for proteins in the D^2P^2 data set, it was first necessary to match an ENSEMBL protein id (from the EMSEMBL database, http://www.ensembl.org/index.html) to each of the genes identified in the RNAseq experiment. This was done by matching entries in the RNAseq data with entries in the org.Hs.eg.db annotation database from which fields for ENSEMBL protein id (ENSEMBLPROT) and gene name (SYMBOL) were extracted and appended to the RNAseq data using ENTREZID as a common key. 18,686 of 23,445 entries in the RNAseq data set were matched and also had identical gene names. This set was used in the further analysis. The annotation for the vast majority of the non-matched genes indicated that they represented non-protein-coding genes, putative protein encoding genes or pseudogenes. 1009 of the 1050 adhesion regulated genes were matched to an ENSEMBL protein id and a control set of 17,612 genes that were not shown to be regulated by adhesion were uniquely matched. Entries for which "SEQID" in the D^2P^2 IDR data matched the ENSEMBL protein id in sets or subsets of the adhesion-regulated genes or non-regulated genes were used for analysis of the sets or subsets. Only D^2P^2 entries for IDRs $\geq$ 30 amino acid residues were used and data for the 9 different IDR predictors were extracted from the database and analysed separately.

Differences in IDR number between test sets and control sets were evaluated statistically by *z*-scores and associated *p*-values calculated from the measured test value compared to the mean of

1000 control values, calculated from 1000 re-samples (with replacement) randomly selected from the control data. The size of the control re-samples was the same as the size of the test set. Differences in IDR length between test sets and control sets were evaluated statistically using a Mann–Whitney test, a non-parametric test appropriate for non-normally distributed data. p-values were adjusted for multiple testing using the false discovery rate method.

Author Contributions: Conceptualization, G.A. and A.P.W.; Methodology, G.A. and A.P.W.; Formal Analysis, G.A. and A.P.W.; Investigation, G.A. and A.P.W.; Data Curation, G.A. and A.P.W.; Writing-Original Draft Preparation, A.P.W.; Writing-Review & Editing, G.A. and A.P.W.; Visualization, G.A. and A.P.W.; Supervision, A.P.W.; Project Administration, A.P.W.; Funding Acquisition, A.P.W.

References

1. Das, R.K.; Ruff, K.M.; Pappu, R.V. Relating sequence encoded information to form and function of intrinsically disordered proteins. *Curr. Opin. Struct. Biol.* **2015**, *32*, 102–112. [CrossRef] [PubMed]
2. Arai, M.; Sugase, K.; Dyson, H.J.; Wright, P.E. Conformational propensities of intrinsically disordered proteins influence the mechanism of binding and folding. *Proc. Natl. Acad. Sci. USA* **2015**, *112*, 9614–9619. [CrossRef] [PubMed]
3. van der Lee, R.; Buljan, M.; Lang, B.; Weatheritt, R.J.; Daughdrill, G.W.; Dunker, A.K.; Fuxreiter, M.; Gough, J.; Gsponer, J.; Jones, D.T.; et al. Classification of intrinsically disordered regions and proteins. *Chem. Rev.* **2014**, *114*, 6589–6631. [CrossRef] [PubMed]
4. Uversky, V.N.; Oldfield, C.J.; Dunker, A.K. Intrinsically disordered proteins in human diseases: Introducing the d2 concept. *Annu. Rev. Biophys.* **2008**, *37*, 215–246. [CrossRef] [PubMed]
5. Marasco, D.; Scognamiglio, P.L. Identification of inhibitors of biological interactions involving intrinsically disordered proteins. *Int. J. Mol. Sci.* **2015**, *16*, 7394–7412. [CrossRef] [PubMed]
6. Arvidsson, G.; Henriksson, J.; Sander, B.; Wright, A.P. Mixed-species rnaseq analysis of human lymphoma cells adhering to mouse stromal cells identifies a core gene set that is also differentially expressed in the lymph node microenvironment of mantle cell lymphoma and chronic lymphocytic leukemia patients. *Haematologica* **2018**, *103*, 666–678. [CrossRef] [PubMed]
7. Medina, D.J.; Goodell, L.; Glod, J.; Gelinas, C.; Rabson, A.B.; Strair, R.K. Mesenchymal stromal cells protect mantle cell lymphoma cells from spontaneous and drug-induced apoptosis through secretion of b-cell activating factor and activation of the canonical and non-canonical nuclear factor kappab pathways. *Haematologica* **2012**, *97*, 1255–1263. [CrossRef] [PubMed]
8. Dyson, H.J.; Wright, P.E. Intrinsically unstructured proteins and their functions. *Nat. Rev. Mol. Cell Biol.* **2005**, *6*, 197–208. [CrossRef] [PubMed]
9. Huang, C.; Rossi, P.; Saio, T.; Kalodimos, C.G. Structural basis for the antifolding activity of a molecular chaperone. *Nature* **2016**, *537*, 202–206. [CrossRef] [PubMed]
10. Scognamiglio, P.L.; Di Natale, C.; Leone, M.; Poletto, M.; Vitagliano, L.; Tell, G.; Marasco, D. G-quadruplex DNA recognition by nucleophosmin: New insights from protein dissection. *Biochim. Biophys. Acta* **2014**, *1840*, 2050–2059. [CrossRef] [PubMed]
11. Pauwels, K.; Lebrun, P.; Tompa, P. To be disordered or not to be disordered: Is that still a question for proteins in the cell? *Cell. Mol. Life Sci.* **2017**, *74*, 3185–3204. [CrossRef] [PubMed]
12. Shin, Y.; Brangwynne, C.P. Liquid phase condensation in cell physiology and disease. *Science* **2017**, *357*, eaaf4382. [CrossRef] [PubMed]
13. Chong, P.A.; Forman-Kay, J.D. Liquid-liquid phase separation in cellular signaling systems. *Curr. Opin. Struct. Biol.* **2016**, *41*, 180–186. [CrossRef] [PubMed]
14. Darling, A.L.; Liu, Y.; Oldfield, C.J.; Uversky, V.N. Intrinsically disordered proteome of human membrane-less organelles. *Proteomics* **2018**, *18*, e1700193. [CrossRef] [PubMed]
15. Mitrea, D.M.; Kriwacki, R.W. Phase separation in biology; functional organization of a higher order. *Cell Commun. Signal.* **2016**, *14*, 1. [CrossRef] [PubMed]
16. Toretsky, J.A.; Wright, P.E. Assemblages: Functional units formed by cellular phase separation. *J. Cell. Biol.* **2014**, *206*, 579–588. [CrossRef] [PubMed]

17. Ping, X.L.; Sun, B.F.; Wang, L.; Xiao, W.; Yang, X.; Wang, W.J.; Adhikari, S.; Shi, Y.; Lv, Y.; Chen, Y.S.; et al. Mammalian wtap is a regulatory subunit of the rna n6-methyladenosine methyltransferase. *Cell Res.* **2014**, *24*, 177–189. [CrossRef] [PubMed]
18. Nagy, Z.; Kalousi, A.; Furst, A.; Koch, M.; Fischer, B.; Soutoglou, E. Tankyrases promote homologous recombination and check point activation in response to dsbs. *PLoS Genet.* **2016**, *12*, e1005791. [CrossRef] [PubMed]
19. Booth, D.G.; Earnshaw, W.C. Ki-67 and the chromosome periphery compartment in mitosis. *Trends Cell. Biol.* **2017**, *27*, 906–916. [CrossRef] [PubMed]
20. Feric, M.; Vaidya, N.; Harmon, T.S.; Mitrea, D.M.; Zhu, L.; Richardson, T.M.; Kriwacki, R.W.; Pappu, R.V.; Brangwynne, C.P. Coexisting liquid phases underlie nucleolar subcompartments. *Cell* **2016**, *165*, 1686–1697. [CrossRef] [PubMed]
21. Katzenberger, T.; Petzoldt, C.; Holler, S.; Mader, U.; Kalla, J.; Adam, P.; Ott, M.M.; Muller-Hermelink, H.K.; Rosenwald, A.; Ott, G. The ki67 proliferation index is a quantitative indicator of clinical risk in mantle cell lymphoma. *Blood* **2006**, *107*, 3407. [CrossRef] [PubMed]
22. Fuxreiter, M.; Tompa, P. Fuzzy complexes: A more stochastic view of protein function. *Adv. Exp. Med. Biol.* **2012**, *725*, 1–14. [PubMed]
23. Lee, S.H.; Kim, D.H.; Han, J.J.; Cha, E.J.; Lim, J.E.; Cho, Y.J.; Lee, C.; Han, K.H. Understanding pre-structured motifs (presmos) in intrinsically unfolded proteins. *Curr. Protein Pept. Sci.* **2012**, *13*, 34–54. [CrossRef] [PubMed]
24. Yao, B.; Li, S.; Chan, E.K. Function of gw182 and gw bodies in sirna and mirna pathways. *Adv. Exp. Med. Biol.* **2013**, *768*, 71–96. [PubMed]
25. Kurtova, A.V.; Balakrishnan, K.; Chen, R.; Ding, W.; Schnabl, S.; Quiroga, M.P.; Sivina, M.; Wierda, W.G.; Estrov, Z.; Keating, M.J.; et al. Diverse marrow stromal cells protect cll cells from spontaneous and drug-induced apoptosis: Development of a reliable and reproducible system to assess stromal cell adhesion-mediated drug resistance. *Blood* **2009**, *114*, 4441–4450. [CrossRef] [PubMed]
26. Lwin, T.; Hazlehurst, L.A.; Dessureault, S.; Lai, R.; Bai, W.; Sotomayor, E.; Moscinski, L.C.; Dalton, W.S.; Tao, J. Cell adhesion induces p27kip1-associated cell-cycle arrest through down-regulation of the scfskp2 ubiquitin ligase pathway in mantle-cell and other non-hodgkin b-cell lymphomas. *Blood* **2007**, *110*, 1631–1638. [CrossRef] [PubMed]
27. Sormanni, P.; Camilloni, C.; Fariselli, P.; Vendruscolo, M. The s2d method: Simultaneous sequence-based prediction of the statistical populations of ordered and disordered regions in proteins. *J. Mol. Biol.* **2015**, *427*, 982–996. [CrossRef] [PubMed]
28. Cilia, E.; Pancsa, R.; Tompa, P.; Lenaerts, T.; Vranken, W.F. The dynamine webserver: Predicting protein dynamics from sequence. *Nucleic Acids Res.* **2014**, *42*, W264–W270. [CrossRef] [PubMed]
29. Meszaros, B.; Erdos, G.; Dosztanyi, Z. Iupred2a: Context-dependent prediction of protein disorder as a function of redox state and protein binding. *Nucleic Acids Res.* **2018**, *46*, W329–W337. [CrossRef] [PubMed]
30. Disfani, F.M.; Hsu, W.L.; Mizianty, M.J.; Oldfield, C.J.; Xue, B.; Dunker, A.K.; Uversky, V.N.; Kurgan, L. Morfpred, a computational tool for sequence-based prediction and characterization of short disorder-to-order transitioning binding regions in proteins. *Bioinformatics* **2012**, *28*, i75–i83. [CrossRef] [PubMed]
31. Vernon, R.M.; Chong, P.A.; Tsang, B.; Kim, T.H.; Bah, A.; Farber, P.; Lin, H.; Forman-Kay, J.D. Pi-pi contacts are an overlooked protein feature relevant to phase separation. *eLife* **2018**, *7*, e31486. [CrossRef] [PubMed]
32. Chong, S.-H.; Lee, C.; Kang, G.; Park, M.; Ham, S. Structural and thermodynamic investigations on the aggregation and folding of acylphosphatase by molecular dynamics simulations and solvation free energy analysis. *J. Am. Chem. Soc.* **2011**, *133*, 7075–7083. [CrossRef] [PubMed]
33. Konrat, R. The protein meta-structure: A novel concept for chemical and molecular biology. *Cell. Mol. Life Sci.* **2009**, *66*, 3625–3639. [CrossRef] [PubMed]
34. Oates, M.E.; Romero, P.; Ishida, T.; Ghalwash, M.; Mizianty, M.J.; Xue, B.; Dosztányi, Z.; Uversky, V.N.; Obradovic, Z.; Kurgan, L.; et al. D^2p^2: Database of disordered protein predictions. *Nucleic Acids Res.* **2013**, *41*, D508–D516. [CrossRef] [PubMed]
35. Piovesan, D.; Tabaro, F.; Paladin, L.; Necci, M.; Micetic, I.; Camilloni, C.; Davey, N.; Dosztanyi, Z.; Meszaros, B.; Monzon, A.M.; et al. Mobidb 3.0: More annotations for intrinsic disorder, conformational diversity and interactions in proteins. *Nucleic Acids Res.* **2018**, *46*, D471–D476. [CrossRef] [PubMed]

36. R Core Team. *R: A Language and Environment for Statistical Computing*; R Foundation for Statistical Computing: Vienna, Austria, 2008.
37. Dowle, M.; Srinivasan, M. Data.Table: Extension of 'Data.Frame'. R Package Version 1.10.4-3. 2017. Available online: https://cran.r-project.org/package=data.table (accessed on 4 October 2018).
38. Gross, J.; Ligges, U. Nortest: Tests for Normality. R Package Version 1.0-4. 2015. Available online: https://cran.r-project.org/package=nortest (accessed on 4 October 2018).
39. Ren, K.; Russell, K. Formattable: Create 'Formattable' Data Structures. R Package Version 0.2.0.1. 2016. Available online: https://cran.r-project.org/package=formattable (accessed on 4 October 2018).
40. Carlson, M. Org.Hs.Eg.Db: Genome Wide Annotation for Human. R Package Version 3.5.0. 2017. Available online: http://bioconductor.org/packages/org.Hs.eg.db/ (accessed on 4 October 2018).

15

Deciphering RNA-Recognition Patterns of Intrinsically Disordered Proteins

Ambuj Srivastava [1], Shandar Ahmad [2] and M. Michael Gromiha [1,*]

1 Department of Biotechnology, Bhupat and Jyoti Mehta School of Biosciences, Indian Institute of Technology Madras, Chennai 600 036, Tamilnadu, India; ambuj.88.in@gmail.com

2 School of Computational and Integrative Sciences, Jawaharlal Nehru University, New Delhi 110 067, India; shandar@jnu.ac.in

* Correspondence: gromiha@iitm.ac.in

Abstract: Intrinsically disordered regions (IDRs) and protein (IDPs) are highly flexible owing to their lack of well-defined structures. A subset of such proteins interacts with various substrates; including RNA; frequently adopting regular structures in the final complex. In this work; we have analysed a dataset of protein–RNA complexes undergoing disorder-to-order transition (DOT) upon binding. We found that DOT regions are generally small in size (less than 3 residues) for RNA binding proteins. Like structured proteins; positively charged residues are found to interact with RNA molecules; indicating the dominance of electrostatic and cation-π interactions. However, a comparison of binding frequency shows that interface hydrophobic and aromatic residues have more interactions in only DOT regions than in a protein. Further; DOT regions have significantly higher exposure to water than their structured counterparts. Interactions of DOT regions with RNA increase the sheet formation with minor changes in helix forming residues. We have computed the interaction energy for amino acids–nucleotide pairs; which showed the preference of His–G; Asn–U and Ser–U at for the interface of DOT regions. This study provides insights to understand protein–RNA interactions and the results could also be used for developing a tool for identifying DOT regions in RNA binding proteins.

Keywords: intrinsically disorder proteins; disorder-to-order regions; protein–RNA interactions; unstructured proteins

---

1. Introduction

Intrinsically disordered proteins lack stable three-dimensional structures under physiological conditions and are known to perform important roles in several processes including signalling, enzymatic activity, and gene regulation [1,2]. To perform these functions, disordered regions interact with protein, RNA, DNA, and other small molecules to gain ordered structures [3,4]. Experimentally, interactions mediated by IDRs can be observed using NMR and X-ray crystallography. However, because of poor resolution, problems in crystallization, and high time and resource consumption, computational methods are necessary to identify disorder-mediated interactions [5,6].

Several methods have been developed for understanding the disorderness of proteins using sequence or structural information [7–10]. In addition, the transition of disorder-to-order regions in protein–protein interactions (PPI) is well studied experimentally and computationally [8–12]. For example, LMO4, a putative breast oncoprotein, interacts with various tandem LIM-domain containing proteins mediated by disordered regions [13]. BRCA1, a tumour suppressor protein, helps in binding with multiple protein and DNA partners by its central disorder region of ~1500 amino acids [14]. Recently, Papadakos et al. [15] showed that inducing intrinsic disorder in high-affinity protein–protein interactions reduces the affinity of binding.

Many proteins contain disordered regions and some of the regions attained ordered structures after binding to their cognate substrates, which are also known as MoRF (Molecular Recognition Features) segments [16,17]. Sugase et al. [18] have shown that folding and binding of IDPs or IDRs are coupled processes. Furthermore, binding partners are also shown to influence affinity and kinetics of binding. The flexibility of IDPs helps them to bind with multiple partners and have co-operative interactions [19]. Although induced fit and conformational selection processes are proposed explanations for the coupling of folding and binding, the exact model which is preferred by IDPs is not known [11,20].

The dynamics of the RNA molecule makes it more amenable to interact with disorder-mediated protein–RNA interactions [21]. The recognition of the protein–RNA complex has been experimentally studied using EMSA, yeast-3-hybrid assay, pull-down assay and CLIP [22,23]. On the other hand, plenty of tools have been developed to identify binding sites in RNA-binding proteins [24–32]. All these methods use the information in their sequence to compute the feature and/or evaluate the performance. Recently, Peng and Kurgan [33] developed a webserver for prediction of disorder-mediated interactions in RNA, DNA and protein–protein complexes. However, the knowledge for understanding the mechanisms or factors responsible for binding of disordered region with RNA has not yet been completely explored.

In this work, we constructed a dataset for protein–RNA complexes (provided in supplementary information), which are involved in disorder-to-order transitions. Utilizing the dataset, we analyzed the number and size of DOT regions in protein–RNA complexes, preference of residues involved in binding in DOT regions, secondary structure, solvent accessibility, pair preference at the interface, preference in different secondary structures of RNA, and interaction energy between protein and RNA DOT and non-DOT regions at the interface.

2. Results and Discussion

Our dataset contains a total of 23,452 and 2412 residues in non-ribosomal and ribosomal protein–RNA complexes. Among them, 1175 (5%) and 155 (6.4%) residues are found to be in DOT regions in non-ribosomal and ribosomal complexes, respectively. The residues binding with RNA are obtained by using 3.5 and 6 Å distance cut-offs and similar trends are obtained. Therefore, we have presented the results with 3.5 Å and those for 6 Å are shown in supplementary material.

2.1. Number of DOT Regions in Protein–RNA Complexes and Length of DOT Regions

The variation in the number of DOT regions in non-ribosomal and ribosomal complexes is shown in Figure 1. We observed that most of the complexes have less than three DOT regions (88% in non-ribosomal and 100% in ribosomal complexes). Most non-ribosomal proteins have one DOT region, whereas ribosomal proteins have mostly two or more DOT regions. In addition, at most eight DOT regions per complex are found in our dataset.

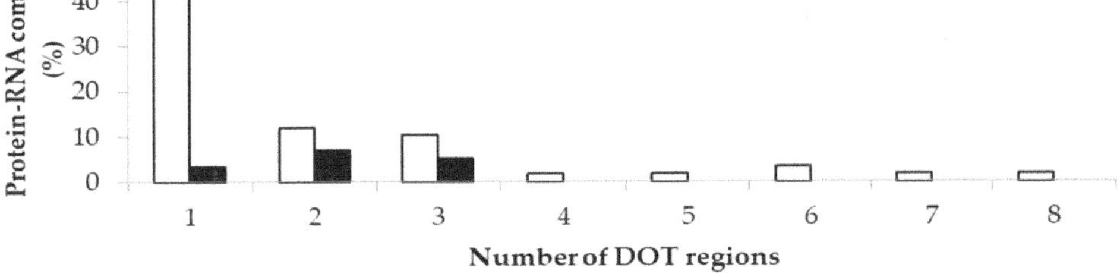

Figure 1. Percentage of protein–RNA complexes containing different number of DOT regions.

Further, we analysed the length of each DOT region in non-ribosomal and ribosomal protein–RNA complexes, which shows that most DOT regions are short, as shown in Figure 2. In both non-ribosomal and ribosomal complexes, more than 70% of DOT regions have three to 10 residues and very few (only 5) regions have a length of more than 50 residues. This leads to a speculation that only a small conformational change might be required for bringing shape complementarity in protein–RNA complexes and these small DOT regions help in obtaining the same.

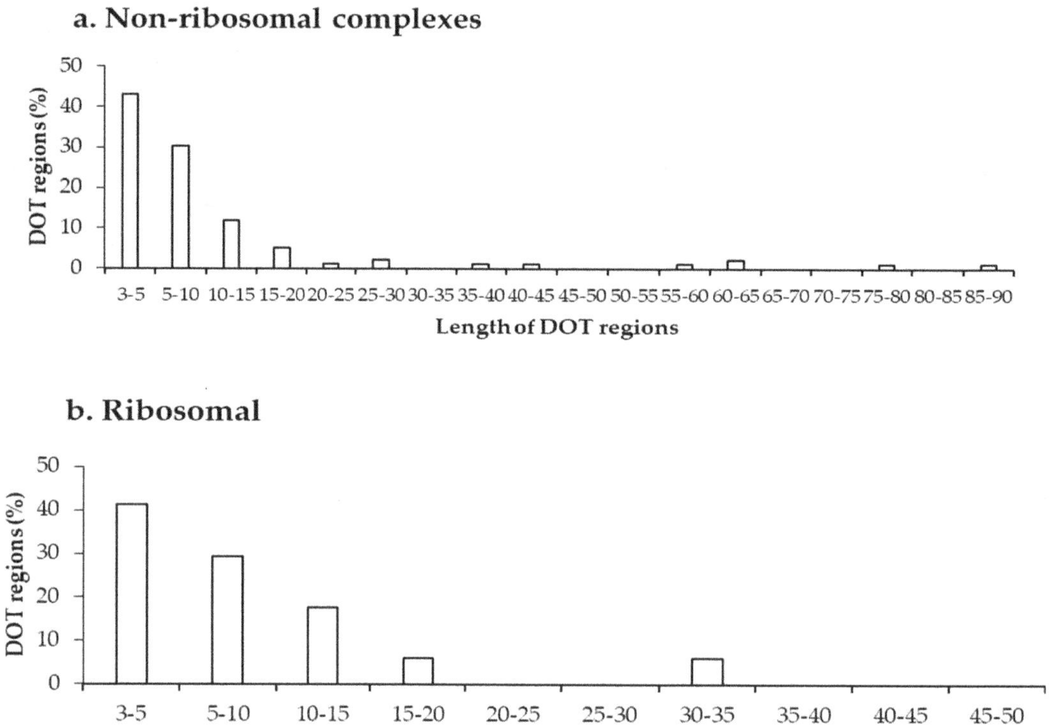

Figure 2. Length distribution of DOT regions in protein–RNA complexes in (**a**) non-ribosomal and (**b**) ribosomal complexes.

2.2. *Binding Frequency of Residues at DOT Regions*

The binding frequencies of residues in DOT regions using 3.5 Å (NR3.5 and RB3.5) and 6 Å (NR6 and RB6) distance cut-offs are shown in Figure 3 and Figure S2, respectively. We observed that among all positively charged residues (Arg, Lys and His), Arg and Lys have high preference for binding in both NR3.5 (Figure 3a) and NR6 (Figure S2a) datasets. Interestingly, only eight and 13 among 20 residues are observed in binding DOT regions at RB3.5 (Figure 3b) and RB6 (Figure S2b) datasets, and Arg has the highest frequency of binding. Cys, Met, and Trp in DOT regions are not involved in binding with RNA, whereas in ordered complexes 0.97%, 4.52%, and 5.54% of Cys, Met, and Trp are involved in binding, respectively. The comparison of binding site residues in DOT regions and the whole protein showed an expected presence of 1.5% and 2.7% of Met and Trp, respectively, in the interface of the DOT region. These results showed that the non-occurrence of Met and Trp at the interface of the DOT regions is statistically significant.

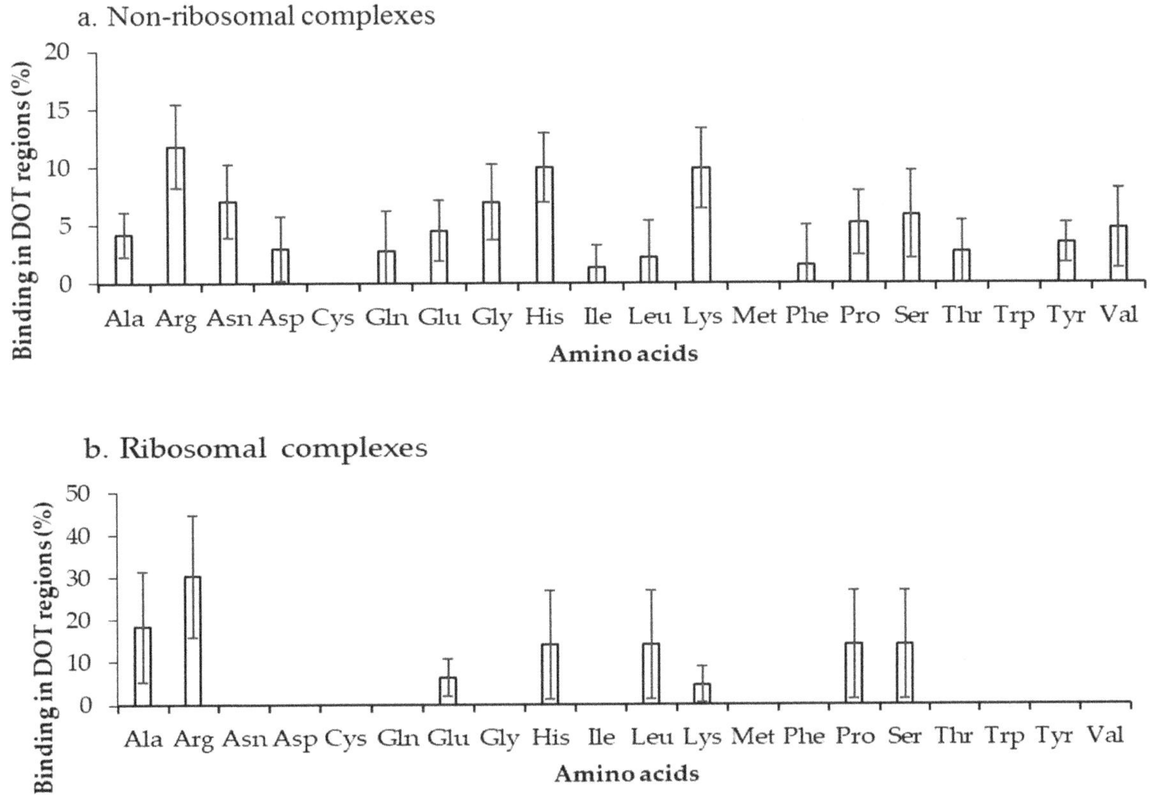

Figure 3. Amino acid frequency of binding in the DOT region for (**a**) non-ribosomal and (**b**) ribosomal protein–RNA complexes.

We have computed the preference of binding of residues in DOT regions by dividing the number of residues in DOT regions with the total number of binding residues, and the results are presented in Figure 4 and Figure S3 for 3.5 Å (NR3.5 and RB3.5 datasets) and 6 Å (NR6 and RB6 datasets), respectively. In Figure 4a, high frequency of Arg, Gly, Lys, and Ser (*z*-score > 1) is observed for the NR3.5 dataset, which suggests that these residues are more probable to contact DOT regions with respect to all residues in contact with RNA. However, for the NR6 dataset (Figure S3a), the result is only consistent for Lys, and two other residues (Glu and Pro) show high binding frequency. In ribosomal protein complexes with 3.5 Å and 6 Å, Ala & Glu, and Glu & Tyr have high frequencies, respectively (Figure 4b and Figure S3b).

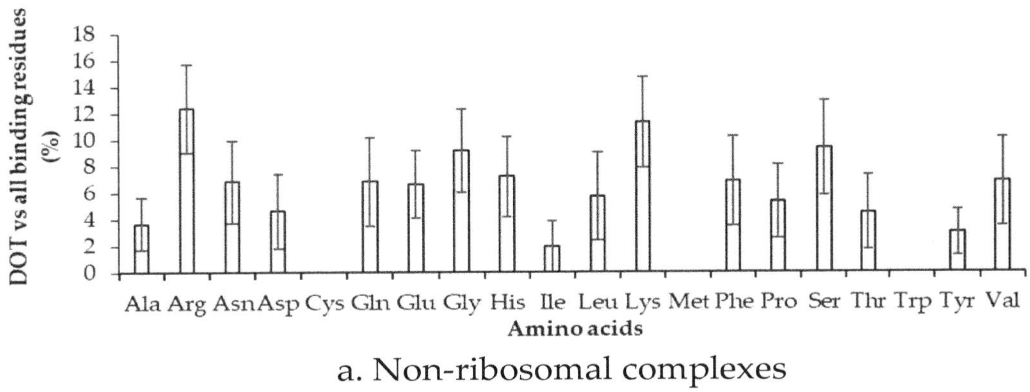

Figure 4. *Cont.*

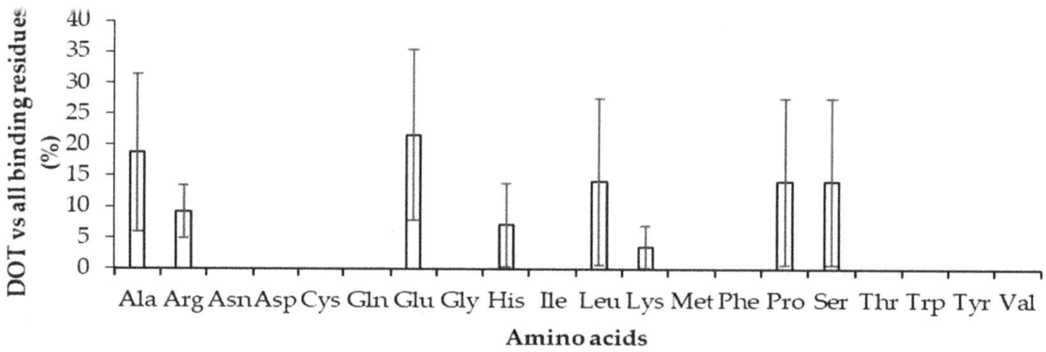

b. Ribosomal complexes

Figure 4. Frequency of DOT regions by contact residues for (**a**) non-ribosomal and (**b**) ribosomal complexes.

2.3. Binding Propensity of Residues at DOT Region

Propensity is calculated by normalizing the binding frequency of residues in DOT regions with the overall frequency of the respective residues to be in a protein, using Equation (3). This can measure the bias in binding of residues in DOT regions, independent of their count in DOT regions. We have calculated the propensity of amino acids to be in DOT regions using distance cut-offs of 3.5 Å and 6 Å and the results are shown in Figure 5 and Figure S4, respectively. In the NR3.5 (Figure 5a) dataset, His, Arg, Asn, Gln, Phe, and Tyr have high propensity of binding, whereas in ribosomal proteins (RB3.5 dataset; Figure 5b), only His showed a high propensity. In the NR6 (Figure S4a), His has high propensity, whereas Asn, His and Tyr have high propensity in the RB6 (Figure S4b) dataset. Similarly, high propensity for binding is observed for positively charged residues along with Tyr and Phe in protein–RNA complexes [34]. On the other hand, among all charged residues only Arg has high tendency to bind with DOT regions in protein–protein complexes [35]. Furthermore, non-specific interactions occurred frequently in protein–protein complexes, which is not a common trend in the binding residues of DOT regions in protein–RNA complexes. Therefore, we can infer that the preferred residues at DOT regions are specific in protein–RNA complexes and, especially, charged interactions are important in DOT regions for binding with RNA.

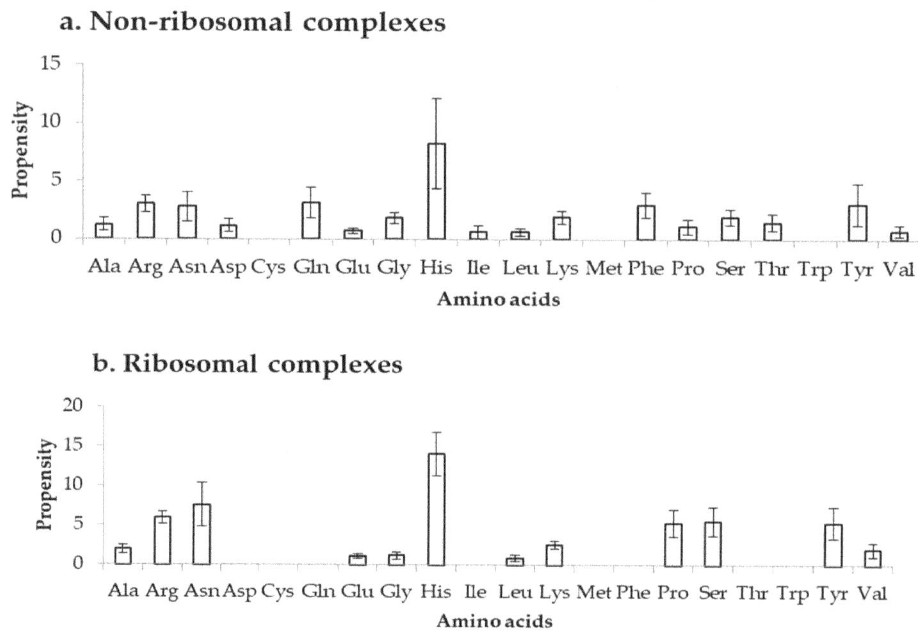

Figure 5. Propensity for amino acids in (**a**) non-ribosomal and (**b**) ribosomal complexes.

2.4. Comparison of Frequency of Binding in the DOT Region and Other Residues of a Protein

To estimate the difference between binding in DOT regions and other part of proteins, we calculated the binding frequency of amino acids in these regions, as shown in Figure 6 and Figure S5. Amino acids significantly differ in their binding with RNA in DOT regions and in the complete protein (p-value for the mean is less than 0.01). In non-ribosomal proteins, when the 3.5 Å cut-off is considered, nonpolar and aromatic residues mostly have high frequency values in the DOT regions than in the overall protein. All the frequencies are observed to be significant when statistical analysis is performed for the bootstrapped sample of the frequencies (p-value is less than 0.01). Residues such as His, Phe, and Leu are found to have a more than 3-fold increase in the frequency of binding in the DOT regions than in other parts of the proteins. A similar trend is observed in the NR6 dataset (Figure S5).

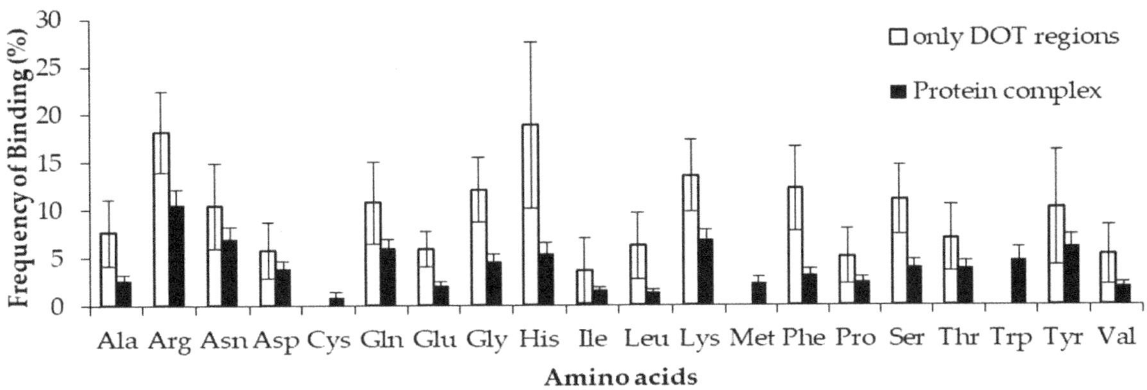

Figure 6. Binding frequency of each amino acid in the DOT region and in the overall protein for non-ribosomal complexes using the 3.5 Å cut-off.

2.5. Amino Acid Contact Frequency with Nucleotides

We have also analysed amino acid contacts with each nucleotide in non-ribosomal complexes using 3.5 Å and 6 Å distance cut-offs for contacting residues and the results are shown in Figure 7 and Figure S6. In the 3.5 Å distance criterion, Arg and Lys have a high frequency to bind with nucleotides. Arg and Lys are observed to have the most and least binding frequencies with Guanine and Uracil, respectively. Whereas in the 6 Å criterion, almost the same frequency of binding is observed for Arg and Lys with Adenine, Guanine and Cytosine nucleotides; least binding was observed in the Uracil nucleotide. When compared with the results presented for ordered protein–DNA and protein–RNA complexes in our earlier works, Arg, Lys, Trp, and Tyr were favoured by RNA and Arg was selected by DNA-binding proteins together with Guanine in DNA and Uracil in RNA–protein complexes [36].

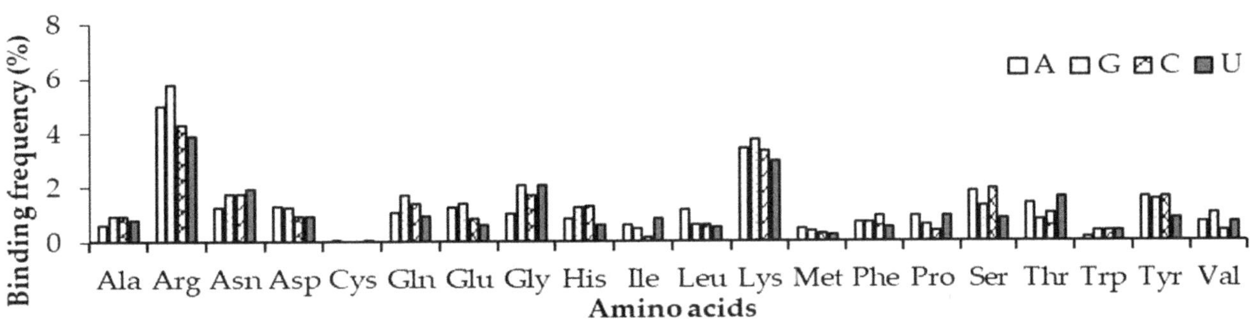

Figure 7. Normalized amino acid nucleotide contact frequency in non-ribosomal protein–RNA complexes at 3.5 Å.

2.6. Secondary Structure of DOT and RNA-Interacting DOT Residues

The secondary structures of DOT residues are quantified to study the bias of residues to have a specific secondary structure in binding and non-binding regions and data are presented in Table 1 and Table S1 for NR3.5 and NR6 datasets, respectively. In the NR3.5 dataset, all the DOT residues have lower and higher preference in sheet (15%) and other structure class (8%), respectively. Interestingly, in DOT residues, binding with RNA molecules, strand-forming residues have a higher preference (15.2%) as compared to helical (8.6%) and other regions (8.9%).

Table 1. Secondary structure of all DOT residues and residues binding with RNA in DOT regions in the NR3.5 dataset.

Secondary Structure	Number of Binding Residues in DOT Regions (N_{idt})	Number of Residues in DOT Region (N_d)	Relative Binding in DOT Regions (%)
Helix	25 (22.12)	288 (24.51)	8.6
Sheet	22 (19.47)	145 (12.34)	15.2
Others (coil, turn, bend)	66 (58.41)	742 (63.15)	8.9

Percentage is mentioned in the parenthesis. Relative binding in DOT regions are calculated by $N_{idt}/N_d \times 100$.

2.7. Relative Solvent Accessibility of DOT Residues

The spatial arrangement of DOT residues is further explored by solvent accessibility calculation and the result is shown in Table 2. Comparison of RASA of DOT regions and complete protein–RNA complex revealed that in DOT regions, solvent accessibility of every amino acid is more than that of other amino acids of a protein. As expected, charged residues have low fold difference (1.18 to 1.28) in RASA in DOT regions and the complete protein. However, most hydrophobic residues (Ala, Cys, Ile, Leu, Met, Phe, Tyr, and Val) have about 1.8 to 2 folds higher RASA in DOT regions than the complete protein, Met has the highest difference. On the other hand, the mean solvent accessibility of DOT regions of proteins is 44 Å^2, which is similar to the average RASA of binding DOT regions (43 Å^2) of protein–protein complexes [17].

Table 2. Relative average solvent accessibility (RASA) of DOT residues and all residues in non-ribosomal protein–RNA complexes.

Amino Acids	RASA in DOT Regions	RASA in Complete Protein	Fold Difference
Ala	44.743	23.305	1.920
Arg	52.168	40.822	1.278
Asn	63.583	42.721	1.488
Asp	56.552	43.811	1.291
Cys	22.32	11.426	1.953
Gln	47.805	38.988	1.226
Glu	53.688	47.838	1.122
Gly	51.599	35.272	1.463
His	43.229	35.372	1.222
Ile	26.618	14.692	1.812
Leu	32.007	16.374	1.955
Lys	58.529	49.520	1.182
Met	41.13	20.391	2.017
Phe	32.481	17.519	1.854
Pro	56.463	38.230	1.477
Ser	54.574	34.622	1.576
Thr	49.852	31.074	1.604
Trp	21.571	19.029	1.134
Tyr	44.579	24.752	1.801
Val	31.433	17.405	1.806

2.8. Number of Residues in Contact with Nucleotides in the DOT Region and in Entire Protein

Among 1175 residues in DOT regions in our dataset, only 96 (8.17%) and 268 (22.81%) are in contact with nucleotides in the NR3.5 and the NR6 dataset, respectively. Almost all the residues have a similar tendency of binding with nucleotides in proteins, ranging between 20% to 29%, as shown in Table 3 and Table S2. However, the number of nucleotides interacting with DOT residues is somewhat different, that is, the range of interaction is 18 to 33%. The DOT residues are more likely to bind with Guanine (20.4%), followed by Cytosine and Uracil, than to binding with Adenine (13.1%).

Table 3. Number of interaction of nucleotides with DOT residues and with complete protein at 3.5 Å.

Nucleotides	Number of Nucleotide in Contact with DOT Regions (N_{idt})	Number of Nucleotides in Contact with Any Residue of Proteins (N_{prot})	Relative Contact in DOT Regions (%)
A	18 (18.75)	137 (25.66)	13.1
C	26 (27.08)	131 (24.53)	19.8
G	32 (33.33)	157 (29.40)	20.4
U	20 (20.83)	109 (20.41)	18.3

Percentage is mentioned in the parenthesis. Relative contact in DOT regions are calculated by $N_{idt}/N_{prot} \times 100$.

2.9. Secondary Structure of Nucleotides Interacting with DOT Residues

Further, we have classified the nucleotides based on location and contacts with DOT residues and preference of amino acids in a protein and the results are presented in Table 4 and Table S3. Among all secondary structures formed by nucleotides, unpaired bases are most likely to bind with DOT residues. Specifically, we observed that A and U in unpaired regions prefer to interact with DOT residues, whereas C and G in unpaired and base-paired positions interact with DOT residues with a similar preference. G and C also interact with DOT residues in pseudoknot secondary structure, whereas A and U are least likely to exist in pseudoknot form when bound to DOT regions.

Table 4. Preference of nucleotides in different secondary structures to bind with DOT residues.

Nucleotides	Secondary Structure	Number of Nucleotide in Contact with DOT Regions (N_{idt})	Number of Nucleotides in Contact with Any Residue of Proteins (N_{prot})	Relative Contact in DOT Regions (%)
A	Unpaired	12 (12.50)	106 (19.56)	11.01
A	Basepaired	6 (6.25)	30 (5.54)	20.00
A	Pseudoknot	0 (0)	0 (0)	0
C	Unpaired	8 (8.33)	70 (12.92)	11.42
C	Basepaired	17 (17.71)	59 (10.89)	28.81
C	Pseudoknot	1 (1.04)	5 (0.92)	20.00
G	Unpaired	16 (16.67)	87 (16.05)	18.39
G	Basepaired	15 (15.63)	71 (13.10)	21.13
G	Pseudoknot	1 (1.04)	4 (0.74)	25.00
U	Unpaired	15 (15.63)	81 (14.94)	18.51
U	Basepaired	5 (5.21)	29 (5.35)	17.24
U	Pseudoknot	0 (0)	0 (0)	0
All	Unpaired	51 (53.13)	344 (63.47)	14.83
All	Basepaired	43 (44.79)	189 (34.87)	22.75
All	Pseudoknot	2 (2.08)	9 (1.66)	22.22

Percentage is mentioned in parenthesis. Relative contacts in DOT regions are calculated by $N_{idt}/N_{prot} \times 100$.

2.10. Interaction Energy of DOT Residues with Nucleotides

We have computed the interaction energy between amino acids and nucleotides in DOT and ordered regions at the binding interface and the results are presented in Table 5. Most of the amino acids have stronger interactions with nucleotides in ordered regions than DOT regions. However, we noticed that some combinations of amino acid–nucleotide pairs have favourable energy when interacting with DOT regions. For example, Arg, His, Ile, Leu, Val, and Phe interact with G, His, Ser, and Val with C, and Asn, Asp, Gly, Ile, Leu, and Ser with U. In addition, hydrophobic residues Ile, Leu, and Val have more favourable interactions with G at DOT regions than others. Since Arg and Lys are important for protein–RNA complex formation through electrostatic interactions these residues have stronger energies in ordered regions than DOT regions. On the other hand, His in the DOT region has favourable energy with G and C. These differences in energy could be important to understand the interactions between DOT regions and the RNA molecule, which might also be used to distinguish the RNA binding residues of proteins in DOT and other regions.

Table 5. Interaction energy between amino acids and nucleotides in DOT regions.

Amino Acids	A	G	C	U
Ala	−0.62 (−0.55)	−0.34 (−0.57)	−0.49 (−0.53)	−0.55 (−0.64)
Arg	−0.36 (−1.23)	**−1.15** (−0.83)	−0.89 (−0.95)	**−1.06** (−0.98)
Asn	−0.45 (−0.68)	−0.59 (−0.73)	−0.48 (−0.83)	**−1.85** (−0.82)
Asp	−0.75 (−0.74)	−0.39 (−0.79)	−0.19 (−0.56)	**−1.40** (−0.92)
Cys	0.00 (−0.87)	−0.01 (−0.03)	−0.03 (−1.10)	−0.63 (−1.13)
Gln	−0.15 (−0.87)	−0.57 (−0.74)	−0.08 (−0.84)	−0.36 (−0.71)
Glu	−0.72 (−0.80)	−0.41 (−0.64)	−0.43 (−0.62)	−0.68 (−0.59)
Gly	−0.28 (−0.47)	−0.37 (−0.69)	−0.58 (−0.57)	**−1.07** (−0.79)
His	−0.81 (−1.17)	**−2.13** (−1.41)	**−1.53** (−1.21)	−0.70 (−1.01)
Ile	−0.60 (−0.64)	**−1.63** (−0.80)	−0.54 (−0.50)	**−1.33** (−0.76)
Leu	−0.35 (−0.75)	**−1.19** (−0.50)	−0.42 (−0.49)	−0.54 (−0.41)
Lys	−0.74 (−0.76)	−0.86 (−0.83)	−0.66 (−0.90)	−0.83 (−0.83)
Met	−0.64 (−1.05)	−0.07 (−0.75)	−0.16 (−1.03)	−0.83 (−1.19)
Phe	−0.81 (−1.03)	**−1.12** (−0.89)	−0.54 (−1.32)	−0.24 (−1.42)
Pro	−0.88 (−0.83)	−0.60 (−0.88)	−0.62 (−0.91)	−0.69 (−1.00)
Ser	−0.79 (−0.77)	−0.29 (−0.56)	**−1.24** (−0.71)	**−1.41** (−0.68)
Thr	−0.39 (−0.68)	−0.66 (−0.56)	−0.67 (−0.64)	−0.53 (−1.00)
Trp	**−1.15** (−1.10)	0.00 (−1.64)	0.00 (−0.99)	0.00 (−1.34)
Tyr	**−1.53** (−1.36)	−1.11 (−1.42)	−0.63 (−1.05)	−0.16 (−1.09)
Val	−0.38 (−0.70)	−0.66 (−0.53)	−0.64 (−0.53)	−0.52 (−0.76)

Interaction energy for non-DOT residues is mentioned in the parenthesis. Amino acid–nucleotide pairs with favourable interaction energies in DOT regions are shown in bold.

We have compared the interaction energy of amino acid–nucleotide pairs in the interface of DOT and other regions and two typical examples are shown in Figure 8. We noticed a wide range of interactions such as stacking, cation-π, electrostatic, and van der Waals interactions at the interface. Most favourable energy is observed for Asn and His with U (−3.26 kcal/mol) and G (−5.44 kcal/mol), respectively, in DOT regions (Figure 8a). On the other hand, Arg and Phe have favourable energy with A (−8.49 kcal/mol) and C (−4.88 kcal/mol), respectively, in non-DOT regions.

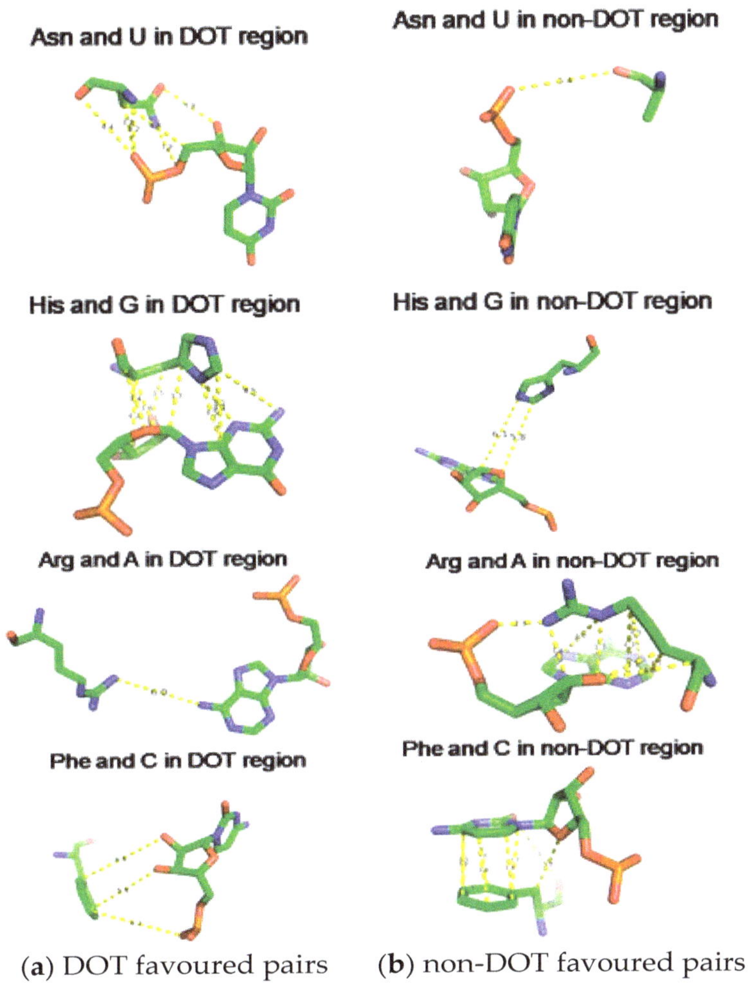

Figure 8. Amino acid showing (**a**) strong interaction in DOT and weak interaction in non-DOT regions and (**b**)weak interaction in DOT and strong interaction in non-DOT regions.

3. Materials and Methods

We adopted the following protocol to obtain a set of protein–RNA complexes with disorder-to-order transition (DOT) regions: (i) Downloaded the protein–RNA complexes from PDB and NDB databases (www.rcsb.org) [37–39]; (ii) Clustered all the protein–RNA complexes with 30% sequence identity cut-off using CD-Hit suite [40]; (iii) Performed BLAST search (using 99% identity cut-off) of protein sequences to obtain free proteins corresponding to each protein–RNA complex [41,42]. The free proteins have the same sequences as the protein part of protein–RNA complexes but crystallized without RNA. Note that free proteins contain unique PDB IDs, which is distinct from the protein–RNA complex; (iv) Disordered residues are obtained from missing residues information in the protein–RNA complex and free protein pairs by locating "REMARK 465" statement in the protein structure file; (v) DOT residues are isolated by comparing the disorder residues of free and protein–RNA complex pairs such that the residue is ordered in the protein–RNA complex but disordered in free protein. Note that only the regions having 3 or more continuous DOT residues are considered. The final dataset contains 101 DOT regions in 52 proteins and complete data are given in supplementary information. The representation of DOT and ordered region in a typical protein–RNA complex (PDB ID: 4H4K) is shown in Figure 9.

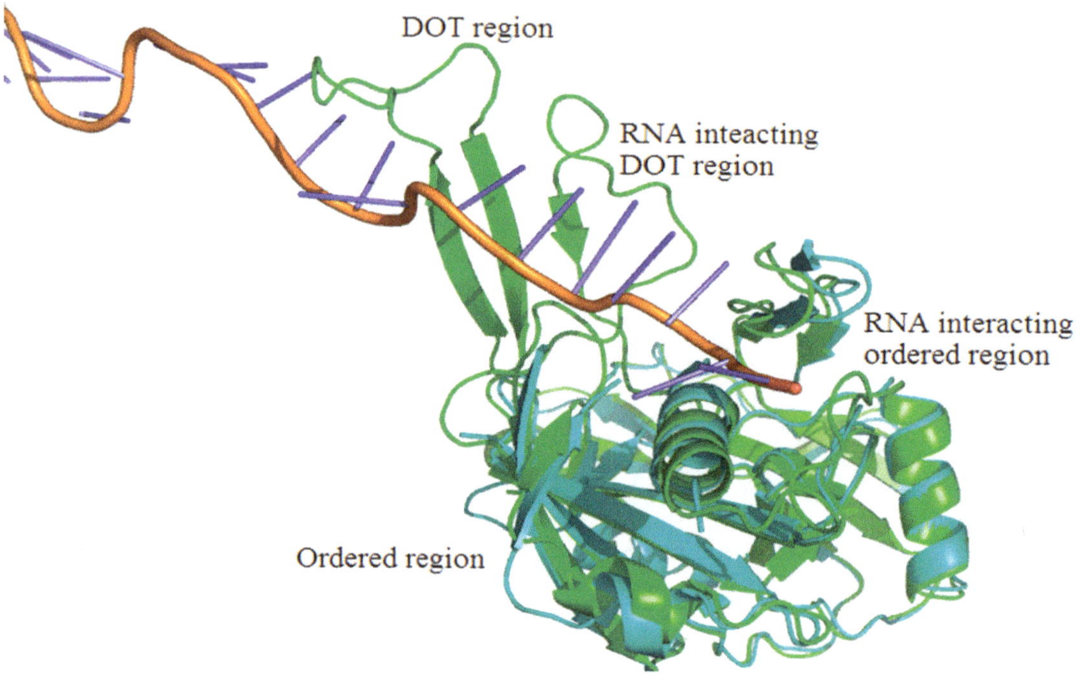

Figure 9. Representation of disorder-to-order mediated interactions. Free protein, RNA, and complex (CRISPR-Cas RNA Silencing Cmr Complex) are shown in cyan, orange and green, respectively. The PDB IDs are 4H4K:A (free protein), 3XIL:I (RNA of protein–RNA complex) and 3XIL:B (RNA-bound protein). The disorder-to-order transition (DOT) region can be clearly seen in green with a missing overlapping region of free protein.

3.1. Number of DOT Regions and Their Lengths

The number of DOT regions and their lengths are obtained by counting the number of non-consecutive and consecutive residues, respectively, using custom build python scripts.

3.2. DOT Residues in Contact with RNA

The residues in contact with RNA molecules are obtained by using distance cut-offs mentioned in literature, that is, 3.5 Å and 6 Å [43–45]. Binding residues in DOT regions are obtained by taking common residues in the DOT dataset and RNA contacting residues. We have classified protein–RNA complexes in non-ribosomal and ribosomal classes because of the difference in their interaction pattern, number of interacting amino acids, and residue bias in them [46]. Therefore, using the type of complex and distance cut-off for interacting residues, we divided protein–RNA complexes into four different datasets: (1) NR3.5: non-ribosomal complex with a contact distance of 3.5 Å; (2) RB3.5: ribosomal complex with a contact distance of 3.5 Å; (3) NR6: non-ribosomal complex with a contact distance of 6 Å; and (4) RB6: ribosomal complex with a contact distance of 6 Å.

We computed the frequency of each DOT residue involved in binding using the Equation (1).

$$\text{Frequency of binding residues in DOT region} = \frac{N_{ib}}{N_{id}} \tag{1}$$

where N_{ib}: number of ith residues binding in the DOT region and N_{id}: number of ith residues in DOT.

Moreover, the differences in the frequency of binding residues in DOT regions and in the protein complexes are obtained.

3.3. Frequency of Binding in DOT and Other Residues

We also computed the frequency of residues binding in DOT regions over all the binding residues by using Equation (2), an error bar is plotted using the bootstrap method by randomly re-sampling an equal sized data with a replacement 1000 times.

$$\text{Frequency of binding by contact residues} = \frac{N_{ibd}}{N_{ib}} \tag{2}$$

where N_{ibd}: number of ith residues binding in DOT region; N_{ib} is number of ith residues binding with RNA in complete protein.

3.4. Propensity of Binding Residues in DOT Region

The normalization of frequency of residues present in DOT regions by individual residue frequency provides the tendency of a residue in DOT regions. Accordingly, propensity values are calculated using the following equation:

$$\text{Propensity(I)} = \frac{N_{ibd}/N_{id}}{N_{ip}/N_p} \tag{3}$$

where Propensity (I): propensity of ith residue; N_{ibd}: number of ith residue binding in DOT region; N_{id}: number of ith residue in DOT regions; N_{ip}: number of ith residue in protein; N_p: number of residues in protein.

3.5. Boot Strap Sampling

To obtain the standard error in frequency and propensity calculations, bootstrap sampling is performed. In this technique all the protein–RNA complexes are sampled randomly and each sample contains complexes equal to the number of protein–RNA complexes. Therefore, each sample will have redundancy of some complexes and will be devoid of some complexes. In this manner, we have created 1000 samples on which the calculations are performed.

3.6. Relative Average Solvent Accessibility (RASA)

The DOT residues buriedness is analysed by the NACCESS [47] program and the RASA of each residue is calculated by using Equation (4).

$$\text{RASA} = \frac{A_{ibd}}{\sum_{i=1}^{n}(A_{ibd})} \tag{4}$$

where A_{ibd}: RASA of ith residue binding with RNA in DOT region; n: number of DOT residues in a protein–RNA complex.

3.7. Secondary Structure of Protein and RNA

Secondary structure of both proteins and RNA molecules are analysed by DSSP and DSSR programs, respectively [48,49]. The DSSR program gives dot bracket notation of secondary structure of RNA as shown in Figure S1, in which "." represents unpaired nucleotide, "(" or ")" represent paired bases, and "{" or "}" or "[" or "]" or "<" or ">" represent pseudoknot bases.

3.8. Binding Preference of Nucleotides for Amino Acids

The binding preference of nucleotide with DOT residues has been calculated by counting the occurrence of nucleotides–amino acid interacting pairs under the distance of 3.5 Å.

3.9. Interaction Energy between Amino Acids and Nucleotides at Binding Interface

The interaction energy of amino acids with nucleotides is computed using van der Waals and coulombs potential using AMBER force field [50]. It is given by

$$\text{Energy} = \sum [\left(\frac{A_{ij}}{r_{ij}{}^{12}} - \frac{B_{ij}}{r_{ij}{}^{6}}\right) + \frac{q_i q_j}{\varepsilon r_{ij}}] \quad (5)$$

where, $A_{ij} = \varepsilon_{ij}{}^* (R_{ij}{}^*)^{12}$ and $B_{ij} = 2\, \varepsilon_{ij}{}^* (R_{ij}{}^*)^6$; $R_{ij}{}^* = (R_i{}^* + R_j{}^*)$; and $\varepsilon_{ij}{}^* = (\varepsilon_i{}^*\, \varepsilon_j{}^*)^{1/2}$; R^* and ε^* van der Waals radius and well depth, respectively, and these parameters are obtained from Gromiha et al. [51]; q_i and q_j is the charge on atom i and j, respectively and R_{ij} is the distance separating atom i and j.

4. Conclusions

The analysis of DOT regions in protein–RNA complexes revealed that in each complex these regions are generally small in size. Electrostatic interactions are found to be important, with the involvement of positively charged residues (Arg, Lys and His) in DOT regions. Among nucleotide–amino acid pairs, guanine–Arg and uracil–Lys pairs are identified to be the most and the least preferred ones at the interface, respectively. Generally, nucleotides prefer to bind DOT regions than other regions of protein. Further, DOT regions are significantly more exposed to solvent than other residues of protein–RNA complexes. Specifically, hydrophobic residues have higher difference in RASA of DOT regions and complete proteins. DOT regions are preferred to form coils, turns, and bends than regular secondary structures such as helices and strands. On the RNA side, DOT residues prefer to bind unpaired A and U and paired regions of C and G. In pseudoknot condition, mostly C and G interact with DOT residues. The interaction energy calculations revealed the types of interactions and preferred amino acid-nucleotide pairs at the interface based on energy.

The frequencies and propensities obtained in the present study could be used for discriminating DOT binding residues from other residues. Further, the location of DOT binding residues based on solvent accessibility and secondary structure of protein and RNA along with energy calculations may help to understand the recognition mechanism.

We obtained the DOT regions by comparing 3D coordinates of the missing residues in protein–RNA complexes and their respective free proteins. This might be an under representation of DOT regions since the structures solved by crystallization often stabilize the residues and reduce the native disorder. Hence, the disordered residues having 3D coordinates in free proteins are not considered. The current study can further be refined with the availability of more numbers of protein–RNA complexes and the improvements in structure determination techniques. In addition, development of disorder specific databases for protein–nucleic acid complexes with large datasets could enhance the confidence level of the result reported in the present study.

Author Contributions: M.M.G. and S.A. conceived the project and designed experiments. A.S. constructed the dataset and performed the analysis. M.M.G, S.A. and A.S took part in discussions. A.S. drafted the manuscript. M.M.G and S.A. edited and refined the manuscript.

Acknowledgments: We thank the Department of Biotechnology, Indian Institute of Technology Madras for computational facilities. A.S. thank Ministry of Human Resource and Development (MHRD) for the fellowship. This project is partially supported by the Council of Scientific & Industrial Research (CSIR), Government of India to M.M.G. and S.A. (grant numbers: 37(1694)/17/EMR-II) and 37(1695)/17/EMRII respectively). S.A. would like to acknowledge a grant from University for potential of excellence (UpoE-II) #270 and support DST-PURSE.

References

1. Habchi, J.; Tompa, P.; Longhi, S.; Uversky, V.N. Introducing protein intrinsic disorder. *Chem. Rev.* **2014**, *114*, 6561–6588. [CrossRef] [PubMed]
2. Fuxreiter, M.; Toóth-Petroóczy, A.; Kraut, D.A.; Matouschek, A.T.; Lim, R.Y.; Xue, B.; Kurgan, L.; Uversky, V.N. Disordered proteinaceous machines. *Chem. Rev.* **2014**, *114*, 6806–6843. [CrossRef] [PubMed]
3. Babu, M.M.; Van der, L.R.; de Groot, N.S.; Gsponer, J. Intrinsically disordered proteins: Regulation and disease. *Curr. Opin. Struct. Biol.* **2011**, *21*, 432–440. [CrossRef] [PubMed]
4. Wright, P.E.; Dyson, H.J. Intrinsically disordered proteins in cellular signaling and regulation. *Nat. Rev. Mol. Cell Biol.* **2015**, *16*, 18–29. [CrossRef] [PubMed]
5. Deller, M.C.; Kong, L.; Rupp, B. Protein stability: A crystallographer's perspective. *Acta Cryst. F* **2016**, *72*, 72–95. [CrossRef] [PubMed]
6. Johnson, D.E.; Xue, B.; Sickmeier, M.D.; Meng, J.; Cortese, M.S.; Oldfield, C.J.; Le Gall, T.; Dunker, A.K.; Uversky, V.N. High-throughput characterization of intrinsic disorder in proteins from the Protein Structure Initiative. *J. Struct. Biol.* **2012**, *180*, 201–215. [CrossRef] [PubMed]
7. Dosztányi, Z.; Csizmok, V.; Tompa, P.; Simon, I. IUPred: Web server for the prediction of intrinsically unstructured regions of proteins based on estimated energy content. *Bioinformatics* **2005**, *21*, 3433–3434. [CrossRef] [PubMed]
8. Jones, D.T.; Cozzetto, D. DISOPRED3: Precise disordered region predictions with annotated protein-binding activity. *Bioinformatics* **2014**, *31*, 857–863. [CrossRef] [PubMed]
9. Disfani, F.M.; Hsu, W.L.; Mizianty, M.J.; Oldfield, C.J.; Xue, B.; Dunker, A.K.; Uversky, V.N.; Kurgan, L. MoRFpred, a computational tool for sequence-based prediction and characterization of short disorder-to-order transitioning binding regions in proteins. *Bioinformatics* **2012**, *28*, i75–i83. [CrossRef] [PubMed]
10. Meng, F.; Uversky, V.N.; Kurgan, L. Comprehensive review of methods for prediction of intrinsic disorder and its molecular functions. *Cell. Mol. Life Sci.* **2017**, *74*, 3069–3090. [CrossRef] [PubMed]
11. Berlow, R.B.; Dyson, H.J.; Wright, P.E. Functional advantages of dynamic protein disorder. *FEBS Lett.* **2015**, *589*, 2433–2440. [CrossRef] [PubMed]
12. Basu, S.; Söderquist, F.; Wallner, B. Proteus: A random forest classifier to predict disorder-to-order transitioning binding regions in intrinsically disordered proteins. *J. Comput. Aided Mol. Des.* **2017**, *31*, 453–466. [CrossRef] [PubMed]
13. Deane, J.E.; Ryan, D.P.; Sunde, M.; Maher, M.J.; Guss, J.M.; Visvader, J.E.; Matthews, J.M. Tandem LIM domains provide synergistic binding in the LMO4: Ldb1 complex. *EMBO J.* **2004**, *23*, 3589–3598. [CrossRef] [PubMed]
14. Mark, W.Y.; Liao, J.C.; Lu, Y.; Ayed, A.; Laister, R.; Szymczyna, B.; Chakrabartty, A.; Arrowsmith, C.H. Characterization of segments from the central region of BRCA1: An intrinsically disordered scaffold for multiple protein–protein and protein–DNA interactions? *J. Mol. Biol.* **2005**, *345*, 275–287. [CrossRef] [PubMed]
15. Papadakos, G.; Sharma, A.; Lancaster, L.; Bowen, R.; Kaminska, R.; Leech, A.P.; Walker, D.; Redfield, C.; Kleanthous, C. Consequences of inducing intrinsic disorder in a high-affinity protein-protein interaction. *J. Am. Chem. Soc.* **2015**, *137*, 5252–5255. [CrossRef] [PubMed]
16. Fukuchi, S.; Amemiya, T.; Sakamoto, S.; Nobe, Y.; Hosoda, K.; Kado, Y.; Murakami, S.D.; Koike, R.; Hiroaki, H.; Ota, M. IDEAL in 2014 illustrates interaction networks composed of intrinsically disordered proteins and their binding partners. *Nucleic Acids Res.* **2014**, *42*, D320–D325. [CrossRef] [PubMed]
17. Vacic, V.; Oldfield, C.J.; Mohan, A.; Radivojac, P.; Cortese, M.S.; Uversky, V.N.; Dunker, A.K. Characterization of molecular recognition features, MoRFs, and their binding partners. *J. Proteome Res.* **2007**, *6*, 2351–2366. [CrossRef] [PubMed]
18. Sugase, K.; Dyson, H.J.; Wright, P.E. Mechanism of coupled folding and binding of an intrinsically disordered protein. *Nature* **2007**, *447*, 1021–1025. [CrossRef] [PubMed]
19. Shammas, S.L.; Travis, A.J.; Clarke, J. Allostery within a transcription coactivator is predominantly mediated through dissociation rate constants. *Proc. Natl. Acad. Sci. USA* **2014**, *111*, 12055–12060. [CrossRef] [PubMed]
20. Shammas, S.L.; Crabtree, M.D.; Dahal, L.; Wicky, B.I.; Clarke, J. Insights into coupled folding and binding mechanisms from kinetic studies. *J. Biol. Chem.* **2016**, *291*, 6689–6695. [CrossRef] [PubMed]

21. Dyson, H.J. Roles of intrinsic disorder in protein–nucleic acid interactions. *Mol. Biosyst.* **2012**, *8*, 97–104. [CrossRef] [PubMed]
22. Dey, B.; Thukral, S.; Krishnan, S.; Chakrobarty, M.; Gupta, S.; Manghani, C.; Rani, V. DNA–protein interactions: Methods for detection and analysis. *Mol. Cell. Biochem.* **2012**, *365*, 279–299. [CrossRef] [PubMed]
23. Popova, V.V.; Kurshakova, M.M.; Kopytova, D.V. Methods to study the RNA-protein interactions. *Mol. Biol.* **2015**, *49*, 472–481. [CrossRef]
24. Walia, R.R.; Caragea, C.; Lewis, B.A.; Towfic, F.; Terribilini, M.; El-Manzalawy, Y.; Dobbs, D.; Honavar, V. Protein–RNA interface residue prediction using machine learning: An assessment of the state of the art. *BMC Bioinform.* **2012**, *13*, 89. [CrossRef] [PubMed]
25. Kumar, M.; Gromiha, M.M.; Raghava, G.P. Prediction of RNA binding sites in a protein using SVM and PSSM profile. *Proteins* **2008**, *71*, 189–194. [CrossRef] [PubMed]
26. Terribilini, M.; Sander, J.D.; Lee, J.H.; Zaback, P.; Jernigan, R.L.; Honavar, V.; Dobbs, D. RNABindR: A server for analyzing and predicting RNA-binding sites in proteins. *Nucleic Acids Res.* **2007**, *35*, W578–W584. [CrossRef] [PubMed]
27. Wang, L.; Brown, S.J. BindN: A web-based tool for efficient prediction of DNA and RNA binding sites in amino acid sequences. *Nucleic Acids Res.* **2006**, *34*, W243–W248. [CrossRef] [PubMed]
28. Zhang, J.; Ma, Z.; Kurgan, L. Comprehensive review and empirical analysis of hallmarks of DNA-, RNA-and protein-binding residues in protein chains. *Brief. Bioinform.* **2017**, 1–19. [CrossRef] [PubMed]
29. Wang, L.; Huang, C.; Yang, M.Q.; Yang, J.Y. BindN+ for accurate prediction of DNA and RNA-binding residues from protein sequence features. *BMC Syst. Biol.* **2010**, *4*, S3. [CrossRef] [PubMed]
30. Yan, J.; Friedrich, S.; Kurgan, L. A comprehensive comparative review of sequence-based predictors of DNA-and RNA-binding residues. *Brief. Bioinform.* **2015**, *17*, 88–105. [CrossRef] [PubMed]
31. Tuszynska, I.; Bujnicki, J.M. DARS-RNP and QUASI-RNP: New statistical potentials for protein–RNA docking. *BMC Bioinform.* **2011**, *12*, 348. [CrossRef] [PubMed]
32. Wang, Y.; Guo, Y.; Pu, X.; Li, M. A sequence-based computational method for prediction of MoRFs. *RSC Adv.* **2017**, *7*, 18937–18945. [CrossRef]
33. Peng, Z.; Kurgan, L. High-throughput prediction of RNA, DNA and protein binding regions mediated by intrinsic disorder. *Nucleic Acids Res.* **2015**, *43*, e121. [CrossRef] [PubMed]
34. Kim, O.T.; Yura, K.; Go, N. Amino acid residue doublet propensity in the protein–RNA interface and its application to RNA interface prediction. *Nucleic Acids Res.* **2006**, *34*, 6450–6460. [CrossRef] [PubMed]
35. Mohan, A.; Oldfield, C.J.; Radivojac, P.; Vacic, V.; Cortese, M.S.; Dunker, A.K.; Uversky, V.N. Analysis of molecular recognition features (MoRFs). *J. Mol. Biol.* **2006**, *362*, 1043–1059. [CrossRef] [PubMed]
36. Fernandez, M.; Kumagai, Y.; Standley, D.M.; Sarai, A.; Mizuguchi, K.; Ahmad, S. Prediction of dinucleotide-specific RNA-binding sites in proteins. *BMC Bioinform.* **2011**, *12*, S5. [CrossRef] [PubMed]
37. Rose, P.W.; Prlić, A.; Altunkaya, A.; Bi, C.; Bradley, A.R.; Christie, C.H.; Costanzo, L.D.; Duarte, J.M.; Dutta, S.; Feng, Z.; Green, R.K. The RCSB protein data bank: Integrative view of protein, gene and 3D structural information. *Nucleic Acids Res.* **2017**, *45*, D271–D281. [PubMed]
38. Berman, H.M.; Olson, W.K.; Beveridge, D.L.; Westbrook, J.; Gelbin, A.; Demeny, T.; Hsieh, S.H.; Srinivasan, A.R.; Schneider, B. The nucleic acid database. A comprehensive relational database of three-dimensional structures of nucleic acids. *Biophys. J.* **1992**, *63*, 751–759. [CrossRef]
39. Coimbatore Narayanan, B.; Westbrook, J.; Ghosh, S.; Petrov, A.I.; Sweeney, B.; Zirbel, C.L.; Leontis, N.B.; Berman, H.M. The Nucleic Acid Database: New features and capabilities. *Nucleic Acids Res.* **2014**, *42*, D114–D122. [CrossRef] [PubMed]
40. Huang, Y.; Niu, B.; Gao, Y.; Fu, L.; Li, W. CD-HIT Suite: A web server for clustering and comparing biological sequences. *Bioinformatics* **2010**, *26*, 680–682. [CrossRef] [PubMed]
41. Boratyn, G.M.; Schäffer, A.A.; Agarwala, R.; Altschul, S.F.; Lipman, D.J.; Madden, T.L. Domain enhanced lookup time accelerated BLAST. *Biol. Direct.* **2012**, *7*, 12. [CrossRef] [PubMed]
42. Altschul, S.F.; Gish, W.; Miller, W.; Myers, E.W.; Lipman, D.J. Basic local alignment search tool. *J. Mol. Biol.* **1990**, *215*, 403–410. [CrossRef]
43. Gromiha, M.M. *Protein Bioinformatics: From Sequence to Function*; Academic Press: Cambridge, MA, USA, 2010.
44. Si, J.; Zhao, R.; Wu, R. An overview of the prediction of protein DNA-binding sites. *Int. J. Mol. Sci.* **2015**, *16*, 5194–5215. [CrossRef] [PubMed]

45. Nagarajan, R.; Gromiha, M.M. Prediction of RNA binding residues: An extensive analysis based on structure and function to select the best predictor. *PLoS ONE* **2014**, *9*, e91140. [CrossRef] [PubMed]
46. Ciriello, G.; Gallina, C.; Guerra, C. Analysis of interactions between ribosomal proteins and RNA structural motifs. *BMC Bioinform.* **2010**, *11*, S41. [CrossRef] [PubMed]
47. NACCESS, V2.1.1. *A Computer Program for Solvent Accessible Area Calculations*; Department of Biochemistry and Molecular Biology, University College London: London, UK, 1993.
48. Kabsch, W.; Sander, C. Dictionary of protein secondary structure: Pattern recognition of hydrogen-bonded and geometrical features. *Biopolymers* **1983**, *22*, 2577–2637. [CrossRef] [PubMed]
49. Lu, X.J.; Bussemaker, H.J.; Olson, W.K. DSSR: An integrated software tool for dissecting the spatial structure of RNA. *Nucleic Acids Res.* **2015**, *43*, e142. [CrossRef] [PubMed]
50. Cornell, W.D.; Cieplak, P.; Bayly, C.I.; Gould, I.R.; Merz, K.M.; Ferguson, D.M.; Spellmeyer, D.C.; Fox, T.; Caldwell, J.W.; Kollman, P.A. A second generation force field for the simulation of proteins, nucleic acids, and organic molecules. *J. Am. Chem. Soc.* **1995**, *117*, 5179–5197. [CrossRef]
51. Gromiha, M.M.; Yokota, K.; Fukui, K. Understanding the recognition mechanism of protein–RNA complexes using energy based approach. *Curr. Protein Pept. Sci.* **2010**, *11*, 629–638. [CrossRef] [PubMed]

16

Disordered Regions of Mixed Lineage Leukemia 4 (MLL4) Protein are Capable of RNA Binding

Beáta Szabó [1], Nikoletta Murvai [1], Rawan Abukhairan [1], Éva Schád [1], József Kardos [2], Bálint Szeder [1], László Buday [1] and Ágnes Tantos [1,*]

1 Institute of Enzymology, Research Centre for Natural Sciences, Hungarian Academy of Sciences, H-1117 Budapest, Hungary; szabo.beata@ttk.mta.hu (B.S.); murvai.nikoletta@ttk.mta.hu (N.M.); rawan.abukhairan@ttk.mta.hu (R.A.); schad.eva@ttk.mta.hu (E.S.); szeder.balint@ttk.mta.hu (B.S.); buday.laszlo@ttk.mta.hu (L.B.)

2 ELTE NAP Neuroimmunology Research Group, Department of Biochemistry, Eötvös Loránd University, H-1117 Budapest, Hungary; kardos@elte.hu

* Correspondence: tantos.agnes@ttk.mta.hu

Abstract: Long non-coding RNAs (lncRNAs) are emerging as important regulators of cellular processes and are extensively involved in the development of different cancers; including leukemias. As one of the accepted methods of lncRNA function is affecting chromatin structure; lncRNA binding has been shown for different chromatin modifiers. Histone lysine methyltransferases (HKMTs) are also subject of lncRNA regulation as demonstrated for example in the case of Polycomb Repressive Complex 2 (PRC2). Mixed Lineage Leukemia (MLL) proteins that catalyze the methylation of H3K4 have been implicated in several different cancers; yet many details of their regulation and targeting remain elusive. In this work we explored the RNA binding capability of two; so far uncharacterized regions of MLL4; with the aim of shedding light to the existence of possible regulatory lncRNA interactions of the protein. We demonstrated that both regions; one that contains a predicted RNA binding sequence and one that does not; are capable of binding to different RNA constructs in vitro. To our knowledge, these findings are the first to indicate that an MLL protein itself is capable of lncRNA binding.

Keywords: MLL proteins; MLL4; lncRNA; HOTAIR; MEG3; leukemia; histone lysine methyltransferase; RNA binding; intrinsically disordered protein

---

1. Introduction

Long non-coding RNAs (lncRNAs) are transcribed RNA molecules longer than 200 nucleotides that do not code for translated proteins. The human genome is estimated to code for about 58,000 lncRNAs [1], that are being more and more recognized as central players in a plethora of biological processes. They can act as flexible scaffolds providing binding platforms for different proteins, they can interfere with other endogenous RNAs acting as microRNA "sponges" and they can modify chromatin state [2], thus regulating the expression of various proteins. LncRNAs have also been shown to play a role in several layers of epigenetic regulation: they are involved in DNA methylation and demethylation, they can modify chromatin conformation through binding to remodelers [3] and many of them interact with histone modifier enzyme complexes such as PRC2, coREST or SMCX [4].

The physiological processes where lncRNA regulation have been suggested involve cell cycle regulation, epithelial mesenchymal transition (EMT) [5], cancer progression [6] and maintenance of cancer stem cells [5], hypoxia [7] and leukemia [8].

Various lncRNAs are shown to have altered expression levels in different leukemias, resulting in a crucial influence on cellular transformation [9], chromosomal translocation [10], apoptosis [11] and on drug resistance [12]. Accumulating evidence regarding the involvement of lncRNAs in leukemic processes prompted the suggestion to use them as prognostic and classification factors. It was found that lncRNA expression has prognostic value in AML patients [13] and multiple pathways were involved in lncRNA expression, including chromosome organization and trans-membrane receptor protein tyrosine kinase signalling pathway.

As lncRNAs are also considered valuable drug targets, it is essential that the molecular details of their functions are uncovered.

Polycomb repressive complex (PRC2) is the most studied histone modifier that relies on lncRNA binding in its function, being able to bind several lncRNAs including HOTAIR, Xist, RepA, Braveheart, MALAT1 and MEG3 [14]. In vitro experiments revealed that not only EZH2, but other PRC2 subunits are also capable of lncRNA binding [15], thus providing a pattern of binding regions distributed along the surface of the complex. Even though there remain open questions regarding the specificity of the RNA binding by PRC2 [16], it is widely accepted that lncRNA binding plays a defining role in PRC2 targeting and the ensuing gene silencing [14]. It is interesting to note that despite the numerous experimental results that show EZH2 to be an RNA binding protein, it cannot be found in databases that list RNA binding proteins, furthermore no RNA binding site is predicted to be located in the region that is shown to be responsible for the RNA-protein interaction [17].

Apart from PRC2, other histone lysine methyltransferases (HKMTs) or HKMT complex components also appear to bind lncRNAs with a relevant physiological outcome.

LncRNA EZR-AS1 enhances EZR expression through recruiting SMYD2 to the upstream region of its promoter region and elevating the activating H3K4 methylation [18].

G9a interacts with lncRNA PARTICLE to regulate MATA2 expression upon mild irradiation [19]. The interaction was shown using ChIP assay and apart from G9a, the PRC2 subunit Suz12 was also pulled down. In a later experiment, it was found that PARTICLE can also interact with DNA methylase DNMT1 and that it increases H3K27 methylation as well as EZH2 expression. It was suggested that PARTICLE may serve as a functional platform that enables the specific targeting of chromatin modifiers, such as PRC2 [20].

WDR5, a component of the MLL1-4 and SET1a/1b complexes was proven to interact with lncRNAs NeST and HOTTIP with an effect on microbial susceptibility through the enhancement of interferon-γ expression [21]. Further investigation of the WDR5-HOTTIP interaction led to the recognition that lncRNA binding by WDR5 is essential in maintaining embryonic stem cell pluripotency [22]. However, not this work nor any previous studies investigated the possibility that the enzymatic component of the methyltransferase complex may also be capable of lncRNA binding.

The family of mammalian MLL (Mixed Lineage Leukemia) proteins consist of Set1a, Set1b and four MLL proteins, MLL1, MLL2, MLL3 and MLL4. They work in COMPASS-like complexes and catalyze H3K4 mono-, di- or tri-methylation, each complex having different specificity and methylase activity [23]. MLL3 and MLL4 are responsible for the monomethylation of H3K4 at enhancer regions [24] and has been linked to a high number of different cancers. Properly functioning MLL3 and MLL4 act as tumor suppressors [23], therefore mutations affecting their activity or stability can result in cancer development. Despite their central role in several types of cancers, many open questions regarding the regulation of the activity and the targeting of the MLL complexes remain unanswered. The exact molecular details of how MLL3 and MLL4 are targeting enhancer regions [23] as well as the specific molecular effects of the interactions of their different regulatory domains [25] are largely unknown. It is also worth noting that the known structured domains represent only 15–21% of the sequences of MLL proteins, leaving the vast majority of these proteins uncharacterized both structurally and functionally.

In a previous work [26] we suggested that the disordered regions of HKMTs may harbor so far unrecognized interaction sites, adding more layers of the regulation of their activity. Based on the observation that many lncRNAs are involved in processes governed by HKMTs, we hypothesized that lncRNA binding might be one of the functions of these regions.

Since multiple evidence point in the direction that leukemic processes are fundamentally affected by lncRNAs and MLL complexes are involved in this regulation, we concentrated on MLL proteins. Taken the analogy of the PRC2 complex, where more than one complex subunits are capable of lncRNA binding, we aimed at testing the ability of MLL4 to bind different RNA molecules.

2. Results

2.1. *In Silico Analysis of the RNA Binding Capacity of MLL Proteins*

As a first step, we mapped the predicted RNA binding motifs on the sequence of four MLL proteins. We used DisoRDPbind, an RNA interaction prediction tool specifically designed to find RNA interaction sites in the disordered regions of proteins. Results shown in Table 1 indicate that all MLL proteins contain several putative RNA interaction motifs in their disordered regions. These regions are found at various positions in the proteins and vary in length from a couple of amino acids to almost a hundred residues, suggesting that RNA binding might be a common feature in MLL proteins.

Table 1. Predicted RNA binding regions in the disordered regions of Mixed Lineage Leukemia (MLL) proteins (aa positions).

MLL1	MLL2	MLL3	MLL4
296–327	84–107	1068–1079	1559–1567
348–408	184–234	1678–1695	3526–3581
415–418	241–244	1701–1709	3899–3983
1155–1194	536–560	1715–1737	4960–5014
1977–1992	783–806	2406–2409	5147–5165
3854–3861	820–828	3052–3073	5227–5251
	1753–1778	3246–3250	
	2600–2616	3394–3427	
	2685–2709	4330–4356	
		4514–4524	
		4586–4625	

A comparison with our earlier studies [26] revealed that two conserved disordered binding sites (residues 3537–3545 and 3560–3567) reside within one of the predicted RNA binding regions (residues 3526–3581, Figure 1A) of MLL4, underlining the reliability of the predictions. This region also harbors several cancer-related point mutations, two of them corresponding to a predicted binding site at positions 3560 (D-N) and 3561 (A-D). All these evidences point to the physiological importance of this protein region, making its structural and functional study worthwhile. ANCHOR prediction [27] shows that within the C-terminal border of the predicted RNA binding region there is a region with a strong tendency of the protein chain to form protein-protein interactions (residues 3597–3613, Figure 1A) that corresponds to a run of 14 glutamine residues. Since polyQ repeats in RNA binding proteins have been linked to protein-RNA droplet formation [28], this raises the intriguing possibility of granule formation potency of this segment. Therefore, we chose to test the RNA binding capacity of the MLL4 region between residues 3500–3630 (Figure 1A). As an internal control, another disordered region with no predicted RNA or protein binding sites was selected between residues 4210–4280 of MLL4 (Figure 1D).

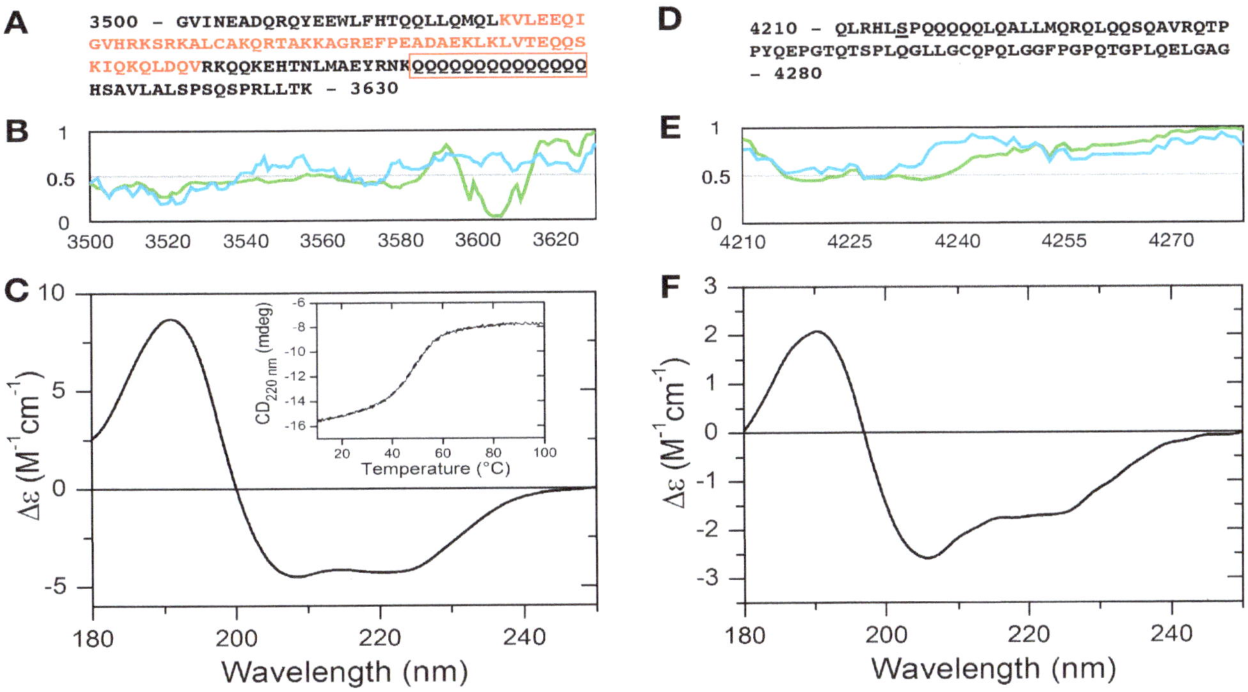

Figure 1. Structural characterization of the MLL4 regions. Sequences of $MLL4_{3500-3630}$ (**A**) and $MLL4_{4210-4280}$ (**D**). Predicted RNA binding region is indicated by red letters and the polyQ stretch is framed with red. IUPRed (blue) and Anchor (green) prediction of $MLL4_{3500-3630}$ (**B**) and $MLL4_{4210-4280}$ (**E**). Residues having an IUPred score above 0.5 are considered to be disordered, while residues with an Anchor score below 0.5 constitute predicted binding sites. Far-UV CD spectra of $MLL4_{3500-3630}$ (**C**) and $MLL4_{4210-4280}$ (**F**). Inset: temperature-dependent changes in the structure of $MLL4_{3500-3630}$ as observed by monitoring the changes in the absorbance at 220 nm.

As for binding RNAs, we opted to test two different lncRNA constructs, both having been reported to play a role in leukemias. The first is HOTAIR, that has the ability to bind EZH2 (PRC2). The 5′ 300 nucleotides of HOTAIR are thought to mediate its binding to PRC2 complex subunits, but the latest annotation in the NCBI database contains an additional 140 bases at the beginning of HOTAIR sequence, compared to the one reported earlier. Therefore, we prepared the longer version ($HOTAIR_{440}$) that encompasses the 300 nucleotides already known to be involved in protein-RNA interactions and also the nucleotides that has not been studied yet. Since there is no information available about the region of MEG3 that is able to bind proteins, we used the full length MEG3 for our experiments.

2.2. Secondary Structure of $MLL4_{3500-3630}$ and $MLL4_{4210-4280}$

Disorder prediction profiles (Figure 1B,E) indicated that both protein regions have a significant disorder tendency. Disorder profile of $MLL4_{3500-3630}$ indicates a rather ambiguous disorder state, with prediction scores fluctuating around the 0.5 limit between ordered and disordered states. This disorder prediction might indicate a disordered region that has an elevated tendency to fold or a relatively unstable folded segment as well. Far-UV CD measurements revealed that $MLL4_{3500-3630}$ has a helical structure in isolation (Figure 1C). The CD spectrum of this region of MLL4 showed a typical alpha helical conformation with a pronounced double minimum at 208 and 220 nm. Secondary structure content calculation using the BeStSel algorithm [29,30] gave an α-helix content of ~36.2%, while another ~36% of the secondary structure content was characterized as "Others", which mainly corresponds to the disordered structure. Thermal unfolding of the observed helical structure was followed by gradually heating the sample to 100 °C while recording the absorbance at 220 nm (Figure 1C inset). The melting curve indicated a cooperative unfolding of the structure with a melting point of 48 °C. The CD spectrum of the thermal denatured state is shown in Supplementary Figure S1, demonstrating a complete loss of structure at high temperatures.

$MLL4_{4210-4280}$ has a more pronounced disorder tendency, as demonstrated by the IUPred profile and is devoid of any predicted ANCHOR binding sites (Figure 1E). Its sequence contains a significant portion of glutamines (Figure 1D), but it does not contain Q stretches longer than 4 residues. Far-UV CD measurements confirmed the disorder predictions, indicating that the protein is mostly disordered in solution, with a considerable α-helical tendency. Secondary structure calculations gave a result of 16% α-helix and ~45% "Others" content, underlining that this segment of MLL4 is not fully disordered and contrary to interaction site predictions, might be involved in molecular recognition.

2.3. RNA Binding of $MLL4_{3500-3630}$ and $MLL4_{4210-4280}$

Microscale thermophoresis measurements were performed to characterize the RNA binding of the expressed protein regions. We used two lncRNA constructs, $HOTAIR_{440}$, a segment of HOTAIR that contains the region involved in binding to EZH2 [31], MEG3, a lncRNA involved in leukemias [32] and a 50 nt long RNA with random nucleotide sequence. Contradicting to the lack of predicted binding sites, $MLL4_{4210-4280}$ showed a relatively strong binding to $HOTAIR_{440}$ with an apparent Kd of 13.05 μM (Figure 2A), while the negative control Thymosin beta 4 (Tβ4) did not bind to the RNA, showing any sign of interaction at only the highest concentrations applied.

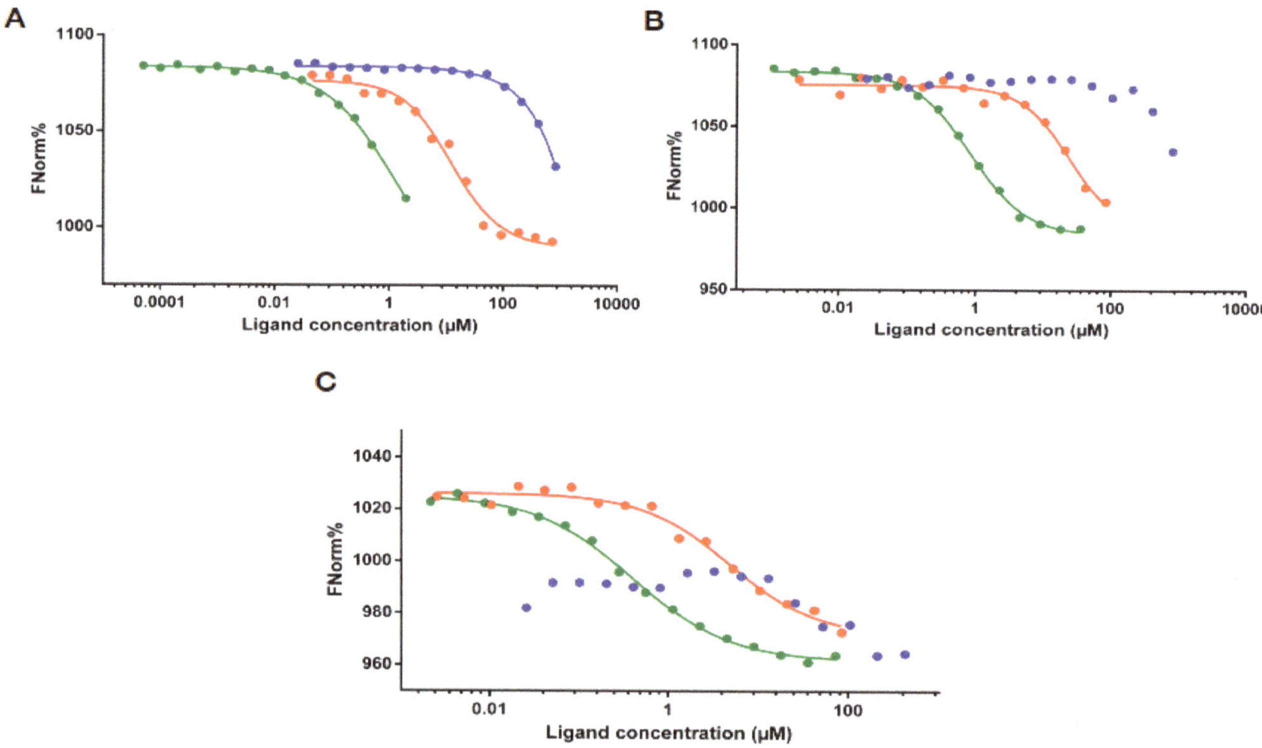

Figure 2. RNA binding detected by microscale thermophoresis. MST binding curves of $MLL4_{3500-3630}$ (green), $MLL4_{4210-4280}$ (red) and thymosin beta 4 (blue) to different RNAs: $HOTAIR_{440}$ (**A**), MEG3 (**B**) and 50 nt RNA (**C**).

In the case of $MLL4_{3500-3630}$, saturation of the reaction could not be reached because of marked aggregation above 1:20 RNA:protein ratio (Supplementary Figure S2) but using the T-jump values of the MST measurement (Supplementary Figure S3) an approximate binding constant of 0.1 μM could be determined. The appearance of large particles in the solution, generally considered to be aggregates, is indicated by a "wavy" MST curve and a randomly fluctuating normalized fluorescent percentage as shown on Supplementary Figures S2 and S5. The observed aggregation was dependent on the RNA species, since it was not seen with either of the other tested RNAs (Figure 3B,C), or with a shorter, 300 nt long HOTAIR construct (Supplementary Figure S4). The $HOTAIR_{300}$ construct overlaps with $HOTAIR_{440}$ in the 3′ 300 nucleotides but lacks the first 140 nucleotides of the latter. This shorter HOTAIR construct bound to $MLL4_{3500–3630}$ with a Kd of 0.97 μM, with no sign of irregular behavior. Centrifugation (15 min at 13,000× *g*) of the samples resulted in the loss of fluorescent signal in a protein concentration-dependent manner (Supplementary Figure S5), indicating a formation of structures containing both RNA and protein. Such phenomenon was not observed with $MLL4_{4210–4280}$, or Tβ4 upon mixing them with $HOTAIR_{440}$, even at significantly higher protein concentrations than $MLL4_{3500–3630}$. Also, $MLL4_{3500–3630}$ did not show aggregation-prone behavior in the absence of RNA.

As we experienced no anomaly in the behavior of $MLL4_{3500–3630}$ when titrated to MEG3, determination of a binding constant was straightforward for this interaction. As shown in Figure 2B, affinity to MEG3 of this region of MLL4 was higher than that of $MLL4_{4210–4280}$. The Kd of $MLL4_{3500–3630}$ binding to MEG3 was calculated to be 0.722 μM, while Kd calculation for $MLL4_{4210–4280}$ was not reliable since saturation of the reaction could not be reached throughout the protein concentration range tested. Tβ4 did not show significant affinity to MEG3, resulting in a failure of binding curve fitting.

To check for any specificity of binding that the expressed MLL4 regions may possess, we also tested a physiologically non-relevant 50 nt RNA construct. Binding curves presented in Figure 2C indicate that both $MLL4_{3500–3630}$ and $MLL4_{4210–4280}$ are capable of binding to this RNA species, but with a remarkably lower affinity than to the lncRNA constructs, while Tβ4 could not bind to it at all. The extended shape of the binding curve and the absence of saturation in the case of both MLL4 constructs indicate weak binding that resulted in an inability to reliably determine the binding constants. Nevertheless, $MLL4_{3500–3630}$ still displayed a stronger affinity towards the RNA than $MLL4_{4210–4280}$.

Electrophoretic Mobility Shift Assay (EMSA) experiments confirmed the findings of the MST measurements (Figure 3) as both MLL4 regions caused a significant change in RNA mobility in the case of $HOTAIR_{440}$ and MEG3 (Figure 3A,B) RNAs. This shift was drastically less pronounced with the 50 nt RNA sample (Figure 3C), resulting only in a minor weakening of the RNA signal in the lane with the highest protein concentration. This observation corresponds to the outcome of the MST experiments, underlining the existence of a certain level of specificity in the RNA recognition by these two MLL4 regions. The negative control Tβ4 failed to cause any visible change in the RNA mobility, indicating a lack of interaction with any of the tested RNAs. Competitive RNA binding (Figure 3, compare the 3rd and 5th lanes) demonstrated that the observed shift in mobility was indeed a result of RNA-protein interaction, since the shift could be prevented at least to some extent by adding excess unlabeled RNA to the reaction mixtures.

The anomalous behavior of the $MLL4_{3500–3630}$:$HOTAIR_{440}$ interaction observed in MST was seen in the EMSA experiments as well, since at high protein:RNA ratios the samples obtained a highly viscous quality and completely remained in the wells during the electrophoretic run. Successful experiments could only be carried out by lowering the applied protein concentration, but the interaction was clearly observable even under these circumstances.

In all of the tested interactions, $MLL4_{3500–3630}$, which contains a predicted RNA binding region presented higher affinities to RNAs than the other MLL4 segment, indicating the validity of the prediction. On the other hand, binding of $MLL4_{4210–4280}$ could also be detected in all cases, raising the possibility of the existence of RNA binding sequences differing from the already described interaction motifs. EZH2, a known RNA binding HKTM also interacts with RNAs through a region [17] that has no recognizable RNA binding sequence, emphasizing our lack of complete knowledge of the sequential determinants of protein-RNA interactions.

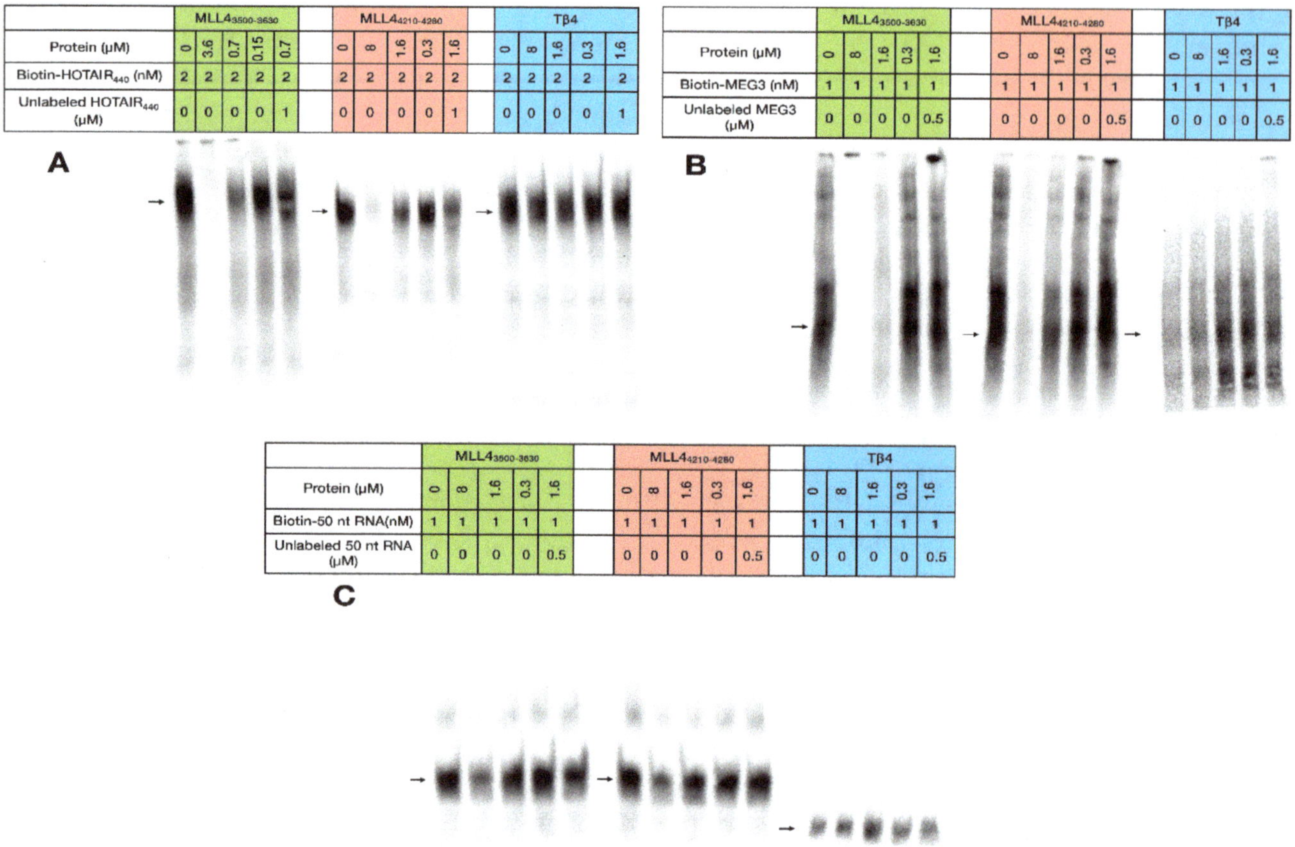

Figure 3. Electrophoretic Mobility Shift Assay. Interaction of MLL4$_{3500-3630}$, MLL4$_{4210-4280}$ and Tβ4 with HOTAIR$_{440}$ (**A**), MEG3 (**B**) and 50 nt RNA (**C**). For easier understanding, the coloring scheme of Figure 2 is followed (MLL4$_{3500-3630}$: green MLL4$_{4210-4280}$: red, Tβ4: blue). Free RNA is indicated by arrows.

3. Discussion

Histone methylation is one of the most studied and best-characterized histone modifications that drive the regulation of complete genetic programs in the cells. However, many details of the regulation and targeting of the enzyme complexes mediating histone methylation remain elusive and a subject of debate [23]. One possible regulatory pathway is represented by the ability of certain HKMT complexes to bind different lncRNAs that serve as a targeting platform, bridging transcription factors and HKMT complexes [20,33] at the promoter regions of target genes. PRC2 is one example where it was shown by multiple experiments that it's binding to different lncRNAs results in different physiological outcomes [34]. lncRNAs are involved in many other processes connected to histone modification and there are examples in the literature of direct interaction between lncRNAs and histone modifier complexes [4,22]. Experimental evidence supports the direct binding of WDR5, a canonical MLL complex subunit, to different lncRNAs in cells [22] indicating the involvement of lncRNAs in the regulation of MLL complexes. Taken the analogy of the PRC2, where multiple subunits are shown to be involved in lncRNA binding (Figure 4A) [15], we hypothesized that MLL proteins might also interact with lncRNAs. This hypothesis was supported by our earlier bioinformatics studies that suggested the existence of several interaction sites in the so far uncharacterized, mostly disordered regions of HKMTs [26] and our prediction presented here that the disordered segments of MLL proteins contain several putative RNA binding sequences. We chose to test the RNA binding capability of one such region of MLL4 that also contains a polyQ stretch and is affected by mutations in different cancers. As an internal control, we also tested a different region of MLL4 that contains no such predicted RNA interaction site.

Our expectation was that the isolated small regions of the MLL4 protein would bind RNAs in a nonspecific manner, such as was observed for the isolated PRC2 complex components [34]. Surprisingly, we found that $MLL4_{4210–4280}$ bound MEG3 stronger than $HOTAIR_{440}$ or the 50 nt random RNA, even though the determination of the exact Kd-s was not successful in all cases.

More interesting was the behavior of the $MLL4_{3500–3630}$ region that showed dramatically different behavior with the different RNAs. Binding to MEG3 gave a Kd of 0.722 μM, while the binding to the 50 nt random RNAs proved to be so weak that a Kd calculation was not successful. Binding to $HOTAIR_{440}$ seemed to be the strongest with an apparent Kd of 0.1 μM, but it led to the aggregation of the protein-RNA complex. The aggregation was dependent on protein-RNA ratio and could be detected through a wide protein concentration range. The same aggregation could not be observed with a shorter HOTAIR construct that consisted of 300 bases (Supplementary Figure S3). The fact that we could not induce such aggregation by the addition of MEG3, which is much longer than $HOTAIR_{440}$, points to specific recognition rather than a side-effect of RNA length. We also observed the aggregation at low protein concentrations, but only in the presence of an appropriate amount of $HOTAIR_{440}$, indicating that the process is not driven by the protein in itself and is not a derivative of sample preparation errors.

It has been recently revealed that many proteins can go through liquid-liquid phase separation when interacting with RNAs, leading to the formation of membraneless organelles that have a significant importance in cellular processes [35]. Experimental evidence supports the involvement of polyQ regions of proteins in the RNA mediated phase separation [28], sometimes in an RNA secondary structure-dependent manner [36]. Since $MLL4_{3500–3630}$ sequence contains 22.9% glutamine residues and a continuous run of 15 glutamines (Figure 1A), it is not unfounded to speculate that this specific region plays a role in the observed anomaly but the fact that it only occurs with one of the tested RNA constructs, indicates that the process is coordinated by the RNA itself. One possibility is that the longer HOTAIR construct contains more than one binding sites for $MLL4_{3500–3630}$, thus facilitating the formation of higher order protein-RNA structures. Alternatively, $HOTAIR_{440}$ may have the ability to form secondary structures not found in $HOTAIR_{300}$ or MEG3, which would also provide an explanation for the different behavior of the three systems. As MLL4 is the only HKMT that contains long polyglutamine repeat stretches [26], phase separation might be a regulatory step specific for this protein. Therefore, it is certainly promising to investigate this peculiar phenomenon in more detail.

Since both tested lncRNAs are implicated in different cancers [5,37,38] involving leukemias, our finding that MLL4 has a capacity to bind them raises the possibility that lncRNAs play a role in MLL/COMPASS complex targeting and regulation to a larger extent than currently recognized.

Although cellular experiments are necessary to prove the validity of the observed interactions, our findings provide the first insights into the structure and function of two regions of MLL4 that have been uncharacterized so far. We were able to show that these regions are capable of RNA binding and may be involved in the lncRNA mediated regulation of the MLL4 complexes. Based on our results, we suggest that and MLL4 complexes utilize different regions on their surface to bind lncRNAs (Figure 4B), similarly to the way PRC2 subunits take part in lncRNA binding. As it was shown that lncRNA binding to WDR5 increases the dwelling time of the protein on the chromatin surface [22], binding of the same RNA to MLL4 might facilitate and accelerate the assembly of a functional methyltransferase complex. Since lncRNAs are large molecules that can adopt various secondary structures and interact with many different partners simultaneously, it is plausible to speculate that a specific and high-affinity interaction can be achieved by the combination of different binding sites distributed along the large surfaces of multi-subunit complexes. Given the central role of histone modifications in gene regulation, it is essential to understand the mechanisms that regulate this process. Mounting evidence supports the involvement of lncRNAs in the coordination of histone modifying enzymes but the exact molecular details of their interactions with proteins are yet to be discovered. Recognizing the importance of the disordered/structurally uncharacterized regions of HKMTs in these interactions might be the first step towards a more complete picture regarding the regulation of histone methylation.

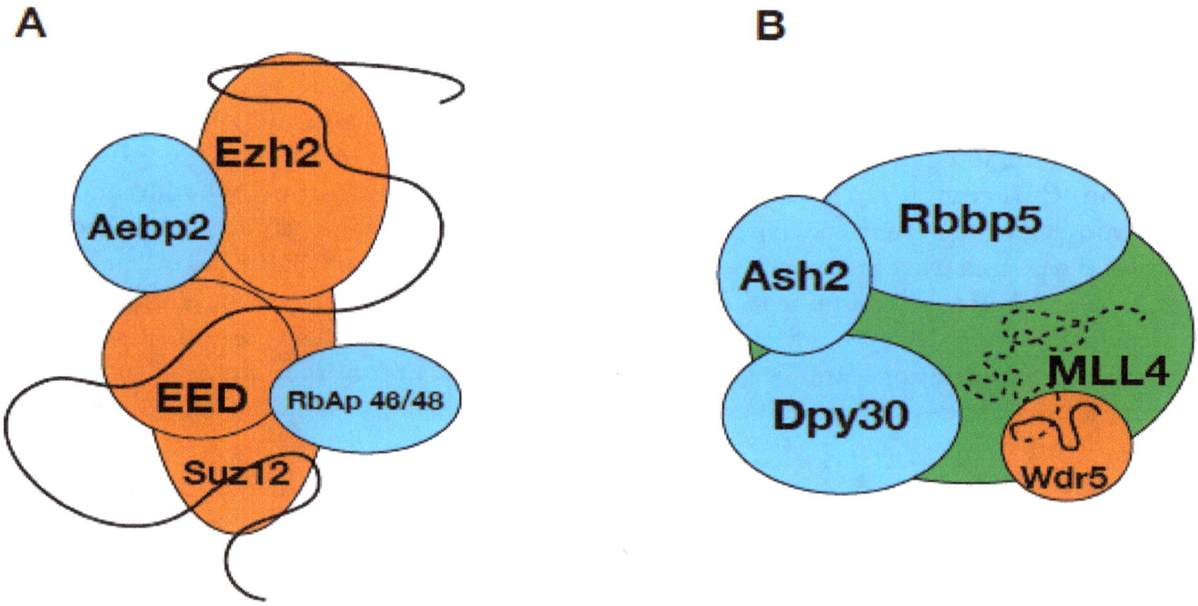

Figure 4. lncRNA binding of PRC2 and MLL4/COMPASS complex. Schematic representation of the PRC2 (**A**) and MLL4/COMPASS (**B**) complexes, where the known RNA binding subunits are shown in orange and the suggested lncRNA binding subunit MLL4 is green. Subunits currently not known to be involved in lncRNA binding are blue and the lncRNA is represented by a black line. Suggested lncRNA-MLL4 interaction is indicated by dashed line.

4. Materials and Methods

4.1. Bioinformatics Analysis

Disorder and disordered binding site predictions were performed with the IUPred2A online prediction tool (https://iupred2a.elte.hu/) [39] which incorporates the IUPred and Anchor predictors. RNA binding regions located in disordered regions were predicted using the DisoRDPbind tool (http://biomine.cs.vcu.edu/servers/DisoRDPbind/) [40]. Cancer-related single nucleotide polymorphisms in the long conserved IDR regions were collected from the BioMuta v2.0 [41] and COSMIC databases [42].

4.2. Accession Numbers

HOTAIR: Gene ID: 100124700
MEG3: Gene ID: 55384
MLL1: Uniprot: Q03164
MLL2: Uniprot: Q9UMN6
MLL3: Uniprot: Q8NEZ4
MLL4: Uniprot: O14686

4.3. Overexpression and Purification of MLL4 Protein Regions

The same methods of protein overexpression and purification were used for both protein constructs, $MLL4_{3500–3630}$ and $MLL4_{4210–4280}$. DNA sequences coding for each protein were cloned into pET22b cloning vector. Induction was done for 4 h at 28 °C by 0.1 M IPTG, cells were pelleted by centrifugation (4000 rpm, 20 min, 4 °C) then lysed by sonication in lysis buffer (50 mM Tris, 200 mM NaCl, 0.5% Triton X-100 pH 8.0 and EDTA-free SIGMAFAST Protease

Inhibitor Cocktail Tablets), cell debris was removed by centrifugation (12,100 rpm, 40 min, 4 °C). The supernatant was filtered through 0.2 μm nitrocellulose filter then purified over HisTrap HP column on an AKTA Explorer system using a gradient elution of two buffers (Buffer A: 20 mM imidazole, 200 mM NaCl, 20 mM Tris. pH 7.5. Buffer B: 1 M imidazole, 200 mM NaCl, 20 mM Tris, pH 7.5). Representative purification results are shown on Supplementary Figure S7. The mostly disordered nature of the $MLL4_{4210-4280}$ region was highlighted by its appearance at a larger size than its actual molecular weight (17 kDa vs. 7 kDa). Elution fractions containing sufficiently pure proteins were dialyzed against distilled water then lyophilized and stored at −20 °C. Lyophilized proteins were dissolved before use in ultrapure water or the appropriate assay buffer. The identity of the purified proteins was confirmed by mass spectrometry.

4.4. RNA Preparation

HOX transcript antisense RNA (HOTAIR):

$HOTAIR_{300}$ (140–440 nt) and $HOTAIR_{440}$ (1–440 nt) DNA sequences cloned into pEX-A128 vector were purchased from Eurofins Genomics (Ebersberg, Germany). After 2 h digestion with EcoRV restriction enzyme at 37 °C, the gel-purified, linearized DNA templates were used to synthesize RNA by in-vitro transcription.

Maternally Expressed 3 (MEG3) lncRNA:

pCI-MEG3 was a gift from Anne Klibanski (Addgene plasmid #44727, Watertown, MA, USA) [43]. Primers to obtain the DNA template for in vitro transcription were as follows:
T7 RNA promoter region followed by:
MEG3 forward primer:
TAATACGACTCACTATAGGGGCAGAGAGGGAGCGCGCCTTGG
MEG3 reverse primer:
GATATCTTTTTGTTAAGACAGGAAACACATTTATTGAGAGC

50 nt RNA:

50 nucleotide RNA was an artificial randomized RNA sequence.
DNA templates were T7 promoter region followed by:
50 nt forward oligo:
TAATACGACTCACTATAGAAGAATGGCCTCGCGGAGGCATGCGTCATGCTAGCGTGCGGGGTACTCTT and
50 nt reverse oligo:
AAGAGTACCCCGCACGCTAGCATGACGCATGCCTCCGCGAGGCCATTCTTCTATAGTGAGTCGTATTA
Transcribed RNA:
GAAGAAUGGCCUCGCGGAGGCAUGCGUCAUGCUAGCGUGCGGGUACUCUU

All primers and oligonucleotides were purchased from Sigma-Aldrich Ltd. (St. Louis, MO, USA).

Tested RNAs were synthesized by in vitro transcription carried out with New England BioLabs HiScribe™ T7 Quick High Yield RNA Synthesis Kit (Ipswich, MA, USA). Fluorescein-labelled, single-stranded RNA probes were generated by using Roche (Basel, Switzerland) Fluorescein RNA Labeling Mix (11685619910) and NEB 10× T7 reaction buffer (#B2041A). After transcription, remaining DNA templates were eliminated with DNaseI treatment. RNA sample purification was carried out using Macherey-Nagel NucleoSpin® RNA Clean-up XS Kit (Düren, Germany). The quality and intactness of the purified transcription products were analysed by native and formaldehyde agarose gel electrophoresis.

Biotinylation of the RNAs was performed using Pierce™ RNA 3′ End Biotinylation Kit (Cat. Number 20160, Thermo Fisher Scientific, Waltham, UK) according to the instructions of the manufacturer. Overnight incubation at 16 °C was applied for the ligation of the biotin label. Final RNA concentrations were determined using NanoDrop™ 1000 Spectrophotometer (Thermo Fisher Scientific, Waltham, UK).

Purified RNAs were stored −80 °C until usage in the presence of RNAINH-RO Roche Protector RNase Inhibitor (20U).

4.5. Far-UV CD Measurements

CD measurements were performed in quartz cells of 0.1 mm pathlengths using a Jasco J-810 (Jasco, Tokyo, Japan) spectropolarimeter. Far-UV CD spectra were recorded in the range of 180–260 nm with a scanning speed of 20 nm/min, bandwidth of 1 nm and integration time of 4 s. 6 scans were accumulated. Thermal denaturation was recorded in a 1 mm cell at 220 nm from 10 to 100 °C with scanning rate of 120 °C/h. The temperature was controlled using a PTE Peltier unit. The thermal denaturation profile was fitted according to the Gibbs-Helmholtz equation assuming a two-state model, which is represented by a sigmoidal curve [44]. CD spectra were quantitatively analyzed by the BeStSel method [29,30].

4.6. Microscale Thermophoresis

RNA-protein binding assays were carried out on a Microscale Thermophoresis system (Monolith NT. 115 from NanoTemper Technologies, München, Germany). Standard treated capillaries (Cat. Number: MO-K002) were used for measurements. Instrument settings are presented in Table 2.

Table 2. Instrument settings for MST.

Title	LED Power (%)	MST Power (%)	Before MST (s)	MST on (s)	After MST (s)	Delay (s)
Round 1	10 or 40	20	5	30	5	25
Round 2	10 or 40	40	5	30	5	25

Normalized fluorescence values after 1.25 s after turning on the IR laser were used as T-jump values.

RNA concentrations were set to give an initial raw fluorescence between 300 and 1000 counts and varied between 30 and 100 nM. All experiments were done at room temperature. DEPC-treated PBS buffer containing 0.05% NP-40 was used as assay buffer.

4.7. Electrophoretic Mobility Shift Assay (EMSA)

LightShift® Chemiluminescent RNA EMSA Kit (Thermo Scientific, Cat. No. 20158, Thermo Fisher Scientific, Waltham, UK) was used for the EMSA experiments. Assay control was performed according to the instructions of the manufacturer with the control reaction provided with the kit. In short, 6.25 nM biotin-labeled IRE RNA was incubated with 2 μg of cytosolic liver extract with or without 1 μM of unlabeled IRE RNA. The result of the assay control is presented on Supplementary Figure S6. Binding, electrophoresis and detection of the tested RNAs with the proteins were carried out following the protocol of the kit. Briefly, proteins of varying concentrations were incubated with 1 or 2 nM of RNAs for 30 min at room temperature, then loaded on 4 or 6% native polyacrylamide gels. RNA was transferred to nitrocellulose membranes using Trans-Blot® Turbo™ Transfer System (Bio-Rad, Hercules, CA, USA) and crosslinked to the membrane by UV-light crosslinking. After proper washing and blocking, biotin labeled RNA was detected by chemiluminescence using Streptavidin-Horseradish Peroxidase Conjugate.

Author Contributions: Conceptualization, A.T.; Data curation, B.S. (Beáta Szabó), E.S. and B.S. (Bálint Szeder); Formal analysis, J.K.; Funding acquisition, J.K., L.B. and A.T.; Investigation, B.S. (Beáta Szabó), N.M., R.A. and J.K.; Methodology, B.S. (Beáta Szabó) and A.T.; Project administration, B.S. (Beáta Szabó) and A.T.; Resources, B.S. (Beáta Szabó) and N.M.; Software, E.S.; Supervision, L.B. and A.T.; Validation, J.K. and L.B.;

Visualization, A.T.; Writing—original draft, A.T.; Writing—review & editing, B.S. (Beáta Szabó), N.M., R.A., E.S., J.K., B.S. (Bálint Szeder) and L.B.

Abbreviations

EMSA	Electrophoretic mobility shift assay
EZH2	Enhancer of zeste homolog 2
HKMT	Histone lysin methyltransferase
HOTAIR	HOX transcript antisense RNA
lncRNA	Long non-coding RNA
MEG3	Maternally Expressed 3
MLL	Mixed lineage leukemia
MST	Microscale thermophoresis
PRC2	Polycomb repressive complex
WDR5	WD repeat-containing protein 5

References

1. Iyer, M.K.; Niknafs, Y.S.; Malik, R.; Singhal, U.; Sahu, A.; Hosono, Y.; Barrette, T.R.; Prensner, J.R.; Evans, J.R.; Zhao, S.; et al. The landscape of long noncoding RNAs in the human transcriptome. *Nat Genet.* **2015**, *47*, 199–208. [CrossRef] [PubMed]
2. Bartonicek, N.; Maag, J.L.V.; Dinger, M.E. Long noncoding RNAs in cancer: Mechanisms of action and technological advancements. *Mol. Cancer* **2016**, *15*, 43. [CrossRef] [PubMed]
3. Yang, F.; Deng, X.; Ma, W.; Berletch, J.B.; Rabaia, N.; Wei, G.; Moore, J.M.; Filippova, G.N.; Xu, J.; Liu, Y.; et al. The lncRNA Firre anchors the inactive X chromosome to the nucleolus by binding CTCF and maintains H3K27me3 methylation. *Genome Biol.* **2015**, *16*, 52. [CrossRef] [PubMed]
4. Khalil, A.M.; Guttman, M.; Huarte, M.; Garber, M.; Raj, A.; Morales, D.R.; Thomas, K.; Presser, A.; Bernstein, B.E.; Van Oudenaarden, A.; et al. Many human large intergenic noncoding RNAs associate with chromatin-modifying complexes and affect gene expression. *Proc. Natl. Acad. Sci. USA* **2009**, *106*, 11667–11672. [CrossRef] [PubMed]
5. Heery, R.; Finn, S.P.; Cuffe, S.; Gray, S.G. Long Non-Coding RNAs: Key Regulators of Epithelial-Mesenchymal Transition, Tumour Drug Resistance and Cancer Stem Cells. *Cancers* **2017**, *9*, 38. [CrossRef] [PubMed]
6. Liu, Z.; Chen, Z.; Fan, R.; Jiang, B.; Chen, X.; Chen, Q.; Nie, F.; Lu, K.; Sun, M. Over-expressed long noncoding RNA HOXA11-AS promotes cell cycle progression and metastasis in gastric cancer. *Mol. Cancer* **2017**, *16*, 82. [CrossRef] [PubMed]
7. Bhan, A.; Deb, P.; Shihabeddin, N.; Ansari, K.I.; Brotto, M.; Mandal, S.S. Histone methylase MLL1 coordinates with HIF and regulate lncRNA HOTAIR expression under hypoxia. *Gene* **2017**, *629*, 16–28. [CrossRef] [PubMed]
8. Chen, S.; Liang, H.; Yang, H.; Zhou, K.; Xu, L.; Liu, J.; Lai, B.; Song, L.; Luo, H.; Peng, J.; et al. Long non-coding RNAs: The novel diagnostic biomarkers for leukemia. *Environ. Toxicol. Pharmacol.* **2017**, *55*, 81–86. [CrossRef] [PubMed]
9. Guo, G.; Kang, Q.; Zhu, X.; Chen, Q.; Wang, X.; Chen, Y.; Ouyang, J.; Zhang, L.; Tan, H.; Chen, R.; et al. A long noncoding RNA critically regulates Bcr-Abl-mediated cellular transformation by acting as a competitive endogenous RNA. *Oncogene* **2015**, *34*, 1768–1779. [CrossRef] [PubMed]
10. Wang, P.; Ren, Z.; Sun, P. Overexpression of the long non-coding RNA MEG3 impairs in vitro glioma cell proliferation. *J. Cell. Biochem.* **2012**, *113*, 1868–1874. [CrossRef] [PubMed]
11. Zhang, L.; Xu, H.-G.; Lu, C. A novel long non-coding RNA T-ALL-R-LncR1 knockdown and Par-4 cooperate to induce cellular apoptosis in T-cell acute lymphoblastic leukemia cells. *Leuk. Lymphoma* **2014**, *55*, 1373–1382. [CrossRef] [PubMed]

12. Zhou, X.; Yuan, P.; Liu, Q.; Liu, Z. LncRNA MEG3 Regulates Imatinib Resistance in Chronic Myeloid Leukemia via Suppressing MicroRNA-21. *Biomol. Ther.* **2017**, *25*, 490–496. [CrossRef] [PubMed]
13. Mer, A.S.; Lindberg, J.; Nilsson, C.; Klevebring, D.; Wang, M.; Grönberg, H.; Lehmann, S.; Rantalainen, M. Expression levels of long non-coding RNAs are prognostic for AML outcome. *J. Hematol. Oncol.* **2018**, *11*, 52. [CrossRef] [PubMed]
14. Davidovich, C.; Cech, T.R. The recruitment of chromatin modifiers by long noncoding RNAs: Lessons from PRC2. *RNA* **2015**, *21*, 2007–2022. [CrossRef] [PubMed]
15. Cifuentes-Rojas, C.; Hernandez, A.J.; Sarma, K.; Lee, J.T. Regulatory interactions between RNA and polycomb repressive complex 2. *Mol. Cell* **2014**, *55*, 171–185. [CrossRef] [PubMed]
16. Davidovich, C.; Wang, X.; Cifuentes-Rojas, C.; Goodrich, K.J.; Gooding, A.R.; Lee, J.T.; Cech, T.R. Toward a Consensus on the Binding Specificity and Promiscuity of PRC2 for RNA. *Mol. Cell* **2015**, *57*, 552–558. [CrossRef] [PubMed]
17. Kaneko, S.; Li, G.; Son, J.; Xu, C.F.; Margueron, R.; Neubert, T.A.; Reinberg, D. Phosphorylation of the PRC2 component Ezh2 is cell cycle-regulated and up-regulates its binding to ncRNA. *Genes Dev.* **2010**, *24*, 2615–2620. [CrossRef] [PubMed]
18. Zhang, X.D.; Huang, G.W.; Xie, Y.H.; He, J.Z.; Guo, J.C.; Xu, X.E.; Liao, L.D.; Xie, Y.M.; Song, Y.M.; Li, E.M.; et al. The interaction of lncRNA EZR-AS1 with SMYD3 maintains overexpression of EZR in ESCC cells. *Nucleic Acids Res.* **2018**, *46*, 1793–1809. [CrossRef] [PubMed]
19. O'Leary, V.B.; Ovsepian, S.V.; Carrascosa, L.G.; Buske, F.A.; Radulovic, V.; Niyazi, M.; Moertl, S.; Trau, M.; Atkinson, M.J.; Anastasov, N. PARTICLE, a Triplex-Forming Long ncRNA, Regulates Locus-Specific Methylation in Response to Low-Dose Irradiation. *Cell Rep.* **2015**, *11*, 474–485. [CrossRef] [PubMed]
20. O'Leary, V.B.; Hain, S.; Maugg, D.; Smida, J.; Azimzadeh, O.; Tapio, S.; Ovsepian, S.V.; Atkinson, M.J. Long non-coding RNA PARTICLE bridges histone and DNA methylation. *Sci. Rep.* **2017**, *7*, 1790. [CrossRef] [PubMed]
21. Gomez, J.A.; Wapinski, O.L.; Yang, Y.W.; Bureau, J.F.; Gopinath, S.; Monack, D.M.; Chang, H.Y.; Brahic, M.; Kirkegaard, K. The NeST Long ncRNA Controls Microbial Susceptibility and Epigenetic Activation of the Interferon-γ Locus. *Cell* **2013**, *152*, 743–754. [CrossRef] [PubMed]
22. Yang, Y.W.; Flynn, R.A.; Chen, Y.; Qu, K.; Wan, B.; Wang, K.C.; Lei, M.; Chang, H.Y. Essential role of lncRNA binding for WDR5 maintenance of active chromatin and embryonic stem cell pluripotency. *Elife* **2014**, *3*, e02046. [CrossRef] [PubMed]
23. Sze, C.C.; Shilatifard, A. MLL3/MLL4/COMPASS Family on Epigenetic Regulation of Enhancer Function and Cancer. *Cold Spring Harb. Perspect. Med.* **2016**, *6*, a026427. [CrossRef] [PubMed]
24. Herz, H.M.; Mohan, M.; Garruss, A.S.; Liang, K.; Takahashi, Y.H.; Mickey, K.; Voets, O.; Verrijzer, C.P.; Shilatifard, A. Enhancer-associated H3K4 monomethylation by Trithorax-related, the Drosophila homolog of mammalian Mll3/Mll4. *Genes Dev.* **2012**, *26*, 2604–2620. [CrossRef] [PubMed]
25. Henikoff, S.; Shilatifard, A. Histone modification: Cause or cog? *Trends Genet.* **2011**, *27*, 389–396. [CrossRef] [PubMed]
26. Lazar, T.; Schad, E.; Szabo, B.; Horvath, T.; Meszaros, A.; Tompa, P.; Tantos, A. Intrinsic protein disorder in histone lysine methylation. *Biol. Dir.* **2016**, *11*, 30. [CrossRef] [PubMed]
27. Mészáros, B.; Simon, I.; Dosztányi, Z. Prediction of Protein Binding Regions in Disordered Proteins. *PLoS Comput. Biol.* **2009**, *5*, e1000376. [CrossRef] [PubMed]
28. Zhang, H.; Elbaum-Garfinkle, S.; Langdon, E.M.; Taylor, N.; Occhipinti, P.; Bridges, A.A.; Brangwynne, C.P.; Gladfelter, A.S. RNA Controls PolyQ Protein Phase Transitions. *Mol. Cell* **2015**, *60*, 220–230. [CrossRef] [PubMed]
29. Micsonai, A.; Wien, F.; Bulyáki, É.; Kun, J.; Moussong, É.; Lee, Y.H.; Goto, Y.; Réfrégiers, M.; Kardos, J. BeStSel: A web server for accurate protein secondary structure prediction and fold recognition from the circular dichroism spectra. *Nucleic Acids Res.* **2018**, *46*, W315–W322. [CrossRef] [PubMed]
30. Micsonai, A.; Wien, F.; Kernya, L.; Lee, Y.H.; Goto, Y.; Réfrégiers, M.; Kardos, J. Accurate secondary structure prediction and fold recognition for circular dichroism spectroscopy. *Proc. Natl. Acad. Sci. USA* **2015**, *112*, E3095–E3103. [CrossRef] [PubMed]
31. Tsai, M.C.; Manor, O.; Wan, Y.; Mosammaparast, N.; Wang, J.K.; Lan, F.; Shi, Y.; Segal, E.; Chang, H.Y. Long noncoding RNA as modular scaffold of histone modification complexes. *Science* **2010**, *329*, 689–693. [CrossRef] [PubMed]

32. Schwarzer, A.; Emmrich, S.; Schmidt, F.; Beck, D.; Ng, M.; Reimer, C.; Adams, F.F.; Grasedieck, S.; Witte, D.; Käbler, S.; et al. The non-coding RNA landscape of human hematopoiesis and leukemia. *Nat. Commun.* **2017**, *8*, 218. [CrossRef] [PubMed]
33. Battistelli, C.; Cicchini, C.; Santangelo, L.; Tramontano, A.; Grassi, L.; Gonzalez, F.J.; de Nonno, V.; Grassi, G.; Amicone, L.; Tripodi, M. The Snail repressor recruits EZH2 to specific genomic sites through the enrollment of the lncRNA HOTAIR in epithelial-to-mesenchymal transition. *Oncogene* **2016**, *36*, 942–955. [CrossRef] [PubMed]
34. Davidovich, C.; Zheng, L.; Goodrich, K.J.; Cech, T.R. Promiscuous RNA binding by Polycomb repressive complex 2. *Nat. Struct. Mol. Biol.* **2013**, *20*, 1250–1257. [CrossRef] [PubMed]
35. Banani, S.F.; Lee, H.O.; Hyman, A.A.; Rosen, M.K. Biomolecular condensates: Organizers of cellular biochemistry. *Nat. Rev. Mol. Cell Biol.* **2017**, *18*, 285–298. [CrossRef] [PubMed]
36. Langdon, E.M.; Qiu, Y.; Niaki, A.G.; McLaughlin, G.A.; Weidmann, C.; Gerbich, T.M.; Smith, J.A.; Crutchley, J.M.; Termini, C.M.; Weeks, K.M.; et al. mRNA structure determines specificity of a polyQ-driven phase separation. *Science* **2018**, *360*, 922–927. [CrossRef] [PubMed]
37. Li, Z.-Y.; Yang, L.; Liu, X.-J.; Wang, X.-Z.; Pan, Y.-X.; Luo, J.-M. The Long Noncoding RNA MEG3 and its Target miR-147 Regulate JAK/STAT Pathway in Advanced Chronic Myeloid Leukemia. *EBioMedicine* **2018**, *34*, 61–75. [CrossRef] [PubMed]
38. Zhang, Y.-Y.; Huang, S.-H.; Zhou, H.-R.; Chen, C.-J.; Tian, L.-H.; Shen, J.-Z. Role of HOTAIR in the diagnosis and prognosis of acute leukemia. *Oncol. Rep.* **2016**, *36*, 3113–3122. [CrossRef] [PubMed]
39. Mészáros, B.; Erdos, G.; Dosztányi, Z. IUPred2A: Context-dependent prediction of protein disorder as a function of redox state and protein binding. *Nucleic Acids Res.* **2018**, *46*, W329–W337. [CrossRef] [PubMed]
40. Peng, Z.; Kurgan, L. High-throughput prediction of RNA, DNA and protein binding regions mediated by intrinsic disorder. *Nucleic Acids Res.* **2015**, *43*, e121. [CrossRef] [PubMed]
41. Wu, T.J.; Shamsaddini, A.; Pan, Y.; Smith, K.; Crichton, D.J.; Simonyan, V.; Mazumder, R. A framework for organizing cancer-related variations from existing databases, publications and NGS data using a High-performance Integrated Virtual Environment (HIVE). *Database* **2014**, *2014*, bau022. [CrossRef] [PubMed]
42. Forbes, S.A.; Beare, D.; Gunasekaran, P.; Leung, K.; Bindal, N.; Boutselakis, H.; Ding, M.; Bamford, S.; Cole, C.; Ward, S.; et al. COSMIC: Exploring the world's knowledge of somatic mutations in human cancer. *Nucleic Acids Res.* **2014**, *43*, D805–D811. [CrossRef] [PubMed]
43. Zhou, Y.; Zhong, Y.; Wang, Y.; Zhang, X.; Batista, D.L.; Gejman, R.; Ansell, P.J.; Zhao, J.; Weng, C.; Klibanski, A. Activation of p53 by MEG3 non-coding RNA. *J. Biol. Chem.* **2007**, *282*, 24731–24742. [CrossRef] [PubMed]
44. Shih, P.; Holland, D.R.; Kirsch, J.F. Thermal stability determinants of chicken egg-white lysozyme core mutants: Hydrophobicity, packing volume, and conserved buried water molecules. *Protein Sci.* **1995**, *4*, 2050–2062. [CrossRef] [PubMed]

17

Decision-Tree Based Meta-Strategy Improved Accuracy of Disorder Prediction and Identified Novel Disordered Residues Inside Binding Motifs

Bi Zhao and Bin Xue *

Department of Cell Biology, Microbiology and Molecular Biology, School of Natural Sciences and Mathematics, College of Arts and Sciences, University of South Florida, Tampa, FL 33620, USA; bizhao@mail.usf.edu

* Correspondence: binxue@usf.edu

Abstract: Using computational techniques to identify intrinsically disordered residues is practical and effective in biological studies. Therefore, designing novel high-accuracy strategies is always preferable when existing strategies have a lot of room for improvement. Among many possibilities, a meta-strategy that integrates the results of multiple individual predictors has been broadly used to improve the overall performance of predictors. Nonetheless, a simple and direct integration of individual predictors may not effectively improve the performance. In this project, dual-threshold two-step significance voting and neural networks were used to integrate the predictive results of four individual predictors, including: DisEMBL, IUPred, VSL2, and ESpritz. The new meta-strategy has improved the prediction performance of intrinsically disordered residues significantly, compared to all four individual predictors and another four recently-designed predictors. The improvement was validated using five-fold cross-validation and in independent test datasets.

Keywords: meta strategy; dual threshold; significance voting; decision tree based artificial neural network; protein intrinsic disorder

---

1. Introduction

Intrinsically disordered proteins (IDPs) and intrinsically disordered regions (IDRs) play critical functions in many biological processes [1–7]. Among all the possible molecule mechanisms for the functions of IDPs/IDRs, a major one is that IDPs/IDRs physically interact with their partners through either conformational search or induced fit [8–10]. Eventually, due to them having structural flexibility, IDPs/IDRs may bind to the partners with low-affinity but high-specificity [11–15], and thus regulate the downstream biological processes. Clearly, to characterize the dynamic process of the interaction and the mechanism of regulation, the exact locations of those intrinsically disordered amino acids (IDAAs) involved in the interaction need to be determined. However, high-accuracy experimental methods for the detection of IDPs/IDRs/IDAAs are time-consuming and cost-inefficient. Besides, high-through experimental identification of disordered residues, although having attracted a lot of attention and approaches have been widely scouted [16], is still challenging and the methods are not currently available.

Consequently, many computational tools have been developed to identify IDPs/IDRs/IDAAs and associated molecular interactions. The Protein Data Bank (PDB) [17], while being used in the majority of cases for the three-dimensional structures of biomolecules, does contain information on residues with missing coordinates. These residues are interpreted as IDAAs. Furthermore, PDB also contains the structure of molecular complexes, which frequently provides information of molecular interactions involving IDPs/IDRs. DisProt [18], which is the first database of IDPs/IDRs, not only collects IDPs/IDRs/IDAAs, but also integrates the information of molecular partners.

Similarly, IDEAL [19], another database of IDPs, incorporates the interaction networks of IDPs in the database. DisBind [20], DIBS [21], and MFIB [22] are three recently developed databases for IDPs/IDRs based molecular interactions. These databases can be used to search for IDAA/IDR/IDP, or to develop computational predictors for various purposes. In fact, both PDB and DisProt are frequently used for the development of disorder predictors. In addition, PDB contains complex structures formed between a short IDR and another protein. Many of these short IDRs are known as MoRFs (Molecular Recognition Features) [23]. MoRFs are the very first type of IDRs found in molecular interaction. Based on this discovery, many MoRF related predictors have developed, such as: MoRF [24], MoRFpred [25], MFSPSSMpred [26], MoRFchibi [27], MoRFPred-plus [28], and OPAL [29], among many others. Furthermore, many other predictors have been developed for the general binding site/regions within IDPs/IDRs, e.g., ANCHOR [30], SLiMpred [31], PepBindPred [32], DISOPRED3 [33], IUPred2A/ANCHOR2 [34], etc.

All these computational tools provide information on protein intrinsic disorder for different aspects. Databases are collections of experimentally observed examples; predictors can be used to analyze novel sequences. Disorder predictors identify the location and, to some extent, the scale of flexibility of IDRs/IDAAs; binding motif predictors spot the location of binding regions; other types of predictors may provide information on various structural features and functional roles. Frequently, the outputs of disorder predictors are used as input for other predictors to improve the prediction accuracy [32,35–40]. Clearly, accurate identification of IDAAs is very important for studies associated with protein structure, intrinsic disorder, interaction, and function. Therefore, improving the prediction accuracy of protein intrinsic disorder predictors is always desirable, though also a real challenge at present time. Furthermore, improving the prediction accuracy of IDAAs has other important impacts on basic science. With more and more IDPs/IDRs being discovered, our knowledge on the actual content of protein intrinsic disorder in nature is still elusive. Part of the reason is that the accuracy of existing computational tools is still not able to meet the requirements. Therefore, developing high-accuracy predictors is still in urgent need. In addition, it could also be expected that when developing new predictors, novel computational strategies could be innovated, and thus, make a much broader impact.

In our previous studies on the development of intrinsic disorder predictors [41,42], as well as studies by many other researchers [43–47], meta-strategy has been demonstrated to have multiple advantages over individual predictors that adopt a single computational strategy in the prediction. One oversimplified but straightforward explanation for the success of meta-strategy is that meta-strategy is able to combine the strengths of all individual predictors, and thus improve the prediction accuracy. Nonetheless, a direct integration of multiple individual predictors may not improve the prediction accuracy significantly [48,49], however, further integration of various data pre-processing techniques will. Data pre-processing, such as angle-shift technique in protein dihedral angle prediction, was used in artificial neural network based predictor and improved the accuracy remarkably [50]. A combination of non-linear transformation and principal component analysis-based dimension reduction together with meta-strategy was used to improve the prediction accuracy of miRNAs [48]. With these proofs, it is expected that other novel techniques can also be used to improve the prediction accuracy of protein intrinsic disorder. In this project, dual-threshold value and two-step significance voting were integrated into a decision-tree based neural network to improve the prediction accuracy of IDAAs.

2. Results

2.1. Prediction Performance of Component Predictors

The ROC (Receiver Operating Characteristic) curves of four component predictors was presented in Figure 1A. The AUC (Area Under the Curve) for DisEMBL, IUPred, VSL2, and ESpritz are 0.78, 0.82, 0.84, and 0.88, respectively. The balanced accuracy (Acc-b) of these four predictors at their default settings are: 68%, 76%, 77%, and 73%, accordingly. In Figure 1B, the overlap and coverage between every two predictors were analyzed for the positive samples (disordered residues) and negative samples (structured residues). Here, overlap stands for the ratio of true-positive (or true-negative) predictions made by both predictors over the total number of positive (or negative) samples, and coverage is defined as the ratio of correct predictions made by either predictors over the total number of samples. Clearly, the overlap of positive samples between predictors normally ranges from ~30% to 50%; however, the number for the overlap between IUpred and VSL2 went up to ~65%. In terms of coverage, the numbers were in the range from ~60% to ~80%. For negative samples, the overlap was from ~70% to ~90%, and the coverage was normally higher than ~90%. The highest values of coverage, as shown by bars at the most right-hand side of both panels, were ~85% and 97% for positive and negative samples, respectively. These two values may outline the theoretical uplimits of combining these four predictors.

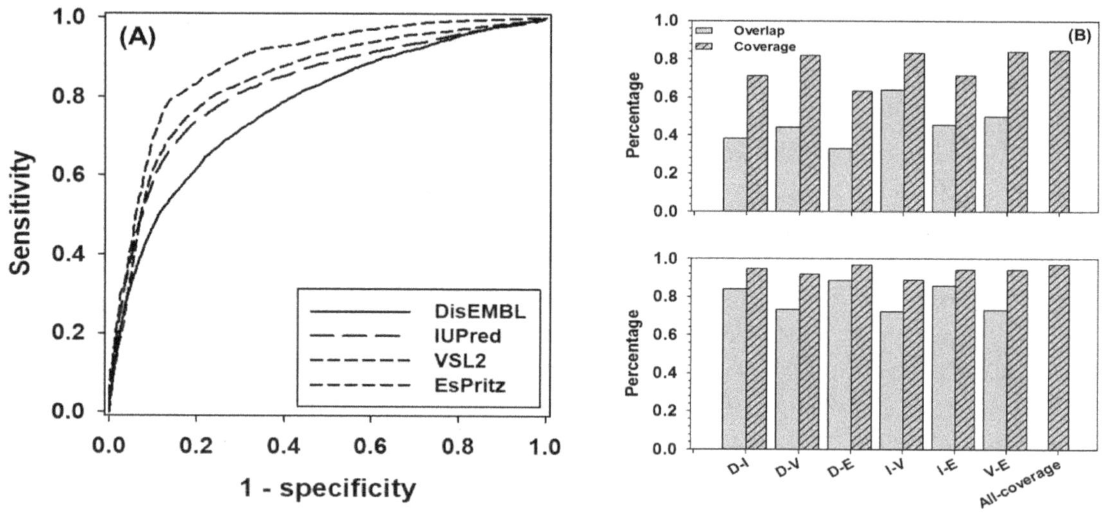

Figure 1. Prediction performance of four component predictors, including DisEMBL, IUPred, VSL2, and ESpritz. (**A**) ROC curves of four component predictors. The ROC curves were obtained by using the default settings of these predictors. (**B**) The pairwise overlap (gray bars) and coverage (dashed bars) for true positive predictions (upper panel) and true negative predictions (lower panel) between each pair of predictors. Axis shows pairs of predictors as follows: D-DisEMBL, I-IUPred, V-VSL2, and E-ESpritz. All-coverage on x-axis stands for the maximum coverage of all predictors.

2.2. Use Information Gain to Choose Threshold Values

To use the new meta-strategy, threshold values of the decision tree need to be determined first. Other than using the method based on the distribution of positive samples and negative samples as a function of prediction score [49], the information gain of all the component predictors in the dataset was analyzed and compared to the distribution of positive samples and negative samples, as shown in Figure 2. The curves of information gain can be roughly characterized by a single-peak distribution, and the location of peaks was, roughly, on the right-hand side of the cross-point where the ratio of positive samples surpassed negative samples. More specifically, the locations of peaks for DisEMBL, IUPred, VSL2, and ESpritz were around 0.5, 0.52, 0.64, and 0.26, respectively. By notation, the locations of the peaks provide a rough estimation of the threshold values, which can be used to maximally partition positive samples and negative samples. Clearly, these values can hardly be determined by the analysis of distribution of positive and negative samples.

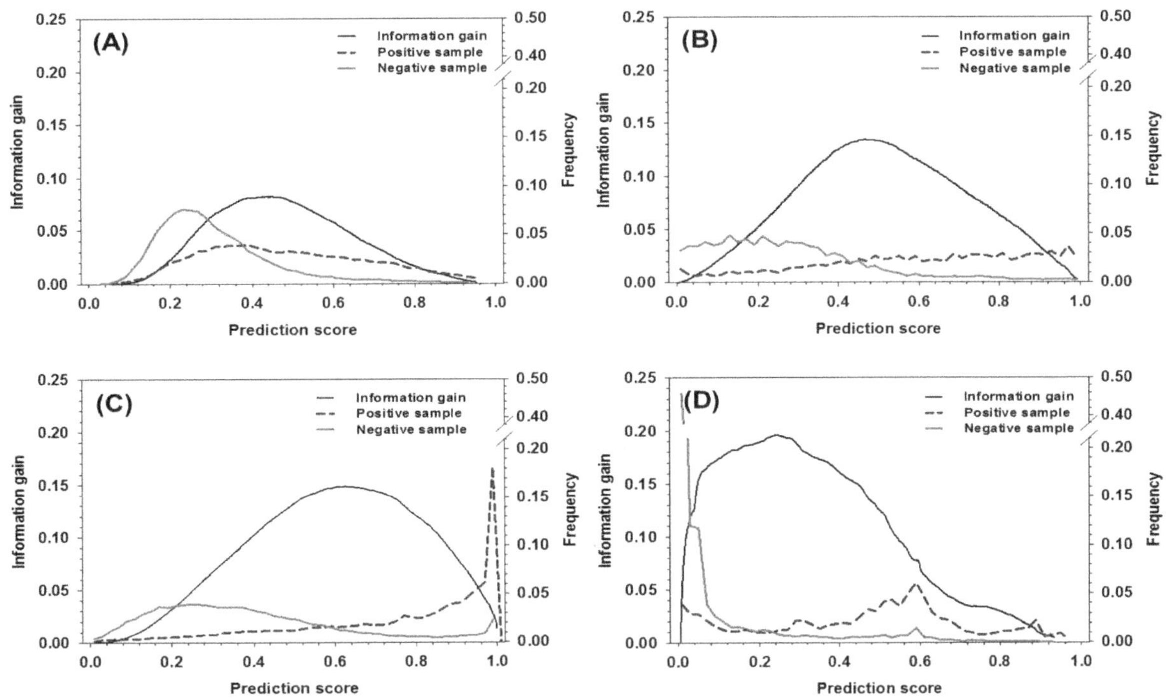

Figure 2. The distribution of information gain, positive sample, and negative samples as a function of prediction score for (**A**) DisEMBL, (**B**) IUPred, (**C**) VSL2, and (**D**) ESpritz. The x-axis shows the prediction score, the y-axis on the left shows the values of information gain, and the y-axis on the right shows the fractions of positive samples and negative samples at different prediction scores, respectively.

2.3. Performance of the New Predictor

Table 1 shows the prediction performance of the new meta-strategy compared to the component predictors, as well as another four recently-developed predictors under five-fold cross-validation. In brief, the performance of new meta-strategy developed in this project was obviously better than others. In terms of accuracy (Acc), balanced accuracy (Acc-b), Matthews Correlation Coefficient (MCC), F1 score, Area Under ROC Curve (AUC_ROC), and Under Precision-Recall Curve (AUC_PR), the new prediction strategy achieved 84.2%, 83.1%, 0,635, 0.744, 0.899, and 0.788, respectively, and was ranked at the first place among all eight different predictors. The new meta-strategy was ranked at the second place on sensitivity (Sens), with one percentage point behind VSL2. With regard to specificity (Spec), the new strategy was inferior to the predictors ESpritz (94%), DisEMBL (91.4%), AUCpreD (90.9%), IUPred2 (87.7%), and IUPred (87.4%). Regardless, it should be noted the Sens values of these predictors are at least 15 percentage points lower than the new meta-strategy.

Table 1. Prediction performance of the new strategy under five-fold cross-validation, in comparison with four component predictors, another four recently-designed predictors.

	DisEMBL	IUPred	VSL2	Espritz	PONDR-FIT	MFDp2	IUPred2	AUCpreD	This Work
Sens	0.440 ± 0.008	0.650 ± 0.003	**0.817 ± 0.004**	0.514 ± 0.009	0.713 ± 0.004	0.777 ± 0.004	0.640 ± 0.003	0.592 ± 0.006	0.807 ± 0.012
Spec	0.914± 0.002	0.874 ± 0.004	0.736 ± 0.003	**0.939 ± 0.002**	0.859 ± 0.004	0.859 ± 0.004	0.877 ± 0.004	0.909 ± 0.002	0.856 ± 0.007
Acc	0.779 ± 0.003	0.810 ± 0.003	0.759 ± 0.002	0.818 ± 0.004	0.817 ± 0.004	0.836 ± 0.003	0.810 ± 0.003	0.819 ± 0.003	**0.842 ± 0.003**
Acc-b	0.677 ± 0.006	0.762 ± 0.002	0.776 ± 0.002	0.726 ± 0.004	0.786 ± 0.003	0.818 ± 0.003	0.759 ± 0.002	0.751 ± 0.003	**0.831 ± 0.004**
MCC	0.410 ± 0.007	0.529 ± 0.006	0.504 ± 0.004	0.521 ± 0.007	0.561 ± 0.007	0.614 ± 0.006	0.526 ± 0.006	0.535 ± 0.006	**0.635 ± 0.006**
F1	0.531 ± 0.006	0.660 ± 0.003	0.658 ± 0.003	0.616 ± 0.006	0.689 ± 0.004	0.729 ± 0.003	0.657 ± 0.004	0.651 ± 0.005	**0.744 ± 0.004**
AUC_ROC	0.776 ± 0.004	0.823 ± 0.001	0.841 ± 0.003	0.886 ± 0.003	0.857 ± 0.003	0.879 ± 0.002	0.822 ± 0.001	0.869 ± 0.003	**0.899 ± 0.004**
AUC_PR	0.607 ± 0.007	0.675 ± 0.007	0.656 ± 0.020	0.752 ± 0.006	0.696 ± 0.004	0.629 ± 0.006	0.657 ± 0.004	0.716 ± 0.007	**0.788 ± 0.010**

Note Bene. The measures of predictor performance include: sensitivity (Sens), specificity (Spec), accuracy (Acc), balanced accuracy (Acc-b), Matthews Correlation Coefficient (MCC), F1 score, Area Under ROC Curve (AUC_ROC), and Area Under Precision-Recall Curve (AUC_PR). The highest value in each of these measures is in bold and highlighted (red).

The performance of all these nine predictors was also assessed using the independent dataset as shown in Table 2. By comparing the data of Tables 1 and 2, it is obvious that although the numbers have fluctuations, the overall levels and trends of all the measures of prediction performance are essentially the same.

Table 2. Prediction performance of all nine predictors in the independent dataset.

	DisEMBL	IUPred	VSL2	Espritz	PONDR-FIT	MFDp2	IUPred2	AUCpreD	This Work
Sens	0.454	0.656	**0.82**	0.529	0.728	0.78	0.647	0.609	0.811 ± 0.007
Spec	0.915	0.872	0.735	**0.932**	0.856	0.857	0.87	0.908	0.856 ± 0.006
Acc	0.784	0.811	0.759	0.818	0.82	0.835	0.811	0.823	**0.844 ± 0.003**
Acc-b	0.684	0.764	0.777	0.731	0.792	0.819	0.761	0.759	**0.834 ± 0.001**
MCC	0.424	0.532	0.507	0.521	0.569	0.615	0.53	0.53	**0.639 ± 0.003**
F1	0.544	0.663	0.659	0.622	0.696	0.729	0.66	0.66	**0.747 ± 0.002**
AUC_ROC	0.779	0.824	0.841	0.888	0.857	0.88	0.822	0.872	**0.9 ± 0.002**
AUC_PR	0.617	0.673	0.642	0.754	0.695	0.622	0.672	0.72	**0.789 ± 0.005**

Note Bene. The new strategy was optimized five times under five-fold cross-validation. Therefore, the performance was also tested in the independent test dataset five times. The results shown in the table is the average of all five times. The highest value in each of these measures is in bold and highlighted (red).

The performance of this new meta-strategy, as well as other predictors, for twenty types of amino acids was analyzed using balanced accuracy in Figure 3. Overall, the new meta-strategy has the highest Acc-b values in fifteen types of residues. The new meta-strategy was also ranked first together with the more recent predictor MFDp2 for residues P and Q. However, the new meta-strategy was ranked at the second position for C, N, and Y, with several percentage points behind MFDp2.

Figure 3. Comparison of balanced accuracy for twenty types of amino acids. The x-axis shows amino acid types in the alphabetic order, while the y-axis shows the value of balanced accuracy. For each type of amino acid, the predictors from left to right are: DisEMBL, IUPred, VSL2, ESpritz, PONDR-FIT, MFDp2, IUPred2A, and AUCpreD.

The balanced accuracies of all predictors for terminal residues were also analyzed in Figure 4. Obviously, the accuracy is location and predictor dependent. For many predictors, the closer to the termini, the lower the accuracy. For N-terminal residues, IUPred, ESpritz, MFDp2, and IUpred2 achieved ~67% balanced accuracy, which was also largely location independent. For DisEMBL, VSL2, and AUCpreD, the balanced accuracies increased gradually from ~55% to ~65% in the window from the 5th to the ~15th residues and then kept similar accuracy afterwards. The newly designed meta-strategy had a lower balanced accuracy of ~52% for the first several residues. The accuracy then increased gradually to ~63% at the 25th residue. PONDR-FIT, a more recently developed predictor, was the least accurate predictor for N-terminal residues, especially in the range from the 10th to the 20th residues where its accuracy was 2–5 points lower than the new strategy. For C-terminal residues, the patterns of accuracy were different from N-terminal residues. First, the balanced accuracy was higher in general than N-terminal residues by several percentage points. Second, although the accuracies of predictors were still either location-independent or location-dependent, the values of accuracies were highly diversified. AUCpreD, MFDp2, IUPred, IUPred2, and ESpritz made location-independent predictions for C-terminal residues, however, the accuracy of these predictors spread from ~74% to 68%, accordingly. DisEMBL, VSL2, and PONDR-FIT's accuracy increased gradually from ~55 to62% at the 5th residue to ~67% at the 20th residue. The accuracy of the newly designed strategy for C-terminal residues was at the lower-end for the first several terminal residues, though increased consistently and achieved the highest balanced accuracy for residues at the ~20th position.

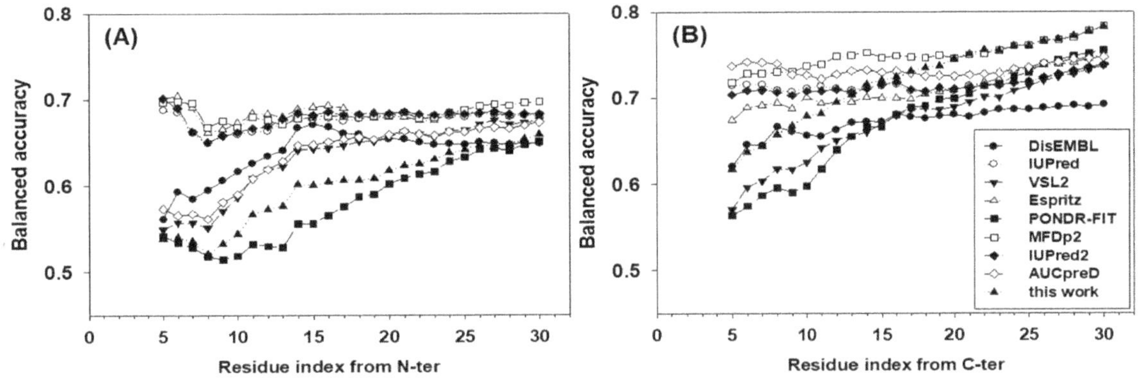

Figure 4. Balanced accuracy of (**A**) N-terminal and (**B**) C-terminal residues. The *x*-axis shows the distance from the first (N-terminal) or the last (C-terminal) residue. The analysis starts at the fifth residue on both N- and C-termini. The *y*-axis shows the value of the balanced accuracy.

With these observations, all the samples were regrouped into three new datasets each containing the first 25 N-terminal residues, the first 25 C-terminal residues, and the middle region, respectively. The meta-strategy was re-trained in three different datasets separately. The prediction performance of all predictors in all three regions under five-fold cross-validation was compared and analyzed in Table 3. Evidently, compared to the results in Figure 4, the prediction accuracy of terminal residues improved substantially. More specifically, the values of improvement of accuracy, balanced accuracy, F1, MCC in N-ter, Mid, and C-ter datasets ranged from 1 to 5 percentage points. For sensitivity and specificity, since many other predictors were trained to maximize either sensitivity or specificity, the new meta-strategy was normally not able to compete with them.

Table 3. Comparison of prediction performance under five-fold cross-validation of eight predictors, as well as the new strategy trained for N-terminal, middle region, and C-terminal residues. The highest value in each of these measures is in bold and highlighted (red).

		DisEMBL	IUPred	VSL2	Espritz	PONDR-FIT	MFDp2	IUPred2	AUCpreD	This Work
N-ter	Sens	0.553 ± 0.009	0.541 ± 0.017	0.782 ± 0.011	0.582 ± 0.010	**0.837 ± 0.004**	0.782 ± 0.009	0.539 ± 0.015	0.748 ± 0.011	0.829 ± 0.023
	Spec	0.741 ± 0.016	0.841 ± 0.020	0.524 ± 0.024	0.789 ± 0.012	0.405 ± 0.020	0.582 ± 0.039	**0.842 ± 0.020**	0.590 ± 0.038	0.572 ± 0.049
	Acc	0.614 ± 0.003	0.639 ± 0.012	0.698 ± 0.005	0.650 ± 0.004	0.697 ± 0.006	0.718 ± 0.012	0.638 ± 0.014	0.697 ± 0.011	**0.746 ± 0.014**
	Acc-b	0.647 ± 0.004	0.691 ± 0.011	0.653 ± 0.010	0.686 ± 0.004	0.621 ± 0.011	0.682 ± 0.020	0.691 ± 0.014	0.669 ± 0.016	**0.701 ± 0.020**
	MCC	0.277 ± 0.009	0.364 ± 0.021	0.308 ± 0.019	0.349 ± 0.009	0.265 ± 0.024	0.361 ± 0.038	0.363 ± 0.027	0.330 ± 0.029	**0.410 ± 0.035**
	F1	0.660 ± 0.007	0.669 ± 0.015	0.778 ± 0.006	0.692 ± 0.008	0.789 ± 0.007	0.789 ± 0.007	0.668 ± 0.014	0.770 ± 0.010	**0.815 ± 0.013**
Middle	Sens	0.387 ± 0.009	0.682 ± 0.005	0.820 ± 0.004	0.481 ± 0.010	**0.663 ± 0.004**	0.777 ± 0.005	0.672 ± 0.005	0.539 ± 0.007	0.807 ± 0.013
	Spec	0.927 ± 0.001	0.877 ± 0.004	0.751 ± 0.004	0.948 ± 0.002	0.888 ± 0.004	0.875 ± 0.005	**0.880 ± 0.004**	0.927 ± 0.001	0.877 ± 0.006
	Acc	0.801 ± 0.004	0.831 ± 0.004	0.767 ± 0.003	0.839 ± 0.004	0.835 ± 0.004	0.852 ± 0.005	0.831 ± 0.004	0.836 ± 0.003	**0.861 ± 0.004**
	Acc-b	0.657 ± 0.005	0.780 ± 0.003	0.786 ± 0.003	0.715 ± 0.005	0.776 ± 0.003	0.826 ± 0.004	0.776 ± 0.003	0.732 ± 0.003	**0.842 ± 0.005**
	MCC	0.376 ± 0.009	0.544 ± 0.005	0.497 ± 0.006	0.506 ± 0.008	0.546 ± 0.006	0.616 ± 0.008	0.540 ± 0.007	0.510 ± 0.006	**0.643 ± 0.008**
	F1	0.477 ± 0.007	0.628 ± 0.003	0.622 ± 0.005	0.583 ± 0.008	0.653 ± 0.004	0.711 ± 0.004	0.650 ± 0.004	0.606 ± 0.005	**0.731 ± 0.006**
C-ter	Sens	0.584 ± 0.014	0.615 ± 0.017	**0.847 ± 0.016**	0.609 ± 0.019	0.828 ± 0.017	0.771 ± 0.016	0.598 ± 0.016	0.681 ± 0.013	0.790 ± 0.018
	Spec	0.787 ± 0.021	0.838 ± 0.023	0.586 ± 0.014	0.857 ± 0.015	0.615 ± 0.017	0.743 ± 0.015	**0.845 ± 0.019**	0.796 ± 0.019	0.769 ± 0.021
	Acc	0.686 ± 0.013	0.727 ± 0.005	0.715 ± 0.007	0.734 ± 0.009	0.720 ± 0.008	0.757 ± 0.007	0.723 ± 0.007	0.739 ± 0.009	**0.780 ± 0.009**
	Acc-b	0.685 ± 0.012	0.726 ± 0.006	0.716 ± 0.007	0.733 ± 0.007	0.721 ± 0.009	0.757 ± 0.007	0.722 ± 0.008	0.739 ± 0.008	**0.780 ± 0.009**
	MCC	0.379 ± 0.026	0.465 ± 0.014	0.448 ± 0.015	0.482 ± 0.015	0.453 ± 0.018	0.514 ± 0.014	0.459 ± 0.018	0.481 ± 0.018	**0.560 ± 0.018**
	F1	0.649 ± 0.012	0.691 ± 0.008	0.747 ± 0.008	0.694 ± 0.009	0.746 ± 0.008	0.759 ± 0.006	0.682 ± 0.012	0.722 ± 0.007	**0.781 ± 0.006**

The performance of all the predictors were then tested in CASP10 dataset and then compared to DISOPRED3, which is one of the two best predictors in CASP10 competition (see Appendix A for more details). In brief, DISOPRED3 and AUCpreD have very similar performance and are better than other predictors on multiple measures, such as specificity, accuracy, MCC, F1, and AUC-ROC. PONDR-FIT achieved the highest balanced accuracy. The new meta-strategy has the highest sensitivity. In addition to the whole dataset analysis, the per-sequence accuracy was also analyzed. The balanced accuracy of PONDR-FIT, MFDp2, AUCpreD, and the new meta-strategy in CASP10 dataset was compared in Figure 5A. All the symbols above the diagonal line represent sequences with higher accuracy when predicted using PONDR-FIT, MFDp2, or AUCpreD, and vice versa. For symbols in the dashed circle, the prediction accuracies of the compared four predictors are all not satisfactory. Symbols in dashed box constitute another group of sequences of which the prediction accuracy of the new meta-strategy is much higher than the other three predictors. For pair-wise comparison between predictors, there are more open circles above the diagonal line, more triangles under the diagonal line, and similar numbers of filled circles on both sides of the diagonal line. Thus, PONDR-FIT (open circles) has better per-sequence prediction performance in the CASP10 dataset. The new meta-strategy and AUCpreD achieved similar results on per-sequence prediction performance. Since the new meta-strategy also made a very low-accuracy prediction on some of the sequences, analyzing the potential reasons could be beneficial. For this purpose, the per-sequence balanced accuracy, fraction of experimentally validated IDAAs per sequence, and the length of each sequence were analyzed in Figure 5B. In this figure, it is apparent that sequences with a very low fraction of experimentally validated IDAAs have very low accuracy. Therefore, the fraction of IDAAs is a critical factor for the performance of the new meta-strategy.

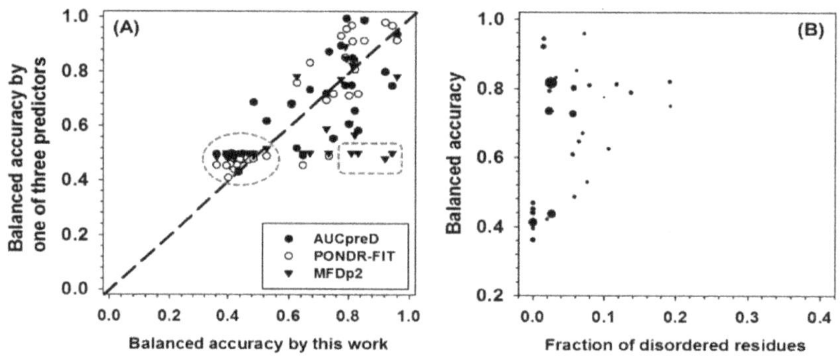

Figure 5. (**A**) Comparison of per-sequence balanced accuracy among AUCpreD (filled circle), PONDR-FIT (open circle), MFDp2 (filled triangle), and this work on sequences in the CASP10 test dataset. The reasons for selecting these predictors are: (1) they are developed in recent years; (2) they have higher performance on some of the accuracy measures; (3) for simplicity of visualization, only four predictors were selected. The x-axis shows the per-sequence balanced accuracy of this work, and the y-axis shows the per-sequence accuracy of the other three predictors. (**B**) Per-sequence balanced accuracy of this work (y-axis) as a function of the fraction of experimentally validated intrinsically disordered amino acids (IDAAs) (x-axis). The size of the symbol is proportional to the length of the sequence.

3. Discussion

Intrinsically disordered proteins play critical roles in biomolecular interaction and signaling; therefore, identifying these residues is crucial for the subsequent analysis and biological studies of the functions and mechanisms. Many experimental techniques have been designed for characterizing these residues. Nonetheless, these techniques are normally time-consuming and/or cost-inefficient. Besides, these techniques may not be appropriate for proteomic studies, although many new approaches are under development [16,51,52]. Therefore, using computational tools to predict intrinsically disordered residues becomes practical, especially for novel protein sequences. Under this situation, using high-accuracy predictors is essential. However, as shown in the previous analysis, the current levels of prediction accuracy of many disordered predictors still have a lot of room for improvement.

There are multiple ways to improve the accuracy of machine learning based techniques. Tuning the list of input features is often the first trial. Recently, deep learning and meta-strategy have also been applied to improve the prediction accuracy. Our previous studies and the studies of other groups [41–47] a direct application of meta-strategy may not lead to the improvement of prediction accuracy, although it has been demonstrated that meta-strategy has many advantages [48]. In these cases, novel data processing techniques are very helpful [48,49]. Therefore, in this project, a dual-threshold was employed; two-step voting with different accuracy stringency was also integrated in the pipeline, based on the analysis of information gain. These techniques eventually contributed remarkably to the improvement of prediction accuracy. The outcomes of this new strategy demonstrate that: (1) integrating lower-accuracy predictors is able to produce higher-accuracy output; (2) the improvement of prediction performance of meta-strategy is significant and impressive, compared to individual predictors and other state-of-the-art predictors, including deep-learning based predictors; (3) the meta-strategy has well-balanced results for sensitivity and specificity, and therefore, is able to achieve higher values on other evaluation quantities, such as F1, MCC, etc.; (4) the meta-strategy provides novel ideas on the renovation of existing predictors.

Many data-processing techniques could be integrated into the meta-strategy. In this project, dual-threshold and two-step significance voting were designed and were critical for the improvement of prediction performance. Dual-threshold refers to true prediction and false prediction having different threshold values. By using dual-threshold, it is possible to control the increase of false positive rate and false negative rate. Two-step voting is a technique to use two sets of threshold values at two steps. At the first step, a set of more stringent threshold values are used, and at the second step less-stringent threshold values are used. In this way, the results from the first step have higher reliability than the second step. Significance-voting is another very useful technique complementary to the well-known majority-voting. When using majority-voting, the number of predictors making true predictions and the number of predictors making false predictions competes to determine the final results. In the application of significance-voting, the Euclidean distance of a prediction score from the corresponding threshold value is calculated, then the sum of distances of predictors making true predictions is compared to that of predictors making false predictions. Clearly, this technique is also beneficial for reducing the prediction error. For majority-voting based strategy, overlap is a critical measurement. However, in significance-voting based predictor, although overlap is still very important, coverage plays a more critical role. In addition, results from majority-voting and from significance-voting predictors have different preferences. Majority-voting is strong in selecting part of the true predictions that have very high confidence. However, significance-voting is able to pick up additional true predictions that cannot be identified by majority-voting.

When selecting individual predictors, overlap and coverage between a pair of predictors or among multiple predictors can be calculated and used to check the similarity of two predictors, and to evaluate whether the combination of these two predictors is able to improve final prediction accuracy. If the two predictors have extremely high overlap and very low coverage, these two predictors are very similar to each other in terms of the predictive results, and vice versa. Evidently, these two types of situations need to be avoided in most cases when selecting the component predictors. Normally, the selected component predictors should have a reasonably level of overlap and a higher level of coverage. The values of coverage also provide an estimation on the maximum values of true-positive and true-negative predictions by combining a pair or several predictors.

It should also be noted that most experimental work aiming at IDAA validation is focused on in vitro approaches, and consequently, the corresponding data analysis and computational strategies are also focused on in vitro data. Regardless, the in vitro foldability of amino acid residues could be very different from in vivo environment [53]. Therefore, novel ideas to develop large-scale in vivo conformational assays are also urgently needed. In fact, novel in vivo labeling strategies of IDAAs have been proposed [53]. It is hopeful that these in vivo techniques or at least the data of in vivo studies will be eventually incorporated into novel predictors of in vivo foldability of IDAA.

4. Materials and Methods

DisProt v7.0 and PDB (Protein DataBank) were combined to build the dataset of disordered residues. DisProt contains over 800 protein sequences, in which the IDAAs/IDRs have been identified using various experimental techniques, such as X-ray, NMR, circular dichroism (CD) spectrometry, proteolysis, etc. For all the DisProt sequences, IDAAs have already been annotated. PDB sequences were extracted using the PISCES server [54]. All the PDB structures in the list have 2.5 angstrom or better resolution and 30% or less sequence identity. Then, 20% of the PDB sequences were randomly selected for further analysis. The missing residues in these PDB sequences were assigned as IDAAs, while all other residues were determined to be structured residues. All the extracted sequences from both DisProt and PDB were further filtered using CD-HIT [55] to remove sequences with 30% or higher sequence identity. Finally, there are 312 protein sequences, containing 30,140 disordered residues and 75,945 structured residues. All the sequences with X-ray structures in CASP10 [56] were also extracted. These sequences were each aligned with all the sequences in the above-mentioned main dataset to check the sequence identity. Only sequences with 30% or lower sequences identity were kept to make the second independent test dataset. This second independent test dataset has 35 sequences.

The infrastructure of the meta-strategy is shown in Figure 6. The prediction results of DisEMBL [57], IUPred [58], VSL2 [59], and ESpritz [60], were used as input. The major reasons for choosing these four predictors are as follows: (1) these predictors were designed using very different strategies. DisEMBL uses artificial neural networks. IUPred uses knowledge-based interaction potential. VSL2 uses neural networks on sequences of different lengths. ESpritz applied bidirectional recursive neural network (BRNN) and was trained separately on N-terminal, C-terminal, and the general sequences; (2) they achieved relatively higher prediction accuracy; (3) these predictors have standalone versions. These four scores were then fed into a decision-tree based artificial neural network (DBann) to make the final prediction. The DBann combines four specific techniques including dual-threshold, significance-voting, two-step selection [49], and two-hidden-layer Artificial Neural Network (ANN). Dual-threshold is a technique using different threshold values for true prediction and false predictions. Significance-voting is complementary to majority-voting by calculating the Euclidean distance of prediction scores to their corresponding threshold values and then comparing the distances of true predictions and false predictions to make selections. For example, when two predictors make true predictions and another two predictors make false predictions, comparing the number of true predictions (N_T) and the number of false predictions (N_P) may have limited usage. In this case, comparing the sum of distances from true thresholds value (d_T) and the total distance from false threshold values (d_F) provides more useful information of the relative significance of true predictions and false predictions. Two-step selection uses two sets of dual-threshold values together with significance-voting as follows: (1) use more stringent values as the first-step threshold values for both true predictions and false predictions; (2) select less stringent values as the second-step threshold values for both true predictions and false predictions; (3) if the numbers of predictors for true prediction and false prediction are equal in the first-step, second-step examination will be performed. If the numbers are still the same, the significance voting will be carried out; (4) based on the results of the above-mentioned comparison, the predictive results of individual predictors will be encoded differently. The encoded predictive results will then be fed into the two-hidden-layer ANN, which is a fully connected ANN and has ten and two nodes in the input and output layers, respectively, as well as twenty nodes in both hidden layers. The activation function for all the nodes is hyperbolic tangent function. In addition, in the output layer, the output was further transformed using a soft matrix function.

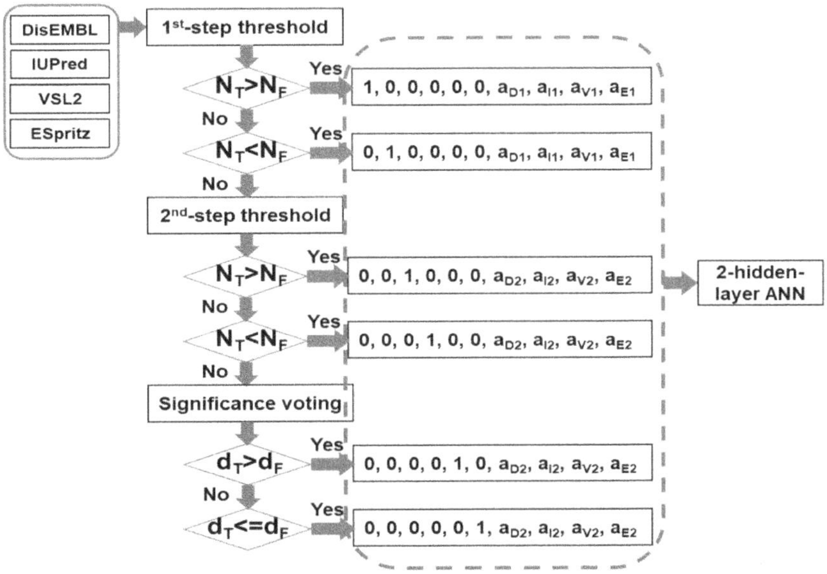

Figure 6. Infrastructure of the new meta-strategy. N_T and N_F are the numbers of predictors making true prediction and false prediction, respectively. "a1" and "a2" are the differences of prediction score from the 1st-step threshold and the 2nd-step threshold values, respectively. The letter subscripts represent DisEMBL (D), IUPred(I), VSL2(V), and ESpritz(E), accordingly. "d_T" and "d_F" are Euclidean distances of prediction scores from their corresponding threshold values for true predictions and false predictions, accordingly.

All the selected sequences were grouped into two datasets. One contains a randomly-selected 20% of all the samples and was set as the independent test dataset, while the other, containing the rest 80% of the samples, was designated as the training and validation dataset. The ratios of positive samples (disordered residues) to negative samples (structured residues) in two datasets are roughly the same. The training and validation dataset was further split into five subsets for five-fold cross-validation. In brief, three out of five subsets were used to train the predictor, the forth subset was used to prevent overfitting, and the last one was used to validate the final prediction performance. By using the different subsets for training, preventing overfitting, and validation, the aforementioned process was repeated five times. The final prediction performance was the average of all five times in the validation subsets. The trained predictors were also evaluated in the independent test dataset.

The performance of predictors was assessed using Sensitivity (Sens), Specificity (Spec), Accuracy (Acc), balanced accuracy (Acc-b, the average of sensitivity and specificity), F1 score (F1), Matthews Correlation Coefficient (MCC), Area Under ROC Curve (AUC, or AUC_ROC), and Area Under precision-recall Curve (AUC_PR) under five-fold cross-validation and in independent datasets. The performance of newly designed predictor was compared to four component predictors (DisEMBL, IUPred, VSL2, and ESpritz), as well as another four recently developed predictors, including PONDR-FIT [42], MFDp2 [61], IUPred2A [34], and AUCpreD [62].

Information Gain (IG) was calculated as a function of predictive score as follows:

$$IG(x) = \sum_{i=1,2} p_i \log_2 p_i - \sum_{j=1,2} f_j(x) \sum_{k=1,2} p_{j,k} \log_2 p_{j,k} \quad (1)$$

In which, p_i is the fraction of positive (i = 1) or negative (i = 2) samples in the dataset; "x" is the threshold prediction score to split the dataset into two groups; $f_j(x)$ is the fraction of samples with prediction score higher than the threshold (j = 1) or the fraction of samples with prediction score lower than the threshold (j = 2); and $p_{j,k}$ refers to the fraction of positive samples (k = 1) or negative samples (k = 2) in the j-th group.

Author Contributions: Conceptualization, B.X.; Methodology, B.X. and B.Z.; Software, B.Z.; Validation, B.Z.; Formal Analysis, B.X. and B.Z.; Writing, B.X.

Acknowledgments: The authors acknowledge with thanks the usage of DisEMBL, IUPred, VSL2, ESpritz, PONDR-FIT, MFDp2, IUPred2A, AUCpreD, and DISOpred3.

Abbreviations

IDP	Intrinsically disordered protein
IDR	Intrinsically disordered region
IDAA	Intrinsically disordered amino acid
ANN	Artificial neural network
IG	Information gain
Sens	Sensitivity
Spec	Specificity
Acc	Accuracy
Acc-b	Balanced accuracy
MCC	Mathew's correlation coefficient
AUC-ROC	Area under ROC curve
AUC-PR	Area under precision-recall curve

Appendix A

Table A1. Comparison of prediction accuracy of in the CASP10 test dataset.

	Sen	Spec	Acc	Acc-b	MCC	F1	AUC_ROC	AUC_PR
Disembl	0.379	0.954	0.929	0.666	0.286	0.318	0.754	0.295
IUPred	0.168	0.958	0.924	0.563	0.122	0.161	0.618	0.175
VSL2	0.612	0.811	0.803	0.712	0.214	0.213	0.774	0.275
Espritz	0.512	0.921	0.903	0.716	0.298	0.316	0.815	0.404
DISOPRED3	0.362	**0.993**	**0.966**	0.678	**0.495**	**0.481**	**0.860**	**0.495**
PONDRFIT	0.586	0.929	0.914	**0.758**	0.362	0.374	0.830	0.358
MFDp2	0.325	0.975	0.947	0.650	0.322	0.349	0.778	0.352
IUPred2	0.164	0.959	0.924	0.561	0.119	0.158	0.616	0.170
AUCpreD	0.425	0.984	**0.960**	0.705	0.465	**0.481**	**0.863**	**0.501**
This work	**0.629**	0.840	0.831	0.734	0.249	0.245	0.793	0.305

Note Bene. DISOPRED3 is one of the two best predictors in CASP10 competition and therefore was used in the comparison. The other one and its web server were not available when the study was carried out. The highest value in each of these measures is in bold and highlighted (red).

References

1. Dunker, A.K.; Silman, I.; Uversky, V.N.; Sussman, J.L. Function and structure of inherently disordered proteins. *Curr. Opin. Struct. Biol.* **2008**, *18*, 756–764. [CrossRef] [PubMed]
2. Uversky, V.N.; Dunker, A.K. Multiparametric analysis of intrinsically disordered proteins: Looking at intrinsic disorder through compound eyes. *Anal. Chem.* **2012**, *84*, 2096–2104. [CrossRef] [PubMed]
3. Csermely, P.; Sandhu, K.S.; Hazai, E.; Hoksza, Z.; Kiss, H.J.; Miozzo, F.; Veres, D.V.; Piazza, F.; Nussinov, R. Disordered proteins and network disorder in network descriptions of protein structure, dynamics and function: Hypotheses and a comprehensive review. *Curr. Protein Pept. Sci.* **2012**, *13*, 19–33. [CrossRef] [PubMed]
4. Tompa, P. Intrinsically disordered proteins: A. 10-year recap. *Trends Biochem. Sci.* **2012**, *37*, 509–516. [CrossRef] [PubMed]

5. Uversky, V.N.; Dave, V.; Iakoucheva, L.M.; Malaney, P.; Metallo, S.J.; Pathak, R.R.; Joerger, A.C. Pathological unfoldomics of uncontrolled chaos: Intrinsically disordered proteins and human diseases. *Chem. Rev.* **2014**, *114*, 6844–6879. [CrossRef] [PubMed]
6. Fuxreiter, M.; Toth-Petroczy, A.; Kraut, D.A.; Matouschek, A.; Lim, R.Y.; Xue, B.; Kurgan, L.; Uversky, V.N. Disordered proteinaceous machines. *Chem. Rev.* **2014**, *114*, 6806–6843. [CrossRef] [PubMed]
7. Wright, P.E.; Dyson, H.J. Intrinsically disordered proteins in cellular signalling and regulation. *Nat. Rev. Mol. Cell Biol.* **2015**, *16*, 18–29. [CrossRef] [PubMed]
8. Follis, A.V.; Hammoudeh, D.I.; Wang, H.; Prochownik, E.V.; Metallo, S.J. Structural rationale for the coupled binding and unfolding of the c-Myc oncoprotein by small molecules. *Chem. Biol.* **2008**, *15*, 1149–1155. [CrossRef] [PubMed]
9. Wright, P.E.; Dyson, H.J. Linking folding and binding. *Curr. Opin. Struct. Biol.* **2009**, *19*, 31–38. [CrossRef] [PubMed]
10. Schulenburg, C.; Hilvert, D. Protein conformational disorder and enzyme catalysis. *Top Curr. Chem.* **2013**, *337*, 41–67. [CrossRef] [PubMed]
11. Dunker, A.K.; Garner, E.; Guilliot, S.; Romero, P.; Albrecht, K.; Hart, J.; Obradovic, Z.; Kissinger, C.; Villafranca, J.E. Protein disorder and the evolution of molecular recognition: Theory, predictions and observations. *Pac. Symp. Biocomput.* **1998**, *3*, 473–484.
12. Uversky, V.N. Intrinsic disorder-based protein interactions and their modulators. *Curr. Pharm. Des.* **2013**, *19*, 4191–4213. [CrossRef] [PubMed]
13. Dogan, J.; Gianni, S.; Jemth, P. The binding mechanisms of intrinsically disordered proteins. *Phys. Chem. Chem. Phys.* **2014**, *16*, 6323–6331. [CrossRef] [PubMed]
14. Liu, Z.; Huang, Y. Advantages of proteins being disordered. *Protein Sci.* **2014**, *23*, 539–550. [CrossRef] [PubMed]
15. Teilum, K.; Olsen, J.G.; Kragelund, B.B. Globular and disordered-the non-identical twins in protein-protein interactions. *Front. Mol. Biosci.* **2015**, *2*, 40. [CrossRef] [PubMed]
16. Minde, D.P.; Dunker, A.K.; Lilley, K.S. Time, space, and disorder in the expanding proteome universe. *Proteomics* **2017**, *17*. [CrossRef] [PubMed]
17. Berman, H.M.; Westbrook, J.; Feng, Z.; Gilliland, G.; Bhat, T.N.; Weissig, H.; Shindyalov, I.N.; Bourne, P.E. The Protein Data Bank. *Nucleic Acids Res.* **2000**, *28*, 235–242. [CrossRef] [PubMed]
18. Piovesan, D.; Tabaro, F.; Micetic, I.; Necci, M.; Quaglia, F.; Oldfield, C.J.; Aspromonte, M.C.; Davey, N.E.; Davidovic, R.; Dosztanyi, Z.; et al. DisProt 7.0: A major update of the database of disordered proteins. *Nucleic Acids Res.* **2017**, *45*, D219–D227. [CrossRef] [PubMed]
19. Fukuchi, S.; Sakamoto, S.; Nobe, Y.; Murakami, S.D.; Amemiya, T.; Hosoda, K.; Koike, R.; Hiroaki, H.; Ota, M. IDEAL: Intrinsically Disordered proteins with Extensive Annotations and Literature. *Nucleic Acids Res.* **2012**, *40*, D507–D511. [CrossRef] [PubMed]
20. Yu, J.F.; Dou, X.H.; Sha, Y.J.; Wang, C.L.; Wang, H.B.; Chen, Y.T.; Zhang, F.; Zhou, Y.; Wang, J.H. DisBind: A database of classified functional binding sites in disordered and structured regions of intrinsically disordered proteins. *BMC Bioinform.* **2017**, *18*, 206. [CrossRef] [PubMed]
21. Schad, E.; Ficho, E.; Pancsa, R.; Simon, I.; Dosztanyi, Z.; Meszaros, B. DIBS: A repository of disordered binding sites mediating interactions with ordered proteins. *Bioinformatics* **2018**, *34*, 535–537. [CrossRef] [PubMed]
22. Ficho, E.; Remenyi, I.; Simon, I.; Meszaros, B. MFIB: A repository of protein complexes with mutual folding induced by binding. *Bioinformatics* **2017**, *33*, 3682–3684. [CrossRef] [PubMed]
23. Cheng, Y.; Oldfield, C.J.; Meng, J.; Romero, P.; Uversky, V.N.; Dunker, A.K. Mining alpha-helix-forming molecular recognition features with cross species sequence alignments. *Biochemistry* **2007**, *46*, 13468–13477. [CrossRef] [PubMed]
24. Malhis, N.; Gsponer, J. Computational identification of MoRFs in protein sequences. *Bioinformatics* **2015**, *31*, 1738–1744. [CrossRef] [PubMed]
25. Disfani, F.M.; Hsu, W.L.; Mizianty, M.J.; Oldfield, C.J.; Xue, B.; Dunker, A.K.; Uversky, V.N.; Kurgan, L. MoRFpred, a computational tool for sequence-based prediction and characterization of short disorder-to-order transitioning binding regions in proteins. *Bioinformatics* **2012**, *28*, i75–i83. [CrossRef] [PubMed]

26. Fang, C.; Noguchi, T.; Tominaga, D.; Yamana, H. MFSPSSMpred: Identifying short disorder-to-order binding regions in disordered proteins based on contextual local evolutionary conservation. *BMC Bioinform.* **2013**, *14*, 300. [CrossRef] [PubMed]
27. Malhis, N.; Jacobson, M.; Gsponer, J. MoRFchibi SYSTEM: Software tools for the identification of MoRFs in protein sequences. *Nucleic Acids Res.* **2016**, *44*, W488–W493. [CrossRef] [PubMed]
28. Sharma, R.; Bayarjargal, M.; Tsunoda, T.; Patil, A.; Sharma, A. MoRFPred-plus: Computational Identification of MoRFs in Protein Sequences using Physicochemical Properties and HMM profiles. *J. Theor. Biol.* **2018**, *437*, 9–16. [CrossRef] [PubMed]
29. Sharma, R.; Raicar, G.; Tsunoda, T.; Patil, A.; Sharma, A. OPAL: Prediction of MoRF regions in intrinsically disordered protein sequences. *Bioinformatics* **2018**, *34*, 1850–1858. [CrossRef] [PubMed]
30. Dosztanyi, Z.; Meszaros, B.; Simon, I. ANCHOR: Web server for predicting protein binding regions in disordered proteins. *Bioinformatics* **2009**, *25*, 2745–2746. [CrossRef] [PubMed]
31. Mooney, C.; Pollastri, G.; Shields, D.C.; Haslam, N.J. Prediction of short linear protein binding regions. *J. Mol. Biol.* **2012**, *415*, 193–204. [CrossRef] [PubMed]
32. Khan, W.; Duffy, F.; Pollastri, G.; Shields, D.C.; Mooney, C. Predicting binding within disordered protein regions to structurally characterised peptide-binding domains. *PLoS ONE* **2013**, *8*, 72838. [CrossRef] [PubMed]
33. Jones, D.T.; Cozzetto, D. DISOPRED3: Precise disordered region predictions with annotated protein-binding activity. *Bioinformatics* **2015**, *31*, 857–863. [CrossRef] [PubMed]
34. Meszaros, B.; Erdos, G.; Dosztanyi, Z. IUPred2A: Context-dependent prediction of protein disorder as a function of redox state and protein binding. *Nucleic Acids Res.* **2018**, *46*, W329–W337. [CrossRef] [PubMed]
35. Li, B.Q.; Cai, Y.D.; Feng, K.Y.; Zhao, G.J. Prediction of protein cleavage site with feature selection by random forest. *PLoS ONE* **2012**, *7*, e45854. [CrossRef] [PubMed]
36. Zhao, X.; Dai, J.; Ning, Q.; Ma, Z.; Yin, M.; Sun, P. Position-specific analysis and prediction of protein pupylation sites based on multiple features. *BioMed Res. Int.* **2013**, *2013*, 109549. [CrossRef] [PubMed]
37. Tretyachenko, V.; Vymetal, J.; Bednarova, L.; Kopecky, V., Jr.; Hofbauerova, K.; Jindrova, H.; Hubalek, M.; Soucek, R.; Konvalinka, J.; Vondrasek, J.; et al. Random protein sequences can form defined secondary structures and are well-tolerated in vivo. *Sci. Rep.* **2017**, *7*, 15449. [CrossRef] [PubMed]
38. Hu, J.; Li, Y.; Zhang, Y.; Yu, D.J. ATPbind: Accurate Protein-ATP Binding Site Prediction by Combining Sequence-Profiling and Structure-Based Comparisons. *J. Chem. Inf. Model.* **2018**, *58*, 501–510. [CrossRef] [PubMed]
39. Basu, S.; Soderquist, F.; Wallner, B. Proteus: A random forest classifier to predict disorder-to-order transitioning binding regions in intrinsically disordered proteins. *J. Comput. Aided Mol. Des.* **2017**, *31*, 453–466. [CrossRef] [PubMed]
40. Klausen, M.S.; Jespersen, M.C.; Nielsen, H.; Jensen, K.K.; Jurtz, V.I.; Soenderby, C.K.; Sommer, M.O.A.; Winther, O.; Nielsen, M.; Petersen, B.; et al. NetSurfP-2.0: Improved prediction of protein structural features by integrated deep learning. *BioRxiv* **2018**. [CrossRef]
41. Xue, B.; Oldfield, C.J.; Dunker, A.K.; Uversky, V.N. CDF it all: Consensus prediction of intrinsically disordered proteins based on various cumulative distribution functions. *FEBS Lett.* **2009**, *583*, 1469–1474. [CrossRef] [PubMed]
42. Xue, B.; Dunbrack, R.L.; Williams, R.W.; Dunker, A.K.; Uversky, V.N. PONDR-FIT: A meta-predictor of intrinsically disordered amino acids. *Biochim. Biophys. Acta* **2010**, *1804*, 996–1010. [CrossRef] [PubMed]
43. Schlessinger, A.; Punta, M.; Yachdav, G.; Kajan, L.; Rost, B. Improved disorder prediction by combination of orthogonal approaches. *PLoS ONE* **2009**, *4*, e4433. [CrossRef] [PubMed]
44. Hirose, S.; Shimizu, K.; Noguchi, T. POODLE-I: Disordered region prediction by integrating POODLE series and structural information predictors based on a workflow approach. *In Silico Biol.* **2010**, *10*, 185–191. [CrossRef] [PubMed]
45. Kozlowski, L.P.; Bujnicki, J.M. MetaDisorder: A meta-server for the prediction of intrinsic disorder in proteins. *BMC Bioinform.* **2012**, *13*, 111. [CrossRef] [PubMed]
46. Huang, Y.J.; Acton, T.B.; Montelione, G.T. DisMeta: A meta server for construct design and optimization. *Methods Mol. Biol.* **2014**, *1091*, 3–16. [CrossRef] [PubMed]

47. Mizianty, M.J.; Stach, W.; Chen, K.; Kedarisetti, K.D.; Disfani, F.M.; Kurgan, L. Improved sequence-based prediction of disordered regions with multilayer fusion of multiple information sources. *Bioinformatics* **2010**, *26*, i489–i496. [CrossRef] [PubMed]
48. Xue, B.; Lipps, D.; Devineni, S. Integrated Strategy Improves the Prediction Accuracy of miRNA in Large Dataset. *PLoS ONE* **2016**, *11*, e0168392. [CrossRef] [PubMed]
49. Zhao, B.; Xue, B. Improving prediction accuracy using decision-tree-based meta-strategy and multi-threshold sequential-voting exemplified by miRNA target prediction. *Genomics* **2017**, *109*, 227–232. [CrossRef] [PubMed]
50. Xue, B.; Dor, O.; Faraggi, E.; Zhou, Y. Real-value prediction of backbone torsion angles. *Proteins* **2008**, *72*, 427–433. [CrossRef] [PubMed]
51. Aebersold, R.; Mann, M. Mass-spectrometric exploration of proteome structure and function. *Nature* **2016**, *537*, 347–355. [CrossRef] [PubMed]
52. Mann, M. Origins of mass spectrometry-based proteomics. *Nat. Rev. Mol. Cell Biol.* **2016**, *17*, 678. [CrossRef] [PubMed]
53. Minde, D.P.; Ramakrishna, M.; Lilley, K.S. Biotinylation by proximity labelling favours unfolded proteins. *BioRxiv* **2018**. [CrossRef]
54. Wang, G.; Dunbrack, R.L., Jr. PISCES: Recent improvements to a PDB sequence culling server. *Nucleic Acids Res.* **2005**, *33*, W94–W98. [CrossRef] [PubMed]
55. Li, W.; Godzik, A. Cd-hit: A fast program for clustering and comparing large sets of protein or nucleotide sequences. *Bioinformatics* **2006**, *22*, 1658–1659. [CrossRef] [PubMed]
56. Monastyrskyy, B.; Kryshtafovych, A.; Moult, J.; Tramontano, A.; Fidelis, K. Assessment of protein disorder region predictions in CASP10. *Proteins* **2014**, *82*, 127–137. [CrossRef] [PubMed]
57. Linding, R.; Jensen, L.J.; Diella, F.; Bork, P.; Gibson, T.J.; Russell, R.B. Protein disorder prediction: Implications for structural proteomics. *Structure* **2003**, *11*, 1453–1459. [CrossRef] [PubMed]
58. Dosztanyi, Z.; Csizmok, V.; Tompa, P.; Simon, I. IUPred: Web server for the prediction of intrinsically unstructured regions of proteins based on estimated energy content. *Bioinformatics* **2005**, *21*, 3433–3434. [CrossRef] [PubMed]
59. Peng, K.; Radivojac, P.; Vucetic, S.; Dunker, A.K.; Obradovic, Z. Length-dependent prediction of protein intrinsic disorder. *BMC Bioinform.* **2006**, *7*, 208. [CrossRef] [PubMed]
60. Walsh, I.; Martin, A.J.; Di Domenico, T.; Tosatto, S.C. ESpritz: Accurate and fast prediction of protein disorder. *Bioinformatics* **2012**, *28*, 503–509. [CrossRef] [PubMed]
61. Mizianty, M.J.; Peng, Z.; Kurgan, L. MFDp2: Accurate predictor of disorder in proteins by fusion of disorder probabilities, content and profiles. *Intrinsically Disord. Proteins* **2013**, *1*, e24428. [CrossRef] [PubMed]
62. Wang, S.; Ma, J.; Xu, J. AUCpreD: Proteome-level protein disorder prediction by AUC-maximized deep convolutional neural fields. *Bioinformatics* **2016**, *32*, i672–i679. [CrossRef] [PubMed]

Permissions

All chapters in this book were first published by MDPI; hereby published with permission under the Creative Commons Attribution License or equivalent. Every chapter published in this book has been scrutinized by our experts. Their significance has been extensively debated. The topics covered herein carry significant findings which will fuel the growth of the discipline. They may even be implemented as practical applications or may be referred to as a beginning point for another development.

The contributors of this book come from diverse backgrounds, making this book a truly international effort. This book will bring forth new frontiers with its revolutionizing research information and detailed analysis of the nascent developments around the world.

We would like to thank all the contributing authors for lending their expertise to make the book truly unique. They have played a crucial role in the development of this book. Without their invaluable contributions this book wouldn't have been possible. They have made vital efforts to compile up to date information on the varied aspects of this subject to make this book a valuable addition to the collection of many professionals and students.

This book was conceptualized with the vision of imparting up-to-date information and advanced data in this field. To ensure the same, a matchless editorial board was set up. Every individual on the board went through rigorous rounds of assessment to prove their worth. After which they invested a large part of their time researching and compiling the most relevant data for our readers.

The editorial board has been involved in producing this book since its inception. They have spent rigorous hours researching and exploring the diverse topics which have resulted in the successful publishing of this book. They have passed on their knowledge of decades through this book. To expedite this challenging task, the publisher supported the team at every step. A small team of assistant editors was also appointed to further simplify the editing procedure and attain best results for the readers.

Apart from the editorial board, the designing team has also invested a significant amount of their time in understanding the subject and creating the most relevant covers. They scrutinized every image to scout for the most suitable representation of the subject and create an appropriate cover for the book.

The publishing team has been an ardent support to the editorial, designing and production team. Their endless efforts to recruit the best for this project, has resulted in the accomplishment of this book. They are a veteran in the field of academics and their pool of knowledge is as vast as their experience in printing. Their expertise and guidance has proved useful at every step. Their uncompromising quality standards have made this book an exceptional effort. Their encouragement from time to time has been an inspiration for everyone.

The publisher and the editorial board hope that this book will prove to be a valuable piece of knowledge for researchers, students, practitioners and scholars across the globe.

List of Contributors

Marta Kolonko and Beata Greb-Markiewicz
Department of Biochemistry, Faculty of Chemistry, Wroclaw University of Science and Technology, Wybrzeże Wyspiańskiego 27, 50-370 Wroclaw, Poland

Maciej Pawel Ciemny
Faculty of Chemistry, Biological and Chemical Research Center, University of Warsaw, Pasteura 1, 02-093 Warsaw, Poland
Faculty of Physics, University of Warsaw, Pasteura 5, 02-093 Warsaw, Poland

Aleksandra Elzbieta Badaczewska-Dawid, Monika Pikuzinska, Andrzej Kolinski and Sebastian Kmiecik
Faculty of Chemistry, Biological and Chemical Research Center, University of Warsaw, Pasteura 1, 02-093 Warsaw, Poland

Rita Pancsa
Research Centre for Natural Sciences of the Hungarian Academy of Sciences, Institute of Enzymology, 1117 Budapest, Hungary

Fruzsina Zsolyomi
Department of Biochemistry and Molecular Biology, Faculty of Medicine, University of Debrecen, 4032 Debrecen, Hungary

Peter Tompa
Research Centre for Natural Sciences of the Hungarian Academy of Sciences, Institute of Enzymology, 1117 Budapest, Hungary
Center for Structural Biology, Flanders Institute for Biotechnology (VIB), Vrije Universiteit Brussel, 1050 Brussels, Belgium

David Alvarez-Ponce
Biology Department, University of Nevada, Reno, NV 89557, USA
Instituto de Biología Molecular y Celular de Plantas, CSIC-UPV, 46022 Valencia, Spain

Mario X. Ruiz-González and Mario A. Fares
Instituto de Biología Molecular y Celular de Plantas, CSIC-UPV, 46022 Valencia, Spain
Smurfit Institute of Genetics, University of Dublin, Trinity College, Dublin 2, Ireland

Francisco Vera-Sirera and Miguel A. Perez-Amador
Instituto de Biología Molecular y Celular de Plantas, CSIC-UPV, 46022 Valencia, Spain

Felix Feyertag
Biology Department, University of Nevada, Reno, NV 89557, USA

Mahdi Muhammad Moosa, Phoebe S. Tsoi, Kyoung-Jae Choi, Allan Chris M. Ferreon and Josephine C. Ferreon
Department of Pharmacology and Chemical Biology, Baylor College of Medicine, Houston, TX 77030, USA

Keiichi Homma, Hiroto Anbo and Satoshi Fukuchi
Department of Life Science and Informatics, Maebashi Institute of Technology, 460-1 Kamisadori-machi, Maebashi-shi 371-0816, Japan

Tamotsu Noguchi
Pharmaceutical Education Research Center, Meiji Pharmaceutical University, 2-522-1 Noshio, Kiyose-shi, Tokyo 204-8588, Japan

Lellys M. Contreras
Center for Environmental Biology and Chemistry Research, Facultad Experimental de Ciencias y Tecnología, Universidad de Carabobo, 2001 Valencia, Venezuela

Paz Sevilla
Facultad de Farmacia, Departamento de Química Física II, Universidad Complutense de Madrid, 28040 Madrid, Spain
Instituto de Estructura de la Materia, IEM-CSIC, Serrano 121, 28006 Madrid, Spain

Ana Cámara-Artigas
Department of Chemistry and Physics, Research Centre CIAIMBITAL, University of Almería- ceiA3, 04120 Almería, Spain

José G. Hernández-Cifre and José García de la Torre
Department of Physical Chemistry, University of Murcia, 30003 Murcia, Spain

Francisco J. Florencio and María Isabel Muro- Pastor
Instituto de Bioquímica Vegetal y Fotosíntesis, CSIC-Universidad de Sevilla, 41092 Seville, Spain

Kálmán Tompa and Mónika Bokor
Institute for Solid State Physics and Optics,Wigner RCP of the HAS, Konkoly-Thege út 29-33, H-1121 Budapest, Hungary

Péter Tompa
Institute of Enzymology, Research Centre for Natural Sciences of the HAS, Magyar Tudósok körút. 27, H-1117 Budapest, Hungary
VIB Center for Structural Biology, Vrije Universiteit Brussel, Building E, Pleinlaan 2, 1050 Brussel, Belgium

Jocelyne Walter, Amandine Barra, Bénédicte Doublet, Nicolas Céré, Justine Charon and Thierry Michon
UMR 1332 Biologie du Fruit et Pathologie, INRA, Université de Bordeaux, CS 20032, 33140 Villenave d'Ornon, France

Ankur Mishra, Wouter Sipma, Erik Van der Giessen and Patrick R. Onck
Zernike Institute for Advanced Materials, University of Groningen, Groningen, 9747 AG, The Netherlands

Liesbeth M. Veenhoff
European Research Institute for the Biology of Ageing, University of Groningen, University Medical Centre, Groningen, 9713 AV, The Netherlands

Allan Chris Ferreon
Department of Pharmacology and Chemical Biology, Baylor College of Medicine, Houston, TX 77030, USA

Antrix Jain
Advanced Technology Cores, Baylor College of Medicine, Houston, TX 77030, USA

Kevin R. MacKenzie
Department of Pharmacology and Chemical Biology, Baylor College of Medicine, Houston, TX 77030, USA
Department of Pathology and Immunology, Baylor College of Medicine, Houston, TX 77030, USA

Sung Yun Jung
Department of Biochemistry and Molecular Biology, Baylor College of Medicine, Houston, TX 77030, USA

Csaba Magyar, Anikó Mentes, Erzsébet Fichó and István Simon
Institute of Enzymology, Research Centre for Natural Sciences, Hungarian Academy of Sciences, Magyar Tudósok krt. 2, H-1117 Budapest, Hungary

Miklós Cserző
Institute of Enzymology, Research Centre for Natural Sciences, Hungarian Academy of Sciences, Magyar Tudósok krt. 2, H-1117 Budapest, Hungary
Department of Physiology, Faculty of Medicine, Semmelweis University, Tűzoltó u. 37-47, H-1094 Budapest, Hungary

José L. Neira
Instituto de Biología Molecular y Celular, Edificio Torregaitán, Universidad Miguel Hernández, Avda. del Ferrocarril s/n, 03202 Elche (Alicante), Spain Instituto de Biocomputación y Física de Sistemas Complejos, Joint Units IQFR-CSIC-BIFI, and GBsC-CSIC-BIFI, Universidad de Zaragoza, 50009 Zaragoza, Spain

A. Marcela Giudici and Felipe Hornos
Instituto de Biología Molecular y Celular, Edificio Torregaitán, Universidad Miguel Hernández, Avda. del Ferrocarril s/n, 03202 Elche (Alicante), Spain

Arantxa Arbe
Centro de Física de Materiales (CFM) (CSIC-UPV/EHU)—Materials Physics Center (MPC), 20018 San Sebastián, Spain

Bruno Rizzuti
CNR-NANOTEC, Licryl-UOS Cosenza and CEMIF. Cal, Department of Physics, University of Calabria, Ponte P. Bucci, 87036 Rende, Italy

Gustav Arvidsson and Anthony P. H. Wright
Clinical Research Center, Department of Laboratory Medicine, Karolinska Institutet, Huddinge SE 141 57, Sweden

Ambuj Srivastava and M. Michael Gromiha
Department of Biotechnology, Bhupat and Jyoti Mehta School of Biosciences, Indian Institute of Technology Madras, Chennai 600 036, Tamilnadu, India

Shandar Ahmad
School of Computational and Integrative Sciences, Jawaharlal Nehru University, New Delhi 110 067, India

Beáta Szabó, Nikoletta Murvai, Rawan Abukhairan, Éva Schád, Bálint Szeder, László Buday and Ágnes Tantos
Institute of Enzymology, Research Centre for Natural Sciences, Hungarian Academy of Sciences, H-1117 Budapest, Hungary

József Kardos
ELTE NAP Neuroimmunology Research Group, Department of Biochemistry, Eötvös Loránd University, H-1117 Budapest, Hungary

Bi Zhao and Bin Xue
Department of Cell Biology, Microbiology and Molecular Biology, School of Natural Sciences and Mathematics, College of Arts and Sciences, University of South Florida, Tampa, FL 33620, USA

Index

www.ingramcontent.com/pod-product-compliance
Lightning Source LLC
LaVergne TN
LVHW080948210926
848485LV00004B/802